Springer Actuarial

Springer Actuarial Lecture Notes

This subseries of Springer Actuarial includes books with the character of lecture notes. Typically these are research monographs on new, cutting-edge developments in actuarial science; sometimes they may be a glimpse of a new field of research activity, or presentations of a new angle in a more classical field.

In the established tradition of Lecture Notes, the timeliness of a manuscript can be more important than its form, which may be informal, preliminary or tentative.

More information about this subseries at http://www.springer.com/series/15682

Michel Denuit · Donatien Hainaut ·
Julien Trufin

Effective Statistical Learning Methods for Actuaries II

Tree-Based Methods and Extensions

Michel Denuit
Institut de Statistique, Biostatistique et
Sciences Actuarielles (ISBA)
Université Catholique Louvain
Louvain-la-Neuve, Belgium

Donatien Hainaut
Institut de Statistique, Biostatistique et
Sciences Actuarielles (ISBA)
Université Catholique Louvain
Louvain-la-Neuve, Belgium

Julien Trufin
Département de Mathématiques
Université Libre de Bruxelles
Brussels, Belgium

ISSN 2523-3262 ISSN 2523-3270 (electronic)
Springer Actuarial
ISSN 2523-3289 ISSN 2523-3297 (electronic)
Springer Actuarial Lecture Notes
ISBN 978-3-030-57555-7 ISBN 978-3-030-57556-4 (eBook)
https://doi.org/10.1007/978-3-030-57556-4

Mathematics Subject Classification: 62P05, 62-XX, 68-XX, 62M45

This Springer imprint is published by the registered company Springer Nature Switzerland AG
The registered company address is: Gewerbestrasse 11, 6330 Cham, Switzerland

Preface

The present material is written for students enrolled in actuarial master programs and practicing actuaries, who would like to gain a better understanding of insurance data analytics. It is built in three volumes, starting from the celebrated Generalized Linear Models, or GLMs and continuing with tree-based methods and neural networks.

This second volume summarizes the state of the art using regression trees and their various combinations such as random forests and boosting trees. This second volume also goes through tools enabling to assess the predictive accuracy of regression models. Throughout this book, we alternate between methodological aspects and numerical illustrations or case studies to demonstrate practical applications of the proposed techniques. The R statistical software has been found convenient to perform the analyses throughout this book. It is a free language and environment for statistical computing and graphics. In addition to our own R code, we have benefited from many R packages contributed by the members of the very active community of R-users. The open-source statistical software R is freely available from `https://www.r-project.org/`.

The technical requirements to understand the material are kept at a reasonable level so that this text is meant for a broad readership. We refrain from proving all results but rather favor an intuitive approach with supportive numerical illustrations, providing the reader with relevant references where all justifications can be found, as well as more advanced material. These references are gathered in a dedicated section at the end of each chapter.

The three authors are professors of actuarial mathematics at the universities of Brussels and Louvain-la-Neuve, Belgium. Together, they accumulate decades of teaching experience related to the topics treated in the three books, in Belgium and throughout Europe and Canada. They are also scientific directors at Detralytics, a consulting office based in Brussels.

Within Detralytics as well as on behalf of actuarial associations, the authors have had the opportunity to teach the material contained in the three volumes of "Effective Statistical Learning Methods for Actuaries" to various audiences of practitioners. The feedback received from the participants to these short courses greatly helped to improve the exposition of the topic. Throughout their contacts with the industry, the authors also implemented these techniques in a variety of consulting and R&D projects. This makes the three volumes of "Effective Statistical Learning Methods for Actuaries" the ideal support for teaching students and CPD events for professionals.

Louvain-la-Neuve, Belgium Michel Denuit
Louvain-la-Neuve, Belgium Donatien Hainaut
Brussels, Belgium Julien Trufin
September 2020

Contents

Chapter 1
Introduction

1.1 The Risk Classification Problem

1.1.1 *Insurance Risk Diversification*

Insurance companies cover risks (that is, random financial losses) by collecting premiums. Premiums are generally paid in advance (hence their name). The pure premium is the amount collected by the insurance company, to be re-distributed as benefits among policyholders and third parties in execution of the contract, without loss nor profit. Under the conditions of validity of the law of large numbers, the pure premium is the expected amount of compensation to be paid by the insurer (sometimes discounted to policy issue in case of long-term liabilities).

The pure premiums are just re-distributed among policyholders to pay for their respective claims, without loss nor profit on average. Hence, they cannot be considered as insurance prices because loadings must be added to face operating costs, in order to ensure solvency, to cover general expenses, to pay commissions to intermediaries, to generate profit for stockholders, not to mention the taxes imposed by the local authorities.

1.1.2 *Why Classifying Risks?*

In practice, most of portfolios are heterogeneous: they mix individuals with different risk levels. Some policyholders tend to report claims more often or to report more expensive claims, on average. In an heterogeneous portfolio with a uniform price list, the financial result of the insurance company depends on the composition of the portfolio.

The modification in the composition of the portfolio may generate losses for the insurer charging a uniform premium to different risk profiles, when competitors

M. Denuit et al., *Effective Statistical Learning Methods for Actuaries II*,
Springer Actuarial, https://doi.org/10.1007/978-3-030-57556-4_1

distinguish premiums according to these profiles. Policyholders who are over-priced by the insurance tariff leave the insurer to enjoy premium discounts offered by the competitors whereas those who appear to have been under-priced remain with the insurer. This change in the portfolio composition generates systematic losses for the insurer applying the uniform tariff. This phenomenon is known as adverse selection, as policyholders are supposed to select the insurance provider offering them the best premium.

This (partly) explains why so many factors are used by insurance companies: insurance companies have to use a rating structure matching the premiums for the risks as closely as the rating structures of competitors. If they do not, they become exposed to the risk of loosing the policyholders who are currently over-priced according to their tariff, breaking the equilibrium between expected losses and collected premiums. This is one of the reasons why the technical pricelist must be as accurate as possible: it is only in this way that the insurer is able to manage its portfolio effectively, by knowing which profiles are over-priced and which ones subsidize the others. In other words, the insurer knows the value of each policy in the portfolio.

1.1.3 The Need for Regression Models

Considering that risk profiles differ inside insurance portfolios, it theoretically suffices to subdivide the entire portfolio into homogeneous risk classes, i.e. groups of policyholders sharing the same risk factors, and to determine an amount of pure premium specific to each risk class. However, if the data are subdivided into risk classes determined by many factors, actuaries often deal with sparsely populated groups of contracts. Therefore, simple averages become useless and regression models are needed.

Regression models predict a response variable from a function of risk factors and parameters. This approach is also referred to as supervised learning. By connecting the different risk profiles, a regression analysis can deal with highly segmented problems resulting from the massive amount of information about the policyholders that has now become available to the insurers.

1.1.4 Observable Versus Hidden Risk Factors

Some risk factors can easily be observed, such as the policyholder's age, gender, marital status or occupation, the type and use of the car, or the place of residence for instance. Other ones can be observed but subject to some effort or cost. This is typically the case with behavioral characteristics reflected in telematics data or information gathered in external databases that can be accessed by the insurer for a fee paid to the provider. But besides these observable factors, there always remain risk factors unknown to the insurer. In motor insurance for instance, these hidden risk factors typically include temper and skills, aggressiveness behind the wheel, respect of the highway code or swiftness of reflexes (even if telematics data now help insurers to figure out these behavioral traits, but only after contract inception).

Henceforth, we denote as $\boldsymbol{X}$ the random vector gathering the observable risk factors used by the insurer. Notice that those risk factors are not necessarily in causal relationship with the response Y. As a consequence, some components of $\boldsymbol{X}$ could become irrelevant if the hidden risk factors influencing the risk (in addition to $\boldsymbol{X}$) in causal relationship with the response Y, denoted $\boldsymbol{X}^+$, would be available.

1.1.5 Insurance Ratemaking Versus Loss Prediction

Consider a response Y and a set of features $X_1, \ldots, X_p$ gathered in the vector $\boldsymbol{X}$. Features are considered here as random variables so that they are denoted by capital letters. This means that we are working with a generic policyholder, taken at random from (and thus representative of) the portfolio under consideration. When it comes to pricing a specific contract, we work conditionally to the realized value of $\boldsymbol{X}$, that is, given $\boldsymbol{X} = \boldsymbol{x}$. The set of all possible features is called the feature space and is denoted $\succ$.

The dependence structure inside the random vector $(Y, X_1, \ldots, X_p)$ is exploited to extract the information contained in $\boldsymbol{X}$ about Y. In actuarial pricing, the aim is to evaluate the pure premium as accurately as possible. This means that the target is the conditional expectation $\mu(\boldsymbol{X}) = \mathrm{E}[Y|\boldsymbol{X}]$ of the response Y (claim number or claim amount) given the available information $\boldsymbol{X}$. Henceforth, $\mu(\boldsymbol{X})$ is referred to as the true (pure) premium.

Notice that the function $\boldsymbol{x} \mapsto \mu(\boldsymbol{x}) = \mathrm{E}[Y|\boldsymbol{X} = \boldsymbol{x}]$ is generally unknown to the actuary, and may exhibit a complex behavior in $\boldsymbol{x}$. This is why this function is approximated by a (working, or actual) premium $\boldsymbol{x} \mapsto \widehat{\mu}(\boldsymbol{x})$ with a relatively simple structure compared to the unknown regression function $\boldsymbol{x} \mapsto \mu(\boldsymbol{x})$.

1.2 Insurance Data

1.2.1 Claim Data

Because of adverse selection, most of actuarial studies are based on insurance-specific data, generally consisting in claims data. Dealing with claim data means that only limited information is available about events that actually occurred. Analyzing insurance data, the actuary draws conclusions about the number of claims filed by policyholders subject to a specific ratemaking mechanism (bonus-malus rules or deductibles, for instance), not about the actual number of accidents. The conclusions of the actuarial analysis are valid only if the existing rules are kept unchanged. The effect of an extension of coverage (decreasing the amount of deductibles, for instance) is extremely difficult to assess.

Also, some policyholders may not report their claims immediately, for various reasons (for instance because they were not aware of the occurrence of the insured event), impacting on the available data. Because of late reporting, the observed number of claims may be smaller than the actual number of claims for recent observation periods. Once reported, claims require some time to be settled. This is especially the case in tort systems, for liability claims. This means that it may take years before the final claim cost is known to the insurer.

The information recorded in the database generally gathers one or several calendar years. The data are as seen from the date of extraction (6 months after the end of the observation period, say). Hence, most of the "small" claims are settled and their final cost is known. However, for the large claims, actuaries can only work with incurred losses (payments made plus reserve, the latter representing a forecast of the final cost still to be paid according to the evaluation made by the claim manager). Incurred losses are routinely used in practice but a better approach would consist in recognizing that actuaries only have partial knowledge about the claim amount.

1.2.2 Frequency-Severity Decomposition

1.2.2.1 Claim Numbers

Even if the actuary wants to model the total claim amount Y generated by a policy of the portfolio over one period (typically, one year), this random variable is generally not the modeling target. Indeed, modeling Y does not allow to study the effect of per-claim deductibles nor bonus-malus rules, for instance. Rather, the total claim amount Y is decomposed into

$$Y = \sum_{k=1}^{N} C_k$$

where

$$\begin{aligned} N &= \text{ number of claims} \\ C_k &= \text{ cost (or severity) of the } k\text{th claim} \\ & \quad C_1, C_2, \ldots \text{ identically distributed} \end{aligned}$$

all these random variables being independent. By convention, the empty sum is zero, that is,

$$N = 0 \Rightarrow Y = 0.$$

The frequency component of Y refers to the number N of claims filed by each policyholder during one period. Considering the number of claims reported by a policyholder in Property and Casualty insurance, the Poisson model is often used as a starting point.

Generally, the different components of the yearly insurance losses Y are modeled separately. Costs may be of different magnitudes, depending on the type of the claim: standard, or attritional claims, with moderate costs versus large claims with much higher costs. If large claims may occur then the mix of these two types of claims is explicitly recognized by

$$C_k = \begin{cases} \text{large claim cost, with probability } p, \\ \text{attritional claim cost, with probability } 1 - p. \end{cases}$$

1.2.2.2 Claim Amounts

Having individual costs for each claim, the actuary often wishes to model their respective amounts (also called claim sizes or claim severities in the actuarial literature). Prior to the analysis, the actuary first needs to exclude possible large claims, keeping only the standard, or attritional ones.

Overall, the modeling of claim amounts is more difficult than claim frequencies. There are several reasons for that. First and foremost, claims sometimes need several years to be settled as explained before. Only estimates of the final cost appear in the insurer's records until the claim is closed. Moreover, the statistics available to fit a model for claim severities are much more scarce, since generally only 5–10% of the policies in the portfolio produced claims. Finally, the unexplained heterogeneity is sometimes more pronounced for costs than for frequencies. The cost of a traffic accident for instance is indeed for the most part beyond the control of a policyholder since the payments of the insurance company are determined by third-party characteristics. The degree of care exercised by a driver mostly influences the number of accidents, but in a much lesser way the cost of these accidents.

1.2.3 Observational Data

Statistical analyzes are conducted with data either from experimental or from observational studies. In the former case, random assignment of individual units (humans or animals, for instance) to the experimental treatments plays a fundamental role to draw conclusions about causal relationships (to demonstrate the usefulness of a new drug, for instance). This is however not the case with insurance data, which consist of observations recorded on past contracts issued by the insurer.

As an example, let us consider motor insurance. The policyholders covered by a given insurance company are generally not a random sample from the entire population of drivers in the country. Each company targets a specific segment of this population (with advertisement campaigns or specific product design, for instance) and attracts particular profiles. This may be due to consumers' perception of insurer's products, sales channels (brokers, agents or direct), not to mention the selection operated by the insurer, screening the applicants before accepting to cover their risks.

In insurance studies, we consider that the portfolio is representative of future policyholders, those who will stay insured by the company or later join the portfolio. The assumption that new policyholders conform with the profiles already in the portfolio needs to be carefully assessed as any change in coverage conditions or in competitors' price lists may attract new profiles with different risk levels (despite they are identical with respect to $\boldsymbol{X}$, they may differ in $\boldsymbol{X}^+$, due to adverse selection against the insurer).

The actuary has always to keep in mind the important difference existing between causal relationships and mere correlations existing among the risk factors and the number of claims or their severity. Such correlations may have been produced by a causal relationship, but could also result from confounding effects. Therefore, the actuary has always to keep in mind that it is generally not possible to disentangle

- a true effect of a risk factor
- from an apparent effect resulting from correlation with hidden characteristics

on the basis of observational data. Also, the effect estimated from portfolio statistics is the dominant one: different stories may apply to different policyholders whereas they are all averaged in the estimates obtained by the actuary.

Notice that correlation with hidden risk factors may even reverse the influence of an available risk factor on the response. This is the case for instance when the feature is negatively correlated with the response but positively correlated with a hidden characteristic, the latter being positively related to the response. The actuary may then observe a positive relationship between this feature and the response, despite the true correlation is negative.

1.2.4 Format of the Data

The data required to perform analyses carried out in this book generally consist of linked policy and claims information at the individual risk level. The appropriate definition of individual risk level varies according to the line of business and the type of study. For instance, an individual risk generally corresponds to a vehicle in motor insurance or to a building in fire insurance.

The database must contain one record for each period of time during which a policy was exposed to the risk of filling out a claim, and during which all risk factors remained unchanged. A new record must be created each time risk factors change, with the previous exposure curtailed at the point of amendment. The policy number then allows the actuary to track the experience of the individual risks over time. Policy cancellations and new business also result in the exposure period to be curtailed. For each record, the database registers policy characteristics together with the number of claims and the total incurred losses. In addition to this policy file, there is a claim file recording all the information about each claim, separately (the link between the two files being made using the policy number). This second file also contains specific features about each claim, such as the presence of bodily injuries, the number of victims, and so on. This second file is interesting to build predictive models for the cost of claims based on the information about the circumstances of each insured event. This allows the insurer to better assess incurred losses.

The information available to perform risk classification is summarized into a set of features x_{ij}, $j = 1, \ldots, p$, available for each policy i. These features may have different formats:

- categorical (such as gender, with two levels, male and female);
- integer-valued, or discrete (such as the number of vehicles in the household);
- continuous (such as policyholder's age).

Categorical covariates may be ordered (when the levels can be ordered in a meaningful way, such as education level) or not (when the levels cannot be ranked, think for instance to marital status, with levels single, married, cohabiting, divorced, or widow, say).

Notice that continuous features are generally available to a finite precision so that they are actually discrete variables with a large number of numerical values.

1.2.5 Data Quality Issues

As in most actuarial textbooks, we assume here that the available data are reliable and accurate. This assumption hides a time-consuming step in every actuarial study, during which data are gathered, checked for consistency, cleaned if needed and sometimes connected to external data bases to increase the volume of information. Setting up the database often takes the most time and does not look very rewarding. Data preparation is however of crucial importance because, as the saying goes, "garbage in, garbage out": there is no hope to get a reliable technical price list from a database suffering many limitations.

Once data have been gathered, it is important to spend enough time on exploratory data analysis. This part of the analysis aims at discovering which features seem to influence the response, as well as subsets of strongly correlated features. This traditional, seemingly old-fashioned view may well conflict with the modern data science approach, where practitioners are sometimes tempted to put all the features in a black-box model without taking the time to even know what they mean. But we firmly believe that such a blind strategy can sometimes lead to disastrous conclusions in insurance pricing so that we strongly advise to dedicate enough time to discover the kind of information recorded in the database under study.

1.3 Exponential Dispersion (ED) Distributions

1.3.1 Frequency and Severity Distributions

Regression models aim to analyze the relationship between a variable whose outcome needs to be predicted and one or more potential explanatory variables. The variable of interest is called the response and is denoted as Y. Insurance analysts typically encounter non-Normal responses such as the number of claims or the claim severities. Actuaries then often select the distribution of the response from the exponential dispersion (or ED) family.

Claim numbers are modeled by means of non-negative integer-valued random variables (often called counting random variables). Such random variables are described by their probability mass function: given a counting random variable Y valued in the set $\{0, 1, 2, \ldots\}$ of non-negative integers, its probability mass function p_Y is defined as

$$y \mapsto p_Y(y) = \mathrm{P}[Y = y], \quad y = 0, 1, 2, \ldots$$

and we set p_Y to zero otherwise. The support $\mathcal{S}$ of Y is defined as the set of all values y such that $p_Y(y) > 0$. Expectation and variance are then respectively given by

$$\mathrm{E}[Y] = \sum_{y=0}^{\infty} y p_Y(y) \text{ and } \mathrm{Var}[Y] = \sum_{y=0}^{\infty} \big(y - \mathrm{E}[Y]\big)^2 p_Y(y).$$

Claim amounts are modeled by non-negative continuous random variables possessing a probability density function. Precisely, the probability density function f_Y of such a random variable Y is defined as

$$y \mapsto f_Y(y) = \frac{\mathrm{d}}{\mathrm{d}y}\mathrm{P}[Y \leq y], \quad y \in (-\infty, \infty).$$

In this case,

$$\mathrm{P}[Y \approx y] = \mathrm{P}\left[y - \frac{\Delta}{2} \leq Y \leq y + \frac{\Delta}{2}\right] \approx f_Y(y)\Delta$$

for sufficiently small $\Delta > 0$, so that f_Y also indicates the region where Y is most likely to fall. In particular, $f_Y = 0$ where Y cannot assume its values. The support $\mathcal{S}$ of Y is then defined as the set of all values y such that $f_Y(y) > 0$. Expectation and variance are then respectively given by

$$\mathrm{E}[Y] = \int_{-\infty}^{\infty} y f_Y(y)\mathrm{d}y \text{ and } \mathrm{Var}[Y] = \int_{-\infty}^{\infty} \big(y - \mathrm{E}[Y]\big)^2 f_Y(y)\mathrm{d}y.$$

1.3.2 From Normal to ED Distributions

The oldest distribution for errors in a regression setting is certainly the Normal distribution, also called Gaussian, or Gauss–Laplace distribution after its inventors. The family of ED distributions in fact extends the nice structure of this probability law to more general errors.

1.3.2.1 Normal Distribution

Recall that a response Y valued in $\mathcal{S} = (-\infty, \infty)$ is Normally distributed with parameters $\mu \in (-\infty, \infty)$ and $\sigma^2 > 0$, denoted as $Y \sim \mathcal{N}or(\mu, \sigma^2)$, if its probability density function f_Y is

$$f_Y(y) = \frac{1}{\sigma\sqrt{2\pi}} \exp\left(-\frac{1}{2\sigma^2}(y-\mu)^2\right), \quad y \in (-\infty, \infty). \tag{1.3.1}$$

Considering (1.3.1), we see that Normally distributed responses can take any real value, positive or negative as $f_Y > 0$ over the whole real line $(-\infty, \infty)$.

Figure 1.1 displays the probability density function (1.3.1) for different parameter values. The $\mathcal{N}or(\mu, \sigma^2)$ probability density function appears to be a symmetric bell-shaped curve centered at μ, with σ^2 controlling the spread of the distribution. The probability density function f_Y being symmetric with respect to μ, positive or negative deviations from the mean μ have the same probability to occur. To be effective, any analysis based on the Normal distribution requires that the probability density function of the data has a shape similar to one of those visible in Fig. 1.1, which is rarely the case in insurance applications.

Notice that the Normal distribution enjoys the convenient convolution stability property, meaning that the sum of independent, Normally distributed random variables remain Normally distributed.

1.3.2.2 ED Distributions

The $\mathcal{N}or(\mu, \sigma^2)$ probability density function can be rewritten in order to be extended to a larger class of probability distributions sharing some convenient properties: the ED family. The idea is as follows. The parameter of interest in insurance pricing is the mean μ involved in pure premium calculations. This is why we isolate components of the Normal probability density function where μ appears. This is done by expanding the square appearing inside the exponential function in (1.3.1), which gives

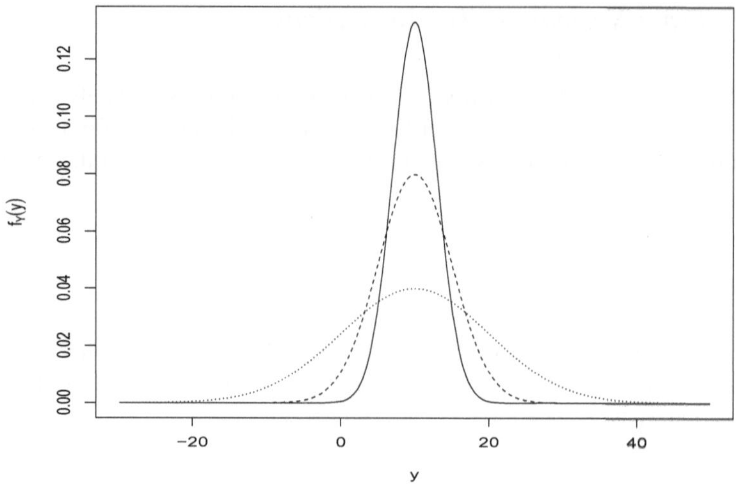

Fig. 1.1 Probability density functions of $\mathcal{N}or(10, 3^2)$ in continuous line, $\mathcal{N}or(10, 5^2)$ in broken line, and $\mathcal{N}or(10, 10^2)$ in dotted line

$$\begin{aligned} f_Y(y) &= \frac{1}{\sigma\sqrt{2\pi}} \exp\left(-\frac{1}{2\sigma^2}(y^2 - 2y\mu + \mu^2)\right) \\ &= \exp\left(\frac{y\mu - \frac{\mu^2}{2}}{\sigma^2}\right) \frac{\exp\left(-\frac{y^2}{2\sigma^2}\right)}{\sigma\sqrt{2\pi}}. \end{aligned} \tag{1.3.2}$$

The second factor appearing in (1.3.2) does not involve μ so that the important component is the first one. We see that it has a very simple form, being the exponential (hence the vocable "exponential" in ED) of a ratio with the variance σ^2, i.e. the dispersion parameter, appearing in the denominator. The numerator appears to be the difference between the product of the response y and the canonical Normal mean parameter μ with a function of μ, only. Notice that the derivative of this second term $\frac{\mu^2}{2}$ is just the mean μ. Such a decomposition allows us to define the whole ED class of distributions as follows.

Definition 1.3.1 Consider a response Y valued in a subset $\mathcal{S}$ of the real line $(-\infty, \infty)$. Its distribution is said to belong to the ED family if Y obeys a probability mass function p_Y or a probability density function f_Y of the form

$$\left.\begin{matrix} p_Y(y) \\ f_Y(y) \end{matrix}\right\} = \exp\left(\frac{y\theta - a(\theta)}{\phi/\nu}\right) c(y, \phi/\nu), \quad y \in \mathcal{S}, \tag{1.3.3}$$

where

$$\begin{aligned}
\theta &= \text{real-valued location parameter, called the canonical parameter}\\
\phi &= \text{positive scale parameter, called the dispersion parameter}\\
\nu &= \text{known positive constant, called the weight}\\
a(\cdot) &= \text{monotonic convex function of } \theta, \text{ called the cumulant function}\\
c(\cdot) &= \text{positive normalizing function.}
\end{aligned}$$

In the majority of actuarial applications, the weight corresponds to some volume measure, hence the notation ν.

The parameters θ and ϕ are essentially location and scale indicators, extending the mean value μ and variance σ^2 to the whole family of ED distributions. Considering (1.3.2), we see that it is indeed of the form (1.3.3) with

$$\begin{aligned}
\theta &= \mu\\
a(\theta) &= \frac{\mu^2}{2} = \frac{\theta^2}{2}\\
\phi &= \sigma^2\\
\nu &= 1\\
c(y, \phi) &= \frac{\exp\left(-\frac{y^2}{2\sigma^2}\right)}{\sigma\sqrt{2\pi}}.
\end{aligned}$$

Remark 1.3.2 Sometimes, (1.3.3) is replaced with the more general form

$$\exp\left(\frac{y\theta - a(\theta)}{b(\phi, \nu)}\right) c(y, \phi, \nu).$$

However, the particular case where ϕ and ν are combined into ϕ/ν, i.e.

$$b(\phi, \nu) = \frac{\phi}{\nu} \text{ and } c(y, \phi, \nu) = c\left(y, \frac{\phi}{\nu}\right)$$

appears to be enough for actuarial applications.

1.3.3 Some ED Distributions

The ED family is convenient for non-life insurance ratemaking. In particular, we show in the following that the Poisson and Gamma distributions, often used by actuaries for modeling claim counts and claim severities, belong to this family. A detailed review of the ED family can be found in Denuit et al. (2019). Thereafter, we only describe in details the Poisson and Gamma distributions.

1.3.3.1 Poisson Distribution

A Poisson-distributed response Y takes its values in $\mathcal{S} = \{0, 1, 2, \ldots\}$ and has probability mass function

$$p_Y(y) = \exp(-\lambda)\frac{\lambda^y}{y!}, \ y = 0, 1, 2, \ldots. \qquad (1.3.4)$$

Having a counting random variable Y, we denote as $Y \sim \mathcal{Poi}(\lambda)$ the fact that Y is Poisson distributed with parameter λ. The parameter λ is often called the rate, in relation to the Poisson process (see below).

The mean and variance of $Y \sim \mathcal{Poi}(\lambda)$ are given by

$$\mathrm{E}[Y] = \lambda \text{ and } \mathrm{Var}[Y] = \lambda. \qquad (1.3.5)$$

Considering (1.3.5), we see that both the mean and variance of the Poisson distribution are equal to λ, a phenomenon termed as equidispersion. The skewness coefficient of the Poisson distribution is

$$\gamma[Y] = \frac{1}{\sqrt{\lambda}}.$$

As λ increases, the Poisson distribution thus becomes more symmetric and is eventually well approximated by a Normal distribution, the approximation turning out to be quite good for $\lambda > 20$. But if $Y \sim \mathcal{Poi}(\lambda)$ then $\sqrt{Y}$ converges much faster to the $\mathcal{Nor}(\lambda, \frac{1}{4})$ distribution. Hence, the square root transformation was often recommended as a variance stabilizing transformation for count data at a time classical methods assuming Normality (and constant variance) were employed.

The shape of the Poisson probability mass function is displayed in the graphs of Fig. 1.2. For small values of λ, we see that the $\mathcal{Poi}(\lambda)$ probability mass function is highly asymmetric. When λ increases, it becomes more symmetric and ultimately looks like the Normal bell curve.

The Poisson distribution enjoys the convenient convolution stability property, i.e.

$$\left.\begin{array}{r} Y_1 \sim \mathcal{Poi}(\lambda_1) \\ Y_2 \sim \mathcal{Poi}(\lambda_2) \\ Y_1 \text{ and } Y_2 \text{ independent} \end{array}\right\} \Rightarrow Y_1 + Y_2 \sim \mathcal{Poi}(\lambda_1 + \lambda_2). \qquad (1.3.6)$$

This property is useful because sometimes the actuary has only access to aggregated data. Assuming that individual data is Poisson distributed, then so is the summed count and Poisson modeling still applies.

In order to establish that the Poisson distribution belongs to the ED family, let us write the $\mathcal{Poi}(\lambda)$ probability mass function (1.3.4) as follows:

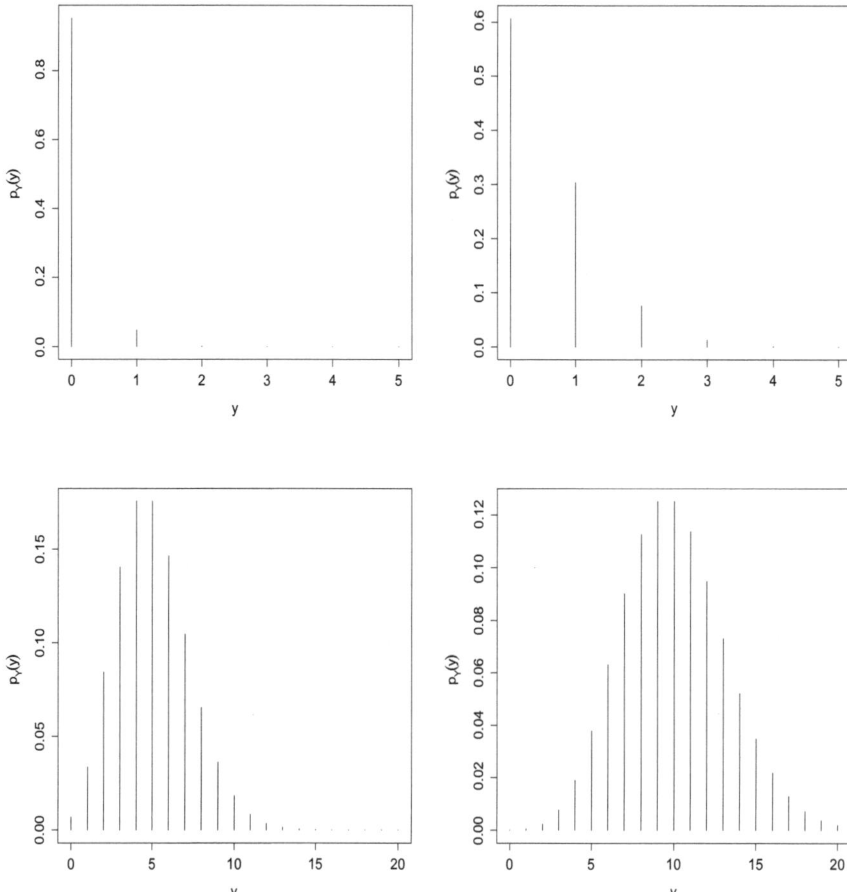

Fig. 1.2 Probability mass functions of $\mathcal{P}oi(\lambda)$ with $\lambda = 0.05, 0.5, 5, 10$ (from upper left to lower right)

$$
\begin{aligned}
p_Y(y) &= \exp(-\lambda)\frac{\lambda^y}{y!} \\
&= \exp\Big(y \ln \lambda - \lambda\Big)\frac{1}{y!}
\end{aligned}
$$

where we recognize the probability mass function (1.3.3) with

$$
\begin{aligned}
\theta &= \ln \lambda \\
a(\theta) &= \lambda = \exp(\theta) \\
\phi &= 1 \\
\nu &= 1 \\
c(y, \phi) &= \frac{1}{y!}.
\end{aligned}
$$

Thus, the Poisson distribution indeed belongs to the ED family.

1.3.3.2 Gamma Distribution

The Gamma distribution is right-skewed, with a sharp peak and a long tail to the right. These characteristics are often visible on empirical distributions of claim amounts. This makes the Gamma distribution a natural candidate for modeling accident benefits paid by the insurer.

Precisely, a random variable Y valued in $\mathcal{S} = (0, \infty)$ is distributed according to the Gamma distribution with parameters $\alpha > 0$ and $\tau > 0$, which will henceforth be denoted as $Y \sim \mathcal{G}am(\alpha, \tau)$, if its probability density function is given by

$$
f_Y(y) = \frac{y^{\alpha-1}\tau^{\alpha}\exp(-\tau y)}{\Gamma(\alpha)}, \quad y > 0, \tag{1.3.7}
$$

where

$$
\Gamma(\alpha) = \int_0^{\infty} x^{\alpha-1}e^{-x}dx.
$$

The parameter α is often called the shape of the Gamma distribution whereas τ is referred to as the scale parameter.

The mean and the variance of $Y \sim \mathcal{G}am(\alpha, \tau)$ are respectively given by

$$
\mathrm{E}[Y] = \frac{\alpha}{\tau} \text{ and } \mathrm{Var}[Y] = \frac{\alpha}{\tau^2} = \frac{1}{\alpha}\big(\mathrm{E}[Y]\big)^2. \tag{1.3.8}
$$

We thus see that the variance is a quadratic function of the mean. The Gamma distribution is useful for modeling a positive, continuous response when the variance grows with the mean but where the coefficient of variation

$$
\mathrm{CV}[Y] = \frac{\sqrt{\mathrm{Var}[Y]}}{\mathrm{E}[Y]} = \frac{1}{\sqrt{\alpha}}
$$

stays constant. As their names suggest, the scale parameter in the Gamma family influences the spread (and incidentally, the location) but not the shape of the distribution, while the shape parameter controls the skewness of the distribution. For $Y \sim \mathcal{G}am(\alpha, \tau)$, we have

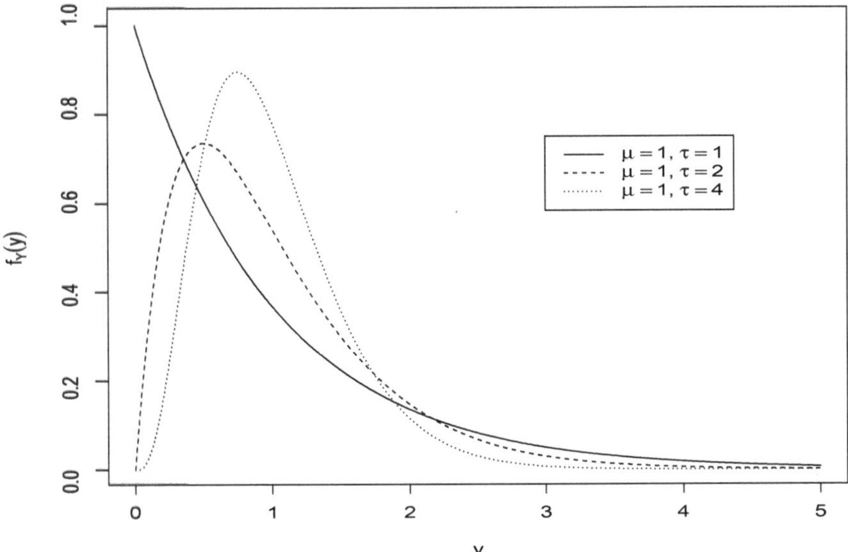

Fig. 1.3 Probability density functions of $\mathcal{G}am(\alpha, \tau)$ with $a = \tau \in \{1, 2, 4\}$

$$\gamma[Y] = \frac{2}{\sqrt{\alpha}}$$

so that the Gamma distribution is positively skewed. As the shape parameter gets larger, the distribution grows more symmetric.

Figure 1.3 displays Gamma probability density functions for different parameter values. Here, we fix the mean $\mu = \frac{\alpha}{\tau}$ to 1 and we take $\tau \in \{1, 2, 4\}$ so that the variance is equal to 1, 0.5, and 0.25. Unlike the Normal distribution (whose probability density function resembles a bell shape centered at μ whatever the variance σ^2), the shape of the Gamma probability density function changes with the parameter α. For $\alpha \leq 1$, the probability density function has a maximum at the origin whereas for $\alpha > 1$ it is unimodal but skewed. The skewness decreases as α increases.

Gamma distributions enjoy the convenient convolution stability property for fixed scale parameter τ. Specifically,

$$\left.\begin{array}{r} Y_1 \sim \mathcal{G}am(\alpha_1, \tau) \\ Y_2 \sim \mathcal{G}am(\alpha_2, \tau) \\ Y_1 \text{ and } Y_2 \text{ independent} \end{array}\right\} \Rightarrow Y_1 + Y_2 \sim \mathcal{G}am(\alpha_1 + \alpha_2, \tau). \tag{1.3.9}$$

When $\alpha = 1$, the Gamma distribution reduces to the Negative Exponential one with probability density function

$$f_Y(y) = \tau \exp(-\tau y), \quad y > 0.$$

In this case, we write $Y \sim \mathcal{E}xp(\tau)$. This distribution enjoys the remarkable memoryless property:

$$\begin{aligned} \mathrm{P}[Y > s + t | Y > s] &= \frac{\exp(-\tau(s+t))}{\exp(-\tau s)} \\ &= \exp(-\tau t) = \mathrm{P}[Y > t]. \end{aligned}$$

When α is a positive integer, the corresponding Gamma distribution function is then given by

$$F(y) = 1 - \sum_{j=0}^{\alpha-1} \exp(-y\tau)\frac{(y\tau)^j}{j!}, \quad y \geq 0.$$

This particular case is referred to as the Erlang distribution. The Erlang distribution corresponds to the distribution of a sum

$$Y = Z_1 + Z_2 + \ldots + Z_\alpha$$

of independent random variables $Z_1, Z_2, \ldots, Z_\alpha$ with common $\mathcal{E}xp(\tau)$ distribution, by virtue of (1.3.9). Hence, when $\alpha = 1$ the Erlang distribution reduces to the Negative Exponential one.

Let us now establish that the Gamma distribution belongs to the ED family. To this end, we let the mean parameter $\mu = \alpha/\tau$ enter the expression of the probability density function and rewrite the $\mathcal{G}am(\alpha, \tau)$ probability density function (1.3.7) as follows:

$$\begin{aligned} f_Y(y) &= \frac{\tau^\alpha}{\Gamma(\alpha)} y^{\alpha-1} \exp(-\tau y) \\ &= \frac{\alpha^\alpha}{\Gamma(\alpha)} \mu^{-\alpha} y^{\alpha-1} \exp\left(-y\frac{\alpha}{\mu}\right) \text{ with } \mu = \frac{\alpha}{\tau} \Leftrightarrow \tau = \frac{\alpha}{\mu} \\ &= \exp\left(\alpha\left(-\frac{y}{\mu} - \ln\mu\right)\right) \frac{\alpha^\alpha}{\Gamma(\alpha)} y^{\alpha-1} \end{aligned}$$

which is well of the form (1.3.3) with

$$\begin{aligned} \theta &= -\frac{1}{\mu} \\ a(\theta) &= \ln\mu = -\ln(-\theta) \\ \phi &= \frac{1}{\alpha} \\ c(y, \phi) &= \frac{\alpha^\alpha}{\Gamma(\alpha)} y^{\alpha-1}. \end{aligned}$$

The Gamma distribution thus belongs to the ED family.

Table 1.1 Examples of ED distributions (with $\nu = 1$)

Distribution	θ	$a(\theta)$	ϕ	$\mu = \mathrm{E}[Y]$	$\mathrm{Var}[Y]$
$\mathcal{Ber}(q)$	$\ln \frac{q}{1-q}$	$\ln(1+\exp(\theta))$	1	q	$\mu(1-\mu)$
$\mathcal{Bin}(m, q)$	$\ln \frac{q}{1-q}$	$m \ln(1+\exp(\theta))$	1	mq	$\mu\left(1-\frac{\mu}{m}\right)$
$\mathcal{Geo}(q)$	$\ln(1-q)$	$-\ln(1-\exp(\theta))$	1	$\frac{1-q}{q}$	$\mu(1+\mu)$
$\mathcal{Pas}(m, q)$	$\ln(1-q)$	$-m\ln(1-\exp(\theta))$	1	$m\frac{1-q}{q}$	$\mu\left(1+\frac{\mu}{m}\right)$
$\mathcal{Poi}(\mu)$	$\ln \mu$	$\exp(\theta)$	1	μ	μ
$\mathcal{Nor}(\mu, \sigma^2)$	μ	$\frac{\theta^2}{2}$	σ^2	μ	ϕ
$\mathcal{Exp}(\mu)$	$-\frac{1}{\mu}$	$-\ln(-\theta)$	1	μ	μ^2
$\mathcal{Gam}(\mu, \alpha)$	$-\frac{1}{\mu}$	$-\ln(-\theta)$	$\frac{1}{\alpha}$	μ	$\phi\mu^2$
$\mathcal{IGau}(\mu, \alpha)$	$-\frac{1}{2\mu^2}$	$-\sqrt{-2\theta}$	$\frac{1}{\alpha}$	μ	$\phi\mu^3$

1.3.3.3 Some Other ED Distributions

In addition to the Normal, Poisson and Gamma distributions, Table 1.1 gives an overview of some other ED distributions that appear to be useful in the analysis of insurance data, namely the Bernoulli distribution $\mathcal{Ber}(q)$ with parameter $q \in (0, 1)$; the Binomial distribution $\mathcal{Bin}(m, q)$ with parameters $m \in \mathbb{N}^+ = \{1, 2, \ldots\}$ and $q \in (0, 1)$; the Geometric distribution $\mathcal{Geo}(q)$ with parameter $q \in (0, 1)$; the Pascal distribution $\mathcal{Pas}(m, q)$ with parameters $m \in \mathbb{N}^+$ and $q \in (0, 1)$; the Negative Exponential distribution $\mathcal{Exp}(\mu)$ with parameter $\mu > 0$ and the Inverse Gaussian distribution $\mathcal{IGau}(\mu, \alpha)$ with parameters $\mu > 0$ and $\alpha > 0$. For each distribution, we list the canonical parameter θ, the cumulant function $a(\cdot)$, and the dispersion parameter ϕ entering the general definition (1.3.3) of ED probability mass or probability density function. We also give the two first moments.

1.3.4 *Mean and Variance*

In the Normal case, the derivative of $a(\theta) = \frac{\theta^2}{2}$ is $\theta = \mu$ so that we recover the mean response from a'. This turns out to be a property generally valid for all ED distributions.

Property 1.3.3 *If the response Y has probability density/mass function of the form* (1.3.3) *then*

$$E[Y] = a'(\theta).$$

The mean response Y then corresponds to the first derivative of the function $a(\cdot)$ involved in (1.3.3). The next result shows that the variance is proportional to the second derivative of $a(\cdot)$

Property 1.3.4 *If the response Y has probability density/mass function of the form* (1.3.3) *then*

$$Var[Y] = \frac{\phi}{\nu}a''(\theta).$$

Notice that increasing the weight thus decreases the variance whereas the variance increases linearly in the dispersion parameter ϕ. The impact of θ on the variance is given by the factor

$$a''(\theta) = \frac{\mathrm{d}}{\mathrm{d}\theta}\mu(\theta)$$

expressing how a change in the canonical parameter θ modifies the expected response. In the Normal case, $a''(\theta) = 1$ and the variance is just constantly equal to $\phi/\nu = \sigma^2/\nu$, not depending on θ. In this case, the mean response does not influence its variance. For the other members of the ED family, a'' is not constant and a change in θ modifies the variance.

An absence of relation between the mean and the variance is only possible for real-valued responses (such as Normally distributed ones, where the variance σ^2 does not depend on the mean μ). Indeed, if Y is non-negative (i.e. $Y \geq 0$) then intuitively the variance of Y tends to zero as the mean of Y tends to zero. That is, the variance is a function of the mean for non-negative responses. The relationship between the mean and variance of an ED distribution is indicated by the variance function $V(\cdot)$.

The variance function $V(\cdot)$ is formally defined as

$$V(\mu) = \frac{\mathrm{d}^2}{\mathrm{d}\theta^2}a(\theta) = \frac{\mathrm{d}}{\mathrm{d}\theta}\mu(\theta).$$

The variance function thus corresponds to the variation in the mean response $\mu(\theta)$ viewed as a function of the canonical parameter θ. In the Normal case, $\mu(\theta) = \theta$ and $V(\mu) = 1$. The other ED distributions have non-constant variance functions. Again, we see that the cumulant function $a(\cdot)$ determines the distributional properties in the ED family.

The variance of the response can thus be written as

$$\mathrm{Var}[Y] = \frac{\phi}{\nu}V(\mu).$$

It is important to keep in mind that the variance function is not the variance of the response, but the function of the mean entering this variance (to be multiplied by ϕ/ν). The variance function is regarded as a function of the mean μ, even if it appears as a function of θ; this is possible by inverting the relationship between θ and μ as we known from Property 1.3.3 that $\mu = \mathrm{E}[Y] = a'(\theta)$. The convexity of $a(\cdot)$ ensures

Table 1.2 Variance functions for some selected ED distributions

Distribution	Variance function $V(\mu)$
$\mathcal{B}er(q)$	$\mu(1-\mu)$
$\mathcal{G}eo(q)$	$\mu(1+\mu)$
$\mathcal{P}oi(\mu)$	μ
$\mathcal{N}or(\mu, \sigma^2)$	1
$\mathcal{G}am(\mu, \alpha)$	μ^2
$\mathcal{IG}au(\mu, \alpha)$	μ^3

that the mean function a' is increasing so that its inverse is well defined. Hence, we can express the canonical parameter in terms of the mean response μ by the relation

$$\theta = (a')^{-1}(\mu).$$

The variance functions corresponding to the usual ED distributions are listed in Table 1.2. In the Poisson case for instance,

$$\begin{aligned} V(\mu) &= \frac{d^2}{d\theta^2} \exp(\theta) \\ &= \exp(\theta) \\ &= \mu \end{aligned}$$

while in the Gamma case,

$$\begin{aligned} V(\mu) &= -\frac{d^2}{d\theta^2} \ln(-\theta) \\ &= -\frac{1}{\theta^2} \\ &= \mu^2. \end{aligned}$$

Notice that

$$V(\mu) = \mu^\xi \text{ with } \xi = \begin{cases} 0 \text{ for the Normal distribution} \\ 1 \text{ for the Poisson distribution} \\ 2 \text{ for the Gamma distribution} \\ 3 \text{ for the Inverse Gaussian distribution.} \end{cases}$$

These members of the ED family thus have power variance functions. The whole family of ED distributions with power variance functions is referred to as the Tweedie class.

1.3.5 Weights

Averaging independent and identically distributed ED responses does not modify their distribution, just the value of the weight.

Property 1.3.5 *If $Y_1, Y_2, \ldots, Y_n$ are independent with common probability mass/density function of the form* (1.3.3), *then their average*

$$\overline{Y} = \frac{1}{n}\sum_{i=1}^{n} Y_i$$

has probability mass/density function

$$\exp\left(\frac{y\theta - a(\theta)}{\phi/(n\nu)}\right) c(y, \phi/(n\nu)).$$

The distribution for $\overline{Y}$ is thus the same as for each Y_i except that the weight ν is replaced with $n\nu$.

Averaging observations is thus accounted for by modifying the weights in the ED family.

In actuarial studies, responses can be ratios with the aggregate exposure or even premiums in the denominator (in case loss ratios are analyzed). The numerator may correspond to individual data, or to grouped data aggregated over a set of homogeneous policies. This means that the size of the group has to be accounted for as the response ratios will tend to be far more volatile in low-volume cells than in high-volume ones. Actuaries generally consider that a large-volume cell is the result of summing smaller independent cells, leading to response variance proportional to the inverse of the volume measure. This implies that weights vary according to the business volume measure.

The next result extends Property 1.3.5 to weighted averages of ED responses.

Property 1.3.6 *Consider independent responses $Y_1, \ldots, Y_n$ obeying ED distributions* (1.3.3) *with common mean μ, dispersion parameter ϕ and specific weights ν_i. Define the total weight*

$$\nu_\bullet = \sum_{i=1}^{n} \nu_i.$$

Then, the weighted average

$$\frac{1}{\nu_\bullet}\sum_{i=1}^{n} \nu_i Y_i$$

still follows an ED distribution (1.3.3) *with mean μ, dispersion parameter ϕ and weight $\nu_\bullet$.*

1.3.6 Exposure-to-Risk

The number of observed events generally depends on a size variable that determines the number of opportunities for the event to occur. This size variable is often the time as the number of claims obviously depends on the length of the coverage period. However, some other choices are possible, such as distance traveled in motor insurance, for instance.

The Poisson process setting is useful when the actuary wants to analyze claims experience from policyholders who have been observed during periods of unequal lengths. Assume that the claims occur according to a Poisson process with rate λ. In this setting, claims occur randomly and independently in time. Denoting as $T_1, T_2, \ldots$ the times between two consecutive events, this means that these random variables are independent and obey the $\mathcal{E}xp(\lambda)$ distribution, the only one enjoying the memoryless property. Hence, the kth claim occurs at time

$$\sum_{j=1}^{k} T_j \sim \mathcal{G}am(k, \lambda)$$

where we recognize the Erlang distribution.

Consider a policyholder covered by the company for a period of length e, that is, the policyholder exposes the insurer to the risk of recording a claim during e time units. Then, the number Y of reported claims is such that

$$\mathrm{P}[Y \geq k] = \mathrm{P}\left[\sum_{j=1}^{k} T_j \leq e\right] = 1 - \sum_{j=0}^{k-1} \exp(-\lambda e)\frac{(\lambda e)^j}{j!}$$

so that it has probability mass function

$$\mathrm{P}[Y = k] = \mathrm{P}[Y \geq k] - \mathrm{P}[Y \geq k+1] = \exp(-\lambda e)\frac{(\lambda e)^k}{k!}, \quad k = 0, 1, \ldots,$$

that is, $Y \sim \mathcal{P}oi(\lambda e)$.

In actuarial studies, the length e of the observation period is generally referred to as the exposure-to-risk (hence the letter e). It allows the analyst to account for the fact that some policies are observed for the whole period whereas others just entered the portfolio or left it soon after the beginning of the observation period. We see that the exposure-to-risk e simply multiplies the annual expected claim frequency λ in the Poisson process model.

1.4 Maximum Likelihood Estimation

1.4.1 Likelihood-Based Statistical Inference

Assuming that the responses are random variables with unknown distribution depending on one (or several) parameter(s) θ, the actuary must draw conclusions about the unknown parameter θ based on available data. Such conclusions are thus subject to sampling errors: another dataset from the same population would have inevitably produced different results. To perform premium calculations, the actuary needs to select a value of θ hopefully close to the true parameter value. Such a value is called an estimate (or pointwise estimation) of the unknown parameter θ. The estimate is distinguished from the model parameters by a hat, which means that an estimate of θ is denoted by $\widehat{\theta}$. This distinction is necessary since it is generally impossible to estimate the true parameter θ without error. Thus, $\widehat{\theta} \neq \theta$ in general. The estimator is itself a random variable as it varies from sample to sample (in a repeated sampling setting, that is, drawing random samples from a given population and computing the estimated value, again and again). Formally, an estimator $\widehat{\theta}$ is a function of the observations $Y_1, Y_2, \ldots, Y_n$, that is,

$$\widehat{\theta} = \widehat{\theta}(Y_1, Y_2, \ldots, Y_n).$$

The corresponding estimate is $\widehat{\theta}(y_1, y_2, \ldots, y_n)$, computed from the realizations of the responses for a particular sample of observations $y_1, y_2, \ldots, y_n$.

In this Sect. 1.4, we do not deal with more complex data structures such as values of responses accompanied by possible explanatory variables, which amounts to consider the case where the features do not bring any information about the response. The fundamental assumption here is that the value of the response is random and thus considered as the realization of a random variable with unknown distribution.

It seems reasonable to require that a good estimate of the unknown parameter θ would be the value of the parameter that maximizes the chance of getting the data that have been recorded in the database (in which the actuary trusts). Maximum likelihood estimation is a method to determine the likely values of the parameters having produced the available responses in a given probability model. Broadly speaking, the parameter values are found such that they maximize the chances that the model produced the data that were actually observed.

1.4.2 Maximum-Likelihood Estimator

The parameter value that makes the observed $y_1, \ldots, y_n$ the most probable is called the maximum-likelihood estimate. Formally, the likelihood function $\mathcal{LF}(\theta)$ is defined as the joint probability mass/density function of the observations. In our case, for independent observations $Y_1, \ldots, Y_n$ obeying the same ED distribution (1.3.3) with unit weights, the likelihood function is given by

$$\mathcal{LF}(\theta) = \prod_{i=1}^{n} \exp\left(\frac{y_i\theta - a(\theta)}{\phi}\right) c(y_i, \phi).$$

The likelihood function $\mathcal{LF}(\theta)$ can be interpreted as the probability or chance of obtaining the actual observations $y_1, \ldots, y_n$ under the parameter θ. It is important to remember that the likelihood is always defined for a given set of observed values $y_1, \ldots, y_n$. In case repeated sampling properties are discussed, the numerical values y_i are replaced with unobserved random variables Y_i.

The maximum-likelihood estimator is the value $\widehat{\theta}$ which maximizes the likelihood function. Equivalently, $\widehat{\theta}$ maximizes the log-likelihood function

$$L(\theta) = \ln \mathcal{LF}(\theta)$$

which is given by

$$\begin{aligned} L(\theta) &= \sum_{i=1}^{n} \left(\frac{y_i\theta - a(\theta)}{\phi} + \ln c(y_i, \phi)\right) \\ &= \frac{n\big(\overline{y}\theta - a(\theta)\big)}{\phi} + \sum_{i=1}^{n} \ln c(y_i, \phi) \end{aligned}$$

where $\overline{y}$ is the sample mean, i.e. the arithmetic average

$$\overline{y} = \frac{1}{n}\sum_{i=1}^{n} y_i$$

of the available observations $y_1, y_2, \ldots, y_n$.

In risk classification, the parameter of interest is the mean μ of the response entering the pure premium calculation. Working with ED distributions, this means that the parameter of interest is θ (that is, the analyst in primarily interested in the mean value of the response), so that ϕ is called a nuisance parameter.

1.4.3 Derivation of the Maximum-Likelihood Estimate

The desired $\widehat{\theta}$ can easily be obtained by solving the likelihood equation

$$\frac{\mathrm{d}}{\mathrm{d}\theta} L(\theta) = 0.$$

This gives

$$
\begin{aligned}
0 &= \frac{\mathrm{d}}{\mathrm{d}\theta} L(\theta) \\
&= \sum_{i=1}^{n} \frac{\mathrm{d}}{\mathrm{d}\theta} \frac{n\big(\overline{y}\theta - a(\theta)\big)}{\phi} \\
&= \frac{n\big(\overline{y} - a'(\theta)\big)}{\phi}
\end{aligned}
$$

so that the maximum-likelihood estimate of θ is the unique root of the equation

$$
\overline{y} = a'(\theta) \Leftrightarrow \widehat{\theta} = (a')^{-1}(\overline{y}).
$$

The solution indeed corresponds to a maximum as the second derivative

$$
\frac{\mathrm{d}^2}{\mathrm{d}\theta^2} L(\theta) = -\frac{na''(\theta)}{\phi} = -\frac{n}{\phi^2}\mathrm{Var}[Y_1]
$$

is always negative. We see that the individual observations $y_1, \ldots, y_n$ are not needed to compute $\widehat{\theta}$ as long as the analyst knows $\overline{y}$ (which thus summarizes all the information contained in the observations about the canonical parameter). Also, we notice that the nuisance parameter ϕ does not show up in the estimation of θ.

1.4.4 Properties of the Maximum-Likelihood Estimators

Properties of estimators are generally inherited from an hypothetical repeated sampling procedure. Assume that we are allowed to draw repeatedly samples from a given population. Averaging the estimates $\widehat{\theta}_1, \widehat{\theta}_2, \ldots$ obtained for each of the samples corresponds to the mathematical expectation as

$$
\mathrm{E}[\widehat{\theta}] = \lim_{k \nearrow \infty} \frac{1}{k} \sum_{j=1}^{k} \widehat{\theta}_j
$$

by the law of large number. In this setting, $\mathrm{Var}[\widehat{\theta}]$ measures the stability of the estimates across these samples.

We now briefly discuss some relevant properties of the maximum-likelihood estimators.

1.4.4.1 Consistency

Let us denote as $\widehat{\theta}_n$ the maximum-likelihood estimator obtained with a sample of size n, that is, with n observations $Y_1, Y_2, \ldots, Y_n$. The maximum-likelihood estimator

$\widehat{\theta}_n = \widehat{\theta}(Y_1, Y_2, \ldots, Y_n)$ is consistent for the parameter θ, namely that

$$\lim_{n \nearrow \infty} \mathrm{P}[|\widehat{\theta}_n - \theta| < \epsilon] = 1 \text{ for all } \epsilon > 0.$$

In our case, in the ED family setting, consistency of the maximum-likelihood estimator is a direct consequence of the law of large numbers. Indeed, as $\overline{Y}$ is known to converge to μ with probability 1, it turns out that $\widehat{\theta} = (a')^{-1}(\overline{Y})$ converges to $(a')^{-1}(\mu) = \theta$.

Consistency is thus an asymptotic property which ensures that, as the sample size gets large, the maximum-likelihood estimator is increasingly likely to fall within a small region around the true value of the parameter. This expresses the idea that $\widehat{\theta}_n$ converges to the true value θ of the parameter as the sample size becomes infinitely large.

1.4.4.2 Invariance

If h is a one-to-one function then $h(\widehat{\theta})$ is the maximum-likelihood estimate for $h(\theta)$, that is,

$$\widehat{h(\theta)} = h(\widehat{\theta})$$

when maximum-likelihood is used for estimation. This ensures that for every distribution in the ED family, the maximum-likelihood estimate fulfills

$$\widehat{\mu} = \overline{y}. \tag{1.4.1}$$

The maximum likelihood estimator $\widehat{\mu}$ is thus unbiased for μ since

$$\mathrm{E}[\widehat{\mu}] = \mathrm{E}[\overline{Y}] = \frac{1}{n}\sum_{i=1}^{n} \mathrm{E}\left[Y_i\right] = \mu.$$

1.4.4.3 Asymptotic Unbiasedness

The maximum-likelihood estimator is asymptotically unbiased, that is,

$$\lim_{n \nearrow \infty} \mathrm{E}[\widehat{\theta}_n] = \theta.$$

The expectation $\mathrm{E}[\widehat{\theta}_n]$ approaches θ as the sample size increases.

1.4.4.4 Minimum Variance

In the class of all estimators, for large samples, the maximum-likelihood estimator $\widehat{\theta}$ has the minimum variance and is therefore the most accurate estimator possible.

We see that many attractive properties of the maximum-likelihood estimation principle hold in large samples. As actuaries generally deal with massive amounts of data, this makes this estimation procedure particularly attractive to conduct insurance studies.

1.4.5 Examples

1.4.5.1 Poisson Distribution

Assume that the responses $Y_1, \ldots, Y_n$ are independent and Poisson distributed with $Y_i \sim \mathcal{P}oi(\lambda e_i)$ for all $i = 1, \ldots, n$, where e_i is the exposure-to-risk for observation i and λ is the annual expected claim frequency that is common to all observations.

We thus have $\mu_i = \lambda e_i$ for all $i = 1, \ldots, n$ so that $\theta_i = \ln \mu_i = \ln e_i + \ln \lambda$ and $a(\theta_i) = \exp(\theta_i) = \lambda e_i$. In this setting, the log-likelihood function writes

$$
\begin{aligned}
L(\lambda) &= \sum_{i=1}^{n} \left(\frac{Y_i \theta_i - a(\theta_i)}{\phi} + \ln c(Y_i, \phi) \right) \\
&= \sum_{i=1}^{n} (Y_i \theta_i - \exp(\theta_i) + \ln c(Y_i, \phi)) \\
&= \sum_{i=1}^{n} (Y_i \ln e_i + Y_i \ln \lambda - \lambda e_i + \ln c(Y_i, \phi)) .
\end{aligned}
$$

Thus, the maximum likelihood estimator $\widehat{\lambda}$ for λ, solution of

$$
\frac{\mathrm{d}}{\mathrm{d}\lambda} L(\lambda) = 0
$$

is given by

$$
\widehat{\lambda} = \frac{\sum_{i=1}^{n} Y_i}{\sum_{i=1}^{n} e_i}.
$$

We have

$$
\mathrm{E}[\widehat{\lambda}] = \mathrm{E}\left[\frac{\sum_{i=1}^{n} Y_i}{\sum_{i=1}^{n} e_i} \right] = \frac{1}{\sum_{i=1}^{n} e_i} \sum_{i=1}^{n} \mathrm{E}\left[Y_i\right] = \lambda,
$$

so that this estimator is unbiased for λ. Furthermore, its variance is given by

$$\mathrm{Var}[\widehat{\lambda}] = \mathrm{Var}\left[\frac{\sum_{i=1}^n Y_i}{\sum_{i=1}^n e_i}\right] = \frac{1}{\left(\sum_{i=1}^n e_i\right)^2}\sum_{i=1}^n \mathrm{Var}\left[Y_i\right] = \frac{\lambda}{\sum_{i=1}^n e_i},$$

which converges to 0 as the denominator goes to infinity. In particular, since $\widehat{\lambda}$ is unbiased and

$$\lim_{n\nearrow\infty} \mathrm{Var}[\widehat{\lambda}] = 0,$$

the maximum likelihood estimator $\widehat{\lambda}$ is consistent for λ.

1.4.5.2 Gamma Distribution

Assume that the responses $Y_1, \ldots, Y_n$ are independent and Gamma distributed with $Y_i \sim \mathcal{G}am(\mu, \alpha\nu_i)$ for all $i = 1, \ldots, n$, where ν_i is the weight for observation i.

The log-likelihood function writes

$$\begin{aligned} L(\mu) &= \sum_{i=1}^n \left(\frac{Y_i\theta - a(\theta)}{\phi/\nu_i} + \ln c(Y_i, \phi/\nu_i)\right) \\ &= \sum_{i=1}^n \left(\frac{-\frac{Y_i}{\mu} - \ln\mu}{\phi/\nu_i} + \ln c(Y_i, \phi/\nu_i)\right). \end{aligned}$$

Taking the derivative of $L(\mu)$ with respect to μ gives

$$\frac{\mathrm{d}}{\mathrm{d}\mu}L(\mu) = \frac{1}{\phi\mu}\left(\frac{\sum_{i=1}^n \nu_i Y_i}{\mu} - \sum_{i=1}^n \nu_i\right).$$

Hence, the maximum likelihood estimator $\widehat{\mu}$ for μ, solution of

$$\frac{\mathrm{d}}{\mathrm{d}\mu}L(\mu) = 0,$$

is given by

$$\widehat{\mu} = \frac{\sum_{i=1}^n \nu_i Y_i}{\sum_{i=1}^n \nu_i}.$$

This estimator is obviously unbiased for μ and

$$\mathrm{Var}[\widehat{\mu}] = \frac{1}{(\sum_{i=1}^n \nu_i)^2}\sum_{i=1}^n \nu_i^2 \mathrm{Var}\left[Y_i\right] = \frac{\mu^2/\alpha}{\sum_{i=1}^n \nu_i},$$

so that $\mathrm{Var}[\widehat{\mu}]$ converges to 0 as $\sum_{i=1}^{n} \nu_i$ becomes infinitely large. The maximum likelihood estimator $\widehat{\mu}$ is then consistent for μ.

1.5 Deviance

In the absence of relation between features and the response, we fit a common mean μ to all observations, such that

$$\widehat{\mu}_i = \overline{y} \text{ for } i = 1, 2, \ldots, n,$$

as we have seen in Sect. 1.4. This model is called the null model and corresponds to the case where the features do not bring any information about the response. In the null model, the data are represented entirely as random variations around the common mean μ. If the null model applies then the data are homogeneous and there is no reason to charge different premium amounts to subgroups of policyholders.

The null model represents one extreme where the data are purely random. Another extreme is the full model which represents the data as being entirely systematic. The model estimate $\widetilde{\mu}_i$ is just the corresponding observation y_i, that is,

$$\widetilde{\mu}_i = y_i \text{ for } i = 1, 2, \ldots, n.$$

Thus, each fitted value is equal to the observation and the full model fits perfectly. However, this model does not extract any structure from the data, but merely repeats the available observations without condensing them.

The deviance, or residual deviance of a regression model $\widehat{\mu}$ is defined as

$$\begin{aligned} D(\widehat{\mu}) &= 2\phi\Big(L_{\text{full}} - L(\widehat{\mu})\Big) \\ &= 2\sum_{i=1}^{n} \nu_i \Big(y_i\big(\widetilde{\theta}_i - \widehat{\theta}_i\big) - a\big(\widetilde{\theta}_i\big) + a\big(\widehat{\theta}_i\big)\Big), \end{aligned} \tag{1.5.1}$$

where L_{full} is the log-likelihood of the full model based on the considered ED distribution and with $\widetilde{\theta}_i = (a')^{-1}(\widetilde{\mu}_i)$ and $\widehat{\theta}_i = (a')^{-1}(\widehat{\mu}_i)$, where $\widehat{\mu}_i = \widehat{\mu}(\boldsymbol{x}_i)$. As the full model gives the highest attainable log-likelihood with the ED distribution under consideration, the difference between the log-likelihood L_{full} of the full model and the log-likelihood $L(\widehat{\mu})$ of the regression model under interest is always positive. The deviance is a measure of distance between a model and the observed data defined by means of the saturated model. It quantifies the variations in the data that are not explained by the model under consideration. A too large value of $D(\widehat{\mu})$ indicates that the model $\widehat{\mu}$ under consideration does not satisfactorily fit the actual data. The larger the deviance, the larger the differences between the actual data and the fitted values. The deviance of the null model is called the null deviance.

Table 1.3 Deviance associated to regression models based on some members of the ED family of distributions

Distribution	Deviance
Binomial	$2\sum_{i=1}^{n}\left(y_i \ln \frac{y_i}{\widehat{\mu}_i} + (n_i - y_i)\ln \frac{n_i - y_i}{n_i - \widehat{\mu}_i}\right)$ where $\widehat{\mu}_i = n_i\widehat{q}_i$
Poisson	$2\sum_{i=1}^{n}\left(y_i \ln \frac{y_i}{\widehat{\mu}_i} - (y_i - \widehat{\mu}_i)\right)$ where $y \ln y = 0$ if $y = 0$
Normal	$\sum_{i=1}^{n}\left(y_i - \widehat{\mu}_i\right)^2$
Gamma	$2\sum_{i=1}^{n}\left(-\ln \frac{y_i}{\widehat{\mu}_i} + \frac{y_i - \widehat{\mu}_i}{\widehat{\mu}_i}\right)$
Inverse Gaussian	$\sum_{i=1}^{n}\frac{\left(y_i - \widehat{\mu}_i\right)^2}{\widehat{\mu}_i^2 y_i}$

Table 1.3 displays the deviance associated to regression models based on some members of the ED family of distributions. Notice that in that table, $y \ln y$ is taken to be 0 when $y = 0$ (its limit as $y \to 0$).

1.6 Actuarial Pricing and Tree-Based Methods

In actuarial pricing, the aim is to evaluate the pure premium as accurately as possible. The target is the conditional expectation $\mu(\boldsymbol{X}) = \mathrm{E}[Y|\boldsymbol{X}]$ of the response Y (claim number or claim amount for instance) given the available information $\boldsymbol{X}$. The function $\boldsymbol{x} \mapsto \mu(\boldsymbol{x}) = \mathrm{E}[Y|\boldsymbol{X} = \boldsymbol{x}]$ is generally unknown to the actuary and is approximated by a working premium $\boldsymbol{x} \mapsto \widehat{\mu}(\boldsymbol{x})$. The goal is to produce the most accurate function $\widehat{\mu}(\boldsymbol{x})$.

Lack of accuracy for $\widehat{\mu}(\boldsymbol{x})$ is defined by the generalization error

$$Err(\widehat{\mu}) = \mathrm{E}\left[L(Y, \widehat{\mu}(\boldsymbol{X}))\right],$$

where $L(.,.)$ is a function measuring the discrepancy between its two arguments, called loss function, and the expected value is over the joint distribution of $(Y, \boldsymbol{X})$. We aim to find a function $\widehat{\mu}(\boldsymbol{x})$ of the features minimizing the generalization error. Notice that the loss function $L(.,.)$ should not be confused with the log-likelihood $L(.)$.

Let

$$\mathcal{L} = \{(y_1, \boldsymbol{x}_1), (y_2, \boldsymbol{x}_2), \ldots, (y_n, \boldsymbol{x}_n)\} \tag{1.6.1}$$

be the set of observations available to the insurer, called learning set. The learning set is often partitioned into a training set

$$\mathcal{D} = \{(y_i, \boldsymbol{x}_i); i \in \mathcal{I}\}$$

and a validation set

$$\overline{\mathcal{D}} = \{(y_i, \boldsymbol{x}_i); i \in \overline{\mathcal{I}}\},$$

with $\mathcal{I} \subset \{1, \ldots, n\}$ labelling the observations in $\mathcal{D}$ and $\overline{\mathcal{I}} = \{1, \ldots, n\} \backslash \mathcal{I}$ labelling the remaining observations of $\mathcal{L}$. The training set is then used to fit the model and the validation set to assess the predictive accuracy of the model.

An approximation $\widehat{\mu}(\boldsymbol{x})$ to $\mu(\boldsymbol{x})$ is obtained by applying a training procedure to the training set. Common training procedures are to restrict $\mu(\boldsymbol{x})$ to be a member of a parametrized class of functions. More specifically, a supervised training procedure assumes that μ belongs to a family of candidate models of restricted structure. Model selection is then defined as finding the best model within this family on the basis of the training set. Of course, the true model is usually not a member of the family under consideration, depending on the restricted structure imposed to the candidate models. However, when these restrictions are flexible enough, some candidate models could be sufficiently close to the true model.

In the GLM setting discussed in Denuit et al. (2019), $\mu(\boldsymbol{x})$ is assumed to be of the form

$$\mu(\boldsymbol{x}) = g^{-1}\left(\text{score}(\boldsymbol{x})\right), \tag{1.6.2}$$

where g is the link function and

$$\text{score}(\boldsymbol{x}) = \beta_0 + \sum_{j=1}^{p} \beta_j x_j. \tag{1.6.3}$$

Estimates $\widehat{\beta}_0, \widehat{\beta}_1, \ldots, \widehat{\beta}_p$ for parameters $\beta_0, \beta_1, \ldots, \beta_p$ are then obtained from the training set by maximum likelihood, so that we get

$$\begin{aligned}\widehat{\mu}(\boldsymbol{x}) &= g^{-1}\left(\widehat{\text{score}}(\boldsymbol{x})\right) \\ &= g^{-1}\left(\widehat{\beta}_0 + \sum_{j=1}^{p} \widehat{\beta}_j x_j\right).\end{aligned} \tag{1.6.4}$$

It is interesting to notice that the maximum likelihood estimates $\widehat{\beta}_0, \widehat{\beta}_1, \ldots, \widehat{\beta}_p$ also minimize the deviance

$$D^{\text{train}}(\widehat{\mu}) = 2\sum_{i \in \mathcal{I}} \nu_i \Big[y_i\big((a')^{-1}(y_i) - (a')^{-1}(\widehat{\mu}_i)\big) - a\big((a')^{-1}(y_i)\big) + a\big((a')^{-1}(\widehat{\mu}_i)\big) \Big] \tag{1.6.5}$$

computed on the training set, also called in-sample deviance. Note that in this book, both notations $D(\widehat{\mu})$ and $D^{\text{train}}(\widehat{\mu})$ are used to designate the in-sample deviance. The appropriate choice for the loss function in the GLM framework is thus given by

$$L(y_i, \widehat{\mu}_i) = 2\nu_i \Big[y_i \big((a')^{-1}(y_i) - (a')^{-1}(\widehat{\mu}_i) \big) - a\big((a')^{-1}(y_i) \big) + a\big((a')^{-1}(\widehat{\mu}_i) \big) \Big], \tag{1.6.6}$$

which makes the training sample estimate of the generalization error

$$\widehat{Err}^{\text{train}}(\widehat{\mu}) = \frac{1}{|\mathcal{I}|} \sum_{i \in \mathcal{I}} L(y_i, \widehat{\mu}_i)$$

correspond to the in-sample deviance (up to the factor $\frac{1}{|\mathcal{I}|}$), that is,

$$\widehat{Err}^{\text{train}}(\widehat{\mu}) = \frac{D^{\text{train}}(\widehat{\mu})}{|\mathcal{I}|}, \tag{1.6.7}$$

where $|\mathcal{I}|$ denotes the number of elements of $\mathcal{I}$. In the following, we extend the choice (1.6.6) for the loss function to the tree-based methods studied in this book. The GLM training procedure thus amounts to estimate scores structured like in (1.6.4) by minimizing the corresponding in-sample deviances (1.6.7) and to select the best model among the GLMs under consideration. Note that selecting the best model among different GLMs on the basis of the deviance computed on the training set will favor the most complex models. In that goal, the Akaike Information Criteria (AIC) is preferred over the in-sample deviance, because it accounts for a measure of model complexity in the penalty. As mentioned in Denuit et al. (2019), comparing different GLMs on the basis of AIC amounts to account for the optimism in the deviance computed on the training set.

In this second volume, we work with models whose scores are linear combinations of regression trees, that is,

$$g(\mu(\boldsymbol{x})) = \text{score}(\boldsymbol{x}) = \sum_{m=1}^{M} \beta_m T_m(\boldsymbol{x}),$$

where M is the number of trees producing the score and $T_m(\boldsymbol{x})$ is the prediction obtained from regression tree $T_m, m = 1, \ldots, M$. The parameters $\beta_1, \ldots, \beta_M$ specify the linear combination used to produce the score. The training procedures studied in this book differ in the way the regression trees are fitted from the training set and in the linear combination used to produce the ensemble.

In Chap. 3, we consider single regression trees for the score. Specifically, we work with $M = 1$ and $\beta_1 = 1$. In this setting, the identity link function is appropriate and implicitly chosen, so that we assume

$$\mu(\boldsymbol{x}) = \text{score}(\boldsymbol{x}) = T_1(\boldsymbol{x}).$$

The estimated score will be in the range of the response, the prediction in a terminal node of the tree will be computed as the (weighted) average of the responses in that node. Note that, contrarily to GLMs for instance for which the form of the score is strongly constrained, the score is here left unspecified and estimated from the training set. Theoretically, any function (here the true model) can be approximated as close as possible by a piecewise constant function (here a regression tree). However, in practice, the training procedure limits the level of accuracy of a regression tree. A large tree might overfit the training set, while a small tree might not capture the important structure of the true model. The size of the tree is thus an important parameter in the training procedure because it controls the score's complexity. Selecting the optimal size of the tree will also be part of the training procedure.

Remark 1.6.1 Because of the high flexibility of the score, a large regression tree is prone to overfit the training set. However, it turns out that the training sample estimate of the generalization error will favor larger trees. To combat this issue, a part of the observations of the training set can only be used to fit trees of different size and the remaining part of the observations of the training set to estimate the generalization error of these trees in order to select the best one. Using the K-fold cross validation estimate of the generalization error, as discussed in the next chapter, is also an alternative to avoid this issue.

In Chap. 4, we work with regression models assumed to be the average of M regression trees, each built on a bootstrap sample of the training set. The goal is to reduce the variance of the predictions obtained from a single tree (slight changes in the training set can drastically change the structure of the tree fitted on it). That is, we suppose $\beta_1 = \beta_2 = \ldots = \beta_M = \frac{1}{M}$, so that we consider models of the form

$$\mu(\boldsymbol{x}) = \text{score}(\boldsymbol{x}) = \frac{1}{M} \sum_{m=1}^{M} T_m(\boldsymbol{x}),$$

where $T_1, \ldots, T_M$ are regression trees that will be estimated on different bootstrap samples of the training set. This training procedure is called bagging trees. The number of trees M and the size of the trees will also be determined during the training procedure. A modification of bagging trees, called random forests, is also studied in Chap. 4. The latter training procedure differs from bagging trees in the way the trees are produced from the bootstrap samples of the training set. Note that in Chap. 4, we use the notation B for the number of trees instead of M, B referring to the number of bootstrap samples produced from the training set.

In Chap. 5, we focus on scores structured as $\text{score}(\boldsymbol{x}) = \sum_{m=1}^{M} T_m(\boldsymbol{x})$, where $T_1, \ldots, T_M$ are relatively small regression trees, called weak learners. Specifically, we suppose $\beta_1 = \ldots = \beta_M = 1$, and the regression trees in the expansion will be sequentially fitted on random subsamples of the training set. This training procedure is called boosting trees. Contrarily to bagging trees and random forests, boosting trees is an iterative training procedure, meaning that the production of the mth regression tree in the ensemble will depend on the previous $m - 1$ trees already fitted. Note that, because of the iterative way the constituent trees are produced, the identity link function may this time lead to predictions that are not in the range of the expected response. Hence, in Chap. 5, we suppose that

$$g(\mu(\boldsymbol{x})) = \text{score}(\boldsymbol{x}) = \sum_{m=1}^{M} T_m(\boldsymbol{x}),$$

where g is an appropriate link function mapping the score to the range of the response and $T_1, \ldots, T_M$ are relatively small regression trees that will be sequentially fitted on random subsamples of the training set. Note that the number of trees M and the size of the trees will also be selected during the training procedure.

1.7 Bibliographic Notes and Further Reading

This chapter closely follows the book of Denuit et al. (2019). Precisely, we summarize the first three chapters of Denuit et al. (2019) with the notions useful for this second volume. We refer the reader to the first three chapters by Denuit et al. (2019) for more details, as well as for an extensive overview of the literature. Section 1.6 gives an overview of the methods used throughout this book. We refer the reader to the bibliographic notes of the next chapters for more details about the corresponding literature.

References

Denuit M, Hainaut D, Trufin J (2019) Effective statistical learning methods for actuaries I: GLMs and extensions. Springer actuarial lecture notes

Friedman J (2001) Greedy function approximation: a gradient boosting machine. Ann Stat 29(5):1189–1232

Chapter 2
Performance Evaluation

2.1 Introduction

In actuarial pricing, the objective is to evaluate the pure premium as accurately as possible. The target is thus the conditional expectation $\mu(\boldsymbol{X}) = \mathrm{E}[Y|\boldsymbol{X}]$ of the response Y (claim number or claim amount for instance) given the available information $\boldsymbol{X}$.

The function $\boldsymbol{x} \mapsto \mu(\boldsymbol{x}) = \mathrm{E}[Y|\boldsymbol{X} = \boldsymbol{x}]$ is generally unknown to the actuary and is approximated by a working premium $\boldsymbol{x} \mapsto \widehat{\mu}(\boldsymbol{x})$. The goal is to produce the most accurate function $\widehat{\mu}(\boldsymbol{x})$. Lack of accuracy for $\widehat{\mu}(\boldsymbol{x})$ is defined by the generalization error. Producing a model $\widehat{\mu}$ whose predictions are as good as possible can be stated as finding a model which minimizes its generalization error. In this chapter, we describe the generalization error used throughout this book for model selection and model assessment.

2.2 Generalization Error

2.2.1 Definition

We denote by

$$\mathcal{L} = \{(y_1, \boldsymbol{x}_1), (y_2, \boldsymbol{x}_2), \ldots, (y_n, \boldsymbol{x}_n)\} \tag{2.2.1}$$

the set of observations available to the insurer. This dataset is called the learning set. We aim to find a model $\widehat{\mu}$ built on the learning set $\mathcal{L}$ (or only on a part of $\mathcal{L}$, called training set, as discussed thereafter) which approximates the best the true model μ.

Lack of accuracy for $\widehat{\mu}$ is defined by the generalization error. The generalization error, also known as expected prediction error, of $\widehat{\mu}$ is defined as follows:

M. Denuit et al., *Effective Statistical Learning Methods for Actuaries II*,
Springer Actuarial, https://doi.org/10.1007/978-3-030-57556-4_2

Definition 2.2.1 The generalization error of the model $\widehat{\mu}$ is

$$Err(\widehat{\mu}) = \mathrm{E}\left[L(Y, \widehat{\mu}(\boldsymbol{X}))\right], \tag{2.2.2}$$

where $L(.,.)$ is a function measuring the discrepancy between its two arguments, called loss function.

The goal is thus to find a function of the covariates which predicts at best the response, that is, which minimizes the generalization error. The model performance is evaluated according to the generalization error which depends on a predefined loss function. The choice of an appropriate loss function in our ED family setting is discussed thereafter.

Notice that the expectation in (2.2.2) is taken over all possible data, that is, with respect to the probability distribution of the random vector $(Y, \boldsymbol{X})$ assumed to be independent of the data used to fit the model.

2.2.2 *Loss Function*

A simple estimate of the generalization error is given by

$$\widehat{Err}(\widehat{\mu}) = \frac{1}{n} \sum_{i=1}^{n} L(y_i, \widehat{\mu}_i) \tag{2.2.3}$$

with $\widehat{\mu}_i = \widehat{\mu}(\boldsymbol{x}_i)$. In the ED family setting, the appropriate choice for the loss function is related to the deviance. It suffices to observe that the regression model $\widehat{\mu}$ maximizing the log-likelihood function $L(\widehat{\mu})$ also minimizes the corresponding deviance $D(\widehat{\mu})$. Specifically, since the deviance $D(\widehat{\mu})$ can be expressed as

$$D(\widehat{\mu}) = 2 \sum_{i=1}^{n} \nu_i \Big[y_i \big((a')^{-1}(y_i) - (a')^{-1}(\widehat{\mu}_i) \big) - a\big((a')^{-1}(y_i) \big) + a\big((a')^{-1}(\widehat{\mu}_i) \big) \Big], \tag{2.2.4}$$

we see from (2.2.3) that the appropriate loss function in our ED family setting is given by

$$L(y_i, \widehat{\mu}_i) = 2\nu_i \Big[y_i \big((a')^{-1}(y_i) - (a')^{-1}(\widehat{\mu}_i) \big) - a\big((a')^{-1}(y_i) \big) + a\big((a')^{-1}(\widehat{\mu}_i) \big) \Big]. \tag{2.2.5}$$

The constant 2 is not necessary and is there to make the loss function match the deviance. Throughout this book, we use the choice (2.2.5) for the loss function.

2.2.3 Estimates

The performance of a model is evaluated throughout the generalization error $Err(\widehat{\mu})$. In practice, we usually do not know the probability distribution from which the observations are drawn, making the direct evaluation of the generalization error $Err(\widehat{\mu})$ not feasible. Hence, the set of observations available to the insurer often constitutes the only data on which the model needs to be fitted and its generalization error estimated.

2.2.3.1 Training Sample Estimate

The learning set

$$\mathcal{L} = \{(y_1, \boldsymbol{x}_1), (y_2, \boldsymbol{x}_2), \ldots, (y_n, \boldsymbol{x}_n)\} \tag{2.2.6}$$

constitutes the only data available to the insurer. When the whole learning set is used to fit the model $\widehat{\mu}$, the generalization error $Err(\widehat{\mu})$ can only be estimated on the same data as the ones used to build the model, that is,

$$\widehat{Err}^{\text{train}}(\widehat{\mu}) = \frac{1}{n}\sum_{i=1}^{n} L(y_i, \widehat{\mu}(\boldsymbol{x}_i)). \tag{2.2.7}$$

This estimate is called the training sample estimate and has been introduced in (2.2.3). In our setting, we thus have

$$\widehat{Err}^{\text{train}}(\widehat{\mu}) = \frac{D(\widehat{\mu})}{n}. \tag{2.2.8}$$

Typically, the training sample estimate (2.2.7) will be less that the true generalization error, because the same data is being used to fit the model and assess its error. A model typically adapts to the data used to train it, and hence the training sample estimate will be an overly optimistic estimate of the generalization error. This is particularly true for tree-based models because of their high flexibility to adapt to the training set: the resulting models are too closely fitted to the training set, which is called overfitting.

2.2.3.2 Validation Sample Estimate

The training sample estimate (2.2.7) directly evaluates the accuracy of the model on the dataset used to build the model. While the training sample estimate is useful to fit the model, the resulting estimate for the generalization error is likely to be very optimistic since the model is precisely built to reduce it. This is of course an issue when we aim to assess the predictive performance of the model, namely its accuracy on new data.

As actuaries generally deal with massive amounts of data, a better approach is to divide the learning set $\mathcal{L}$ into two disjoint sets $\mathcal{D}$ and $\overline{\mathcal{D}}$, called training set and validation set, and to use the training set for fitting the model and the validation set for estimating the generalization error of the model. The learning set is thus partitioned into a training set

$$\mathcal{D} = \{(y_i, \boldsymbol{x}_i); i \in \mathcal{I}\}$$

and a validation set

$$\overline{\mathcal{D}} = \{(y_i, \boldsymbol{x}_i); i \in \overline{\mathcal{I}}\},$$

with $\mathcal{I} \subset \{1, \ldots, n\}$ labelling the observations in $\mathcal{D}$ considered for fitting the model and $\overline{\mathcal{I}} = \{1, \ldots, n\} \backslash \mathcal{I}$ labelling the remaining observations of $\mathcal{L}$ used to assess the predictive accuracy of the model. The validation sample estimate of the generalization error of the model $\widehat{\mu}$ that has been built on the training set $\mathcal{D}$ is then given by

$$\begin{aligned}\widehat{Err}^{\text{val}}(\widehat{\mu}) &= \frac{1}{|\overline{\mathcal{I}}|} \sum_{i \in \overline{\mathcal{I}}} L(y_i, \widehat{\mu}(\boldsymbol{x}_i)) \\ &= \frac{D^{\text{val}}(\widehat{\mu})}{|\overline{\mathcal{I}}|}\end{aligned} \tag{2.2.9}$$

while the training sample estimate (2.2.7) now writes

$$\begin{aligned}\widehat{Err}^{\text{train}}(\widehat{\mu}) &= \frac{1}{|\mathcal{I}|} \sum_{i \in \mathcal{I}} L(y_i, \widehat{\mu}(\boldsymbol{x}_i)) \\ &= \frac{D^{\text{train}}(\widehat{\mu})}{|\mathcal{I}|},\end{aligned} \tag{2.2.10}$$

where we denote by $D^{\text{train}}(\widehat{\mu})$ the deviance computed from the observations of the training set (also called in-sample deviance) and $D^{\text{val}}(\widehat{\mu})$ the deviance computed from the observations of the validation set (also called out-of-sample deviance). As a rule-of-thumb, the training set usually represents 80% of the learning set and the validation set the remaining 20%. Of course, this allocation depends on the problem under consideration. In any case, the splitting of the learning set must be done in a way that observations in the training set can be considered independent from those in the validation set and drawn from the same population. Usually, this is guaranteed by drawing both sets at random from the learning set.

Training and validation sets should be as homogeneous as possible. Creating those two sets by taking simple random samples, as mentioned above, is usually sufficient to guarantee similar data sets. However, in some cases, the distribution of the response can be quite different between the training and validation sets. For instance, consider the annual number of claims in MTPL insurance. Typically, the vast majority of the policyholders makes no claim over the year (say 95%). Some policyholders experience one claim (say 4%) while only a few of them have more than one claim (say 1% with two claims). In such a situation, because the proportions of policyholders with one or two claims are small compared to the proportion of policyholders with no claim, the distribution of the response can be very different between the training and validation sets.

To address this potential issue, random sampling can be applied within subgroups, a subgroup being a set of observations with the same response. In our example, we would thus have three subgroups: a first one made of the observations with no claim (95% of the observations), a second one composed of observations with one claim (4% of the observations) and a third one with remaining observations (1% of the observations). Applying the randomization within these subgroups is called stratified random sampling.

2.2.3.3 K-Fold Cross Validation Estimate

The validation sample estimate $\widehat{Err}^{\text{val}}(\widehat{\mu})$ is an unbiased estimate of $Err(\widehat{\mu})$. It is obtained from a dataset independent of the training set used to fit the model. As such, it constitutes a more reliable estimate of the generalization error to assess the predictive performance of the model.

However, using a validation set reduces the size of the set used to build the model. In actuarial science, it is generally not an issue as we are often in data-rich situations. But sometimes, when the size of the learning set is too small, the K-fold cross validation estimate is preferred over the validation sample estimate.

The K-fold cross validation works as follows. We randomly partition the learning set $\mathcal{L}$ into K roughly equal-size disjoint subsets $\mathcal{L}_1, \ldots, \mathcal{L}_K$ and we label by $\mathcal{I}_k \subset \{1, \ldots, n\}$ the observations in $\mathcal{L}_k$ for all $k = 1, \ldots, K$. For each subset $\mathcal{L}_k$, $k = 1, \ldots, K$, we fit the model on the set of observations $\mathcal{L}\backslash\mathcal{L}_k$, that we denote $\widehat{\mu}_{\mathcal{L}\backslash\mathcal{L}_k}$, and we estimate its generalization error on the remaining data $\mathcal{L}_k$ as

$$\widehat{Err}^{\text{val}}(\widehat{\mu}_{\mathcal{L}\backslash\mathcal{L}_k}) = \frac{1}{|\mathcal{I}_k|} \sum_{i \in \mathcal{I}_k} L(y_i, \widehat{\mu}_{\mathcal{L}\backslash\mathcal{L}_k}(\boldsymbol{x}_i)). \tag{2.2.11}$$

For the model $\widehat{\mu}_{\mathcal{L}\backslash\mathcal{L}_k}$, the set of observations $\mathcal{L}\backslash\mathcal{L}_k$ plays the role of training set while $\mathcal{L}_k$ of validation set. The generalization error can then be estimated as the weighted average of the estimates $\widehat{Err}^{\text{val}}(\widehat{\mu}_{\mathcal{L}\backslash\mathcal{L}_k})$ given in (2.2.11), that is,

$$\begin{aligned} \widehat{Err}^{\text{CV}}(\widehat{\mu}) &= \sum_{k=1}^{K} \frac{|\mathcal{I}_k|}{n} \widehat{Err}^{\text{val}}(\widehat{\mu}_{\mathcal{L}\backslash\mathcal{L}_k}) \\ &= \frac{1}{n} \sum_{k=1}^{K} \sum_{i \in \mathcal{I}_k} L(y_i, \widehat{\mu}_{\mathcal{L}\backslash\mathcal{L}_k}(\boldsymbol{x}_i)). \end{aligned} \tag{2.2.12}$$

The idea behind the K-fold cross validation estimate $\widehat{Err}^{CV}(\widehat{\mu})$ is that each model $\widehat{\mu}_{\mathcal{L}\setminus\mathcal{L}_k}$ should be close enough to the model $\widehat{\mu}$ fitted on the whole learning set $\mathcal{L}$. Therefore, the estimates $\widehat{Err}^{val}(\widehat{\mu}_{\mathcal{L}\setminus\mathcal{L}_k})$ given in (2.2.11), that are unbiased estimates of $Err(\widehat{\mu})$, should also be close enough to $Err(\widehat{\mu})$.

Contrary to the validation sample estimate, the K-fold cross validation estimate uses every observation $(y_i, \boldsymbol{x}_i)$ in the learning set $\mathcal{L}$ for estimating the generalization error. Typically, K is fixed to 10, which appears to be a value that produces stable and reliable estimates. However, the K-fold cross validation estimate is more computationally intensive since it requires to fit K models, while only one model needs to be fitted for computing its validation sample counterpart.

Note that the K partitions can be chosen in a way that makes the subsets $\mathcal{L}_1, \ldots, \mathcal{L}_K$ balanced with respect to the response. Applying stratified random sampling as discussed in Sect. 2.2.3.2 produces folds that have similar distributions for the response.

2.2.3.4 Model Selection and Model Assessment

Model selection consists in choosing the best model among different models produced by a training procedure, say the final model, while model assessment consists in assessing the generalization error of this final model. In practice, we do both model selection and model assessment.

Model assessment should be performed on data that are kept out of the entire training procedure (which includes the fit of the different models together with model selection). Ideally, the training set $\mathcal{D}$ should be entirely dedicated to the training procedure and the validation set $\overline{\mathcal{D}}$ to model assessment.

As part of the training procedure, model selection is thus based on observations from the training set. To guarantee an unbiased estimate of the generalization error for each model under consideration during model selection, a possibility is to divide in its turn the training set into two parts: a part used to fit the models and another (sometimes called test set) to estimate the corresponding generalization errors. Of course, this approach supposes to be in a data-rich situation. If we are in situations where there is insufficient observations in the training set to make this split, another possibility for model selection consists in relying on K-fold cross validation estimates as described above, using the training set $\mathcal{D}$ instead of the entire learning set $\mathcal{L}$.

2.2.4 *Decomposition*

According to Definition 2.2.1, the generalization error $Err(\widehat{\mu})$ of a model $\widehat{\mu}$ is given by

$$Err(\widehat{\mu}) = \mathrm{E}\left[L(Y, \widehat{\mu}(\boldsymbol{X}))\right]. \tag{2.2.13}$$

In the same way, the generalization error of $\widehat{\mu}$ can be defined for a fixed value $\boldsymbol{X} = \boldsymbol{x}$ as

$$Err(\widehat{\mu}(\boldsymbol{x})) = \mathrm{E}\left[L(Y, \widehat{\mu}(\boldsymbol{X}))|\boldsymbol{X} = \boldsymbol{x}\right]. \tag{2.2.14}$$

Notice that averaging the local errors $Err(\widehat{\mu}(\boldsymbol{x}))$ enables to recover the generalization error $Err(\widehat{\mu})$, that is,

$$Err(\widehat{\mu}) = \mathrm{E}\left[Err(\widehat{\mu}(\boldsymbol{X}))\right]. \tag{2.2.15}$$

2.2.4.1 Squared Error Loss

Consider that the loss function is the squared error loss. In our ED family setting, it amounts to assume that the responses are normally distributed. The generalization error of model $\widehat{\mu}$ at $\boldsymbol{X} = \boldsymbol{x}$ becomes

$$\begin{aligned}
Err(\widehat{\mu}(\boldsymbol{x})) &= \mathrm{E}\left[(Y - \widehat{\mu}(\boldsymbol{x}))^2 \middle| \boldsymbol{X} = \boldsymbol{x}\right] \\
&= \mathrm{E}\left[(Y - \mu(\boldsymbol{x}) + \mu(\boldsymbol{x}) - \widehat{\mu}(\boldsymbol{x}))^2 \middle| \boldsymbol{X} = \boldsymbol{x}\right] \\
&= \mathrm{E}\left[(Y - \mu(\boldsymbol{x}))^2 \middle| \boldsymbol{X} = \boldsymbol{x}\right] + \mathrm{E}\left[(\mu(\boldsymbol{x}) - \widehat{\mu}(\boldsymbol{x}))^2 \middle| \boldsymbol{X} = \boldsymbol{x}\right] \\
&\quad +2\,\mathrm{E}\left[(Y - \mu(\boldsymbol{x}))\,(\mu(\boldsymbol{x}) - \widehat{\mu}(\boldsymbol{x})) \middle| \boldsymbol{X} = \boldsymbol{x}\right] \\
&= \mathrm{E}\left[(Y - \mu(\boldsymbol{x}))^2 \middle| \boldsymbol{X} = \boldsymbol{x}\right] + \mathrm{E}\left[(\mu(\boldsymbol{x}) - \widehat{\mu}(\boldsymbol{x}))^2 \middle| \boldsymbol{X} = \boldsymbol{x}\right]
\end{aligned}$$

since

$$\begin{aligned}
&\mathrm{E}\left[(Y - \mu(\boldsymbol{x}))\,(\mu(\boldsymbol{x}) - \widehat{\mu}(\boldsymbol{x})) \middle| \boldsymbol{X} = \boldsymbol{x}\right] \\
&= (\mu(\boldsymbol{x}) - \widehat{\mu}(\boldsymbol{x}))\ \mathrm{E}\left[(Y - \mu(\boldsymbol{x})) \middle| \boldsymbol{X} = \boldsymbol{x}\right] \\
&= (\mu(\boldsymbol{x}) - \widehat{\mu}(\boldsymbol{x}))\ (\mathrm{E}\left[Y|\boldsymbol{X} = \boldsymbol{x}\right] - \mu(\boldsymbol{x})) \\
&= 0
\end{aligned}$$

by definition of $\mu(\boldsymbol{x}) = \mathrm{E}\left[Y|\boldsymbol{X} = \boldsymbol{x}\right]$. So, it comes

$$Err\ (\widehat{\mu}(\boldsymbol{x})) = Err\ (\mu(\boldsymbol{x})) + (\mu(\boldsymbol{x}) - \widehat{\mu}(\boldsymbol{x}))^2. \tag{2.2.16}$$

By (2.2.15), the generalization error $Err(\widehat{\mu})$ thus writes

$$Err\ (\widehat{\mu}) = Err\ (\mu) + \mathrm{E}\left[(\mu(\boldsymbol{X}) - \widehat{\mu}(\boldsymbol{X}))^2\right]. \tag{2.2.17}$$

The generalization error of $\widehat{\mu}$ can be expressed as the sum of two terms, the first one corresponding to the generalization error of the true model μ and the second one representing the estimation error, that is, the discrepancy of $\widehat{\mu}$ from the true model

μ. The further our model from the true one, the larger the generalization error. The generalization error of the true model is called the residual error and is irreducible. Indeed, we have

$$Err\left(\widehat{\mu}\right) \geq Err\left(\mu\right),$$

which means that the smallest generalization error coincides with the one associated to the true model.

2.2.4.2 Poisson Deviance Loss

Consider that the loss function is the Poisson deviance. This choice is appropriate when the responses are assumed to be Poisson distributed, as when examining the number of claims for instance. The generalization error of model $\widehat{\mu}$ at $\boldsymbol{X} = \boldsymbol{x}$ is then given by

$$\begin{aligned}
Err\left(\widehat{\mu}(\boldsymbol{x})\right) &= 2\mathrm{E}\left[Y \ln\left(\frac{Y}{\widehat{\mu}(\boldsymbol{x})}\right) - \left(Y - \widehat{\mu}(\boldsymbol{x})\right) \Big| \boldsymbol{X} = \boldsymbol{x}\right] \\
&= 2\mathrm{E}\left[Y \ln\left(\frac{Y}{\mu(\boldsymbol{x})}\right) - \left(Y - \mu(\boldsymbol{x})\right) \Big| \boldsymbol{X} = \boldsymbol{x}\right] \\
&\quad + 2\mathrm{E}\left[\widehat{\mu}(\boldsymbol{x}) - \mu(\boldsymbol{x}) - Y \ln\left(\frac{\widehat{\mu}(\boldsymbol{x})}{\mu(\boldsymbol{x})}\right) \Big| \boldsymbol{X} = \boldsymbol{x}\right] \\
&= 2\mathrm{E}\left[Y \ln\left(\frac{Y}{\mu(\boldsymbol{x})}\right) - \left(Y - \mu(\boldsymbol{x})\right) \Big| \boldsymbol{X} = \boldsymbol{x}\right] \\
&\quad + 2\left(\widehat{\mu}(\boldsymbol{x}) - \mu(\boldsymbol{x})\right) - 2\mathrm{E}\left[Y | \boldsymbol{X} = \boldsymbol{x}\right] \ln\left(\frac{\widehat{\mu}(\boldsymbol{x})}{\mu(\boldsymbol{x})}\right).
\end{aligned}$$

Replacing $\mathrm{E}\left[Y | \boldsymbol{X} = \boldsymbol{x}\right]$ by $\mu(\boldsymbol{x})$, we get

$$\begin{aligned}
Err\left(\widehat{\mu}(\boldsymbol{x})\right) &= Err\left(\mu(\boldsymbol{x})\right) + 2\left(\widehat{\mu}(\boldsymbol{x}) - \mu(\boldsymbol{x})\right) - 2\mu(\boldsymbol{x}) \ln\left(\frac{\widehat{\mu}(\boldsymbol{x})}{\mu(\boldsymbol{x})}\right) \\
&= Err\left(\mu(\boldsymbol{x})\right) + 2\mu(\boldsymbol{x}) \left(\frac{\widehat{\mu}(\boldsymbol{x})}{\mu(\boldsymbol{x})} - 1 - \ln\left(\frac{\widehat{\mu}(\boldsymbol{x})}{\mu(\boldsymbol{x})}\right)\right).
\end{aligned} \tag{2.2.18}$$

The generalization error $Err(\widehat{\mu})$ thus writes

$$Err\left(\widehat{\mu}\right) = Err\left(\mu\right) + 2\mathrm{E}\left[\mu(\boldsymbol{X}) \left(\frac{\widehat{\mu}(\boldsymbol{X})}{\mu(\boldsymbol{X})} - 1 - \ln\left(\frac{\widehat{\mu}(\boldsymbol{X})}{\mu(\boldsymbol{X})}\right)\right)\right]. \tag{2.2.19}$$

As for the squared error loss, the generalization error of $\widehat{\mu}$ can be decomposed as the sum of the generalization error of the true model and an estimation error $\mathrm{E}[\mathcal{E}^{\mathcal{P}}\left(\widehat{\mu}(\boldsymbol{X})\right)]$, where

$$\mathcal{E}^{\mathcal{P}}\left(\widehat{\mu}(\boldsymbol{x})\right) = 2\mu(\boldsymbol{x})\left(\frac{\widehat{\mu}(\boldsymbol{x})}{\mu(\boldsymbol{x})} - 1 - \ln\left(\frac{\widehat{\mu}(\boldsymbol{x})}{\mu(\boldsymbol{x})}\right)\right).$$

Notice that $\mathcal{E}^{\mathcal{P}}\left(\widehat{\mu}(\boldsymbol{x})\right)$ is always positive because $y \to y - 1 - \ln y$ is positive on $\mathbb{R}^+$, so that we have

$$Err\left(\widehat{\mu}\right) \geq Err\left(\mu\right).$$

2.2.4.3 Gamma Deviance Loss

Consider the Gamma deviance loss. This choice is often made when we study claim severities for instance. The generalization error of model $\widehat{\mu}$ at $\boldsymbol{X} = \boldsymbol{x}$ is then given by

$$\begin{aligned}
Err\left(\widehat{\mu}(\boldsymbol{x})\right) &= 2\mathrm{E}\left[-\ln\left(\frac{Y}{\widehat{\mu}(\boldsymbol{x})}\right) + \frac{Y}{\widehat{\mu}(\boldsymbol{x})} - 1 \middle| \boldsymbol{X} = \boldsymbol{x}\right] \\
&= 2\mathrm{E}\left[-\ln\left(\frac{Y}{\mu(\boldsymbol{x})}\right) + \frac{Y}{\mu(\boldsymbol{x})} - 1 \middle| \boldsymbol{X} = \boldsymbol{x}\right] \\
&\quad +2\mathrm{E}\left[-\ln\left(\frac{Y}{\widehat{\mu}(\boldsymbol{x})}\right) + \ln\left(\frac{Y}{\mu(\boldsymbol{x})}\right) + \frac{Y}{\widehat{\mu}(\boldsymbol{x})} - \frac{Y}{\mu(\boldsymbol{x})} \middle| \boldsymbol{X} = \boldsymbol{x}\right] \\
&= Err\left(\mu(\boldsymbol{x})\right) + 2\mathrm{E}\left[\ln\left(\frac{\widehat{\mu}(\boldsymbol{x})}{\mu(\boldsymbol{x})}\right) + Y\left(\frac{\mu(\boldsymbol{x}) - \widehat{\mu}(\boldsymbol{x})}{\widehat{\mu}(\boldsymbol{x})\mu(\boldsymbol{x})}\right) \middle| \boldsymbol{X} = \boldsymbol{x}\right] \\
&= Err\left(\mu(\boldsymbol{x})\right) + 2\left(\ln\left(\frac{\widehat{\mu}(\boldsymbol{x})}{\mu(\boldsymbol{x})}\right) + \mathrm{E}\left[Y|\boldsymbol{X} = \boldsymbol{x}\right]\left(\frac{\mu(\boldsymbol{x}) - \widehat{\mu}(\boldsymbol{x})}{\widehat{\mu}(\boldsymbol{x})\mu(\boldsymbol{x})}\right)\right) \\
&= Err\left(\mu(\boldsymbol{x})\right) + 2\left(\ln\left(\frac{\widehat{\mu}(\boldsymbol{x})}{\mu(\boldsymbol{x})}\right) + \frac{\mu(\boldsymbol{x}) - \widehat{\mu}(\boldsymbol{x})}{\widehat{\mu}(\boldsymbol{x})}\right) \\
&= Err\left(\mu(\boldsymbol{x})\right) + 2\left(\frac{\mu(\boldsymbol{x})}{\widehat{\mu}(\boldsymbol{x})} - 1 - \ln\left(\frac{\mu(\boldsymbol{x})}{\widehat{\mu}(\boldsymbol{x})}\right)\right) \qquad (2.2.20)
\end{aligned}$$

since $\mathrm{E}\left[Y|\boldsymbol{X} = \boldsymbol{x}\right] = \mu(\boldsymbol{x})$. The generalization error $Err(\widehat{\mu})$ thus writes as the sum of the generalization error of the true model and an estimation error $\mathrm{E}[\mathcal{E}^{\mathcal{G}}\left(\widehat{\mu}(\boldsymbol{X})\right)]$, where

$$\mathcal{E}^{\mathcal{G}}\left(\widehat{\mu}(\boldsymbol{x})\right) = 2\left(\frac{\mu(\boldsymbol{x})}{\widehat{\mu}(\boldsymbol{x})} - 1 - \ln\left(\frac{\mu(\boldsymbol{x})}{\widehat{\mu}(\boldsymbol{x})}\right)\right),$$

that is,

$$Err\left(\widehat{\mu}\right) = Err\left(\mu\right) + \mathrm{E}[\mathcal{E}^{\mathcal{G}}\left(\widehat{\mu}(\boldsymbol{X})\right)]. \qquad (2.2.21)$$

Note that $\mathcal{E}^{\mathcal{G}}\left(\widehat{\mu}(\boldsymbol{x})\right)$ is always positive since we have already noticed in the Poisson case that $y \to y - 1 - \ln y$ is positive on $\mathbb{R}^+$, so that we have

$$Err\left(\widehat{\mu}\right) \geq Err\left(\mu\right).$$

2.3 Expected Generalization Error

The model $\widehat{\mu}$ under consideration is estimated on the training set $\mathcal{D}$ so that it depends on $\mathcal{D}$. To make explicit the dependence on the training set, we use from now on both notations $\widehat{\mu}$ and $\widehat{\mu}_{\mathcal{D}}$ for the model under interest. We assume in a first time there is only one model which corresponds to a given training set, that is, we consider training procedures that are said to be deterministic. Training procedures that can produce different models for a fixed training set are discussed in Sect. 2.4.

The generalization error $Err(\widehat{\mu}_{\mathcal{D}})$ is evaluated conditional on the training set. That is, the model $\widehat{\mu}_{\mathcal{D}}$ under study is first fitted on the training set $\mathcal{D}$ before computing the expectation over all possible observations independently from the training set $\mathcal{D}$. In that sense, the generalization error $Err(\widehat{\mu}_{\mathcal{D}})$ gives an idea of the general accuracy of the training procedure for the particular training set $\mathcal{D}$. In order to study the general behavior of our training procedure, and not only its behavior for a specific training set, it is interesting to evaluate the training procedure on different training sets of the same size.

The training set $\mathcal{D}$ is itself a random variable sampled from a distribution usually unknown in practice, so that the generalization error $Err(\widehat{\mu}_{\mathcal{D}})$ is in its turn a random variable. In order to study the general performance of the training procedure, it is then of interest to take the average of the generalization error $Err\left(\widehat{\mu}_{\mathcal{D}}\right)$ over $\mathcal{D}$, that is, to work with the expected generalization error $\mathrm{E}_{\mathcal{D}}\left[Err\left(\widehat{\mu}_{\mathcal{D}}\right)\right]$ over the models learned from all possible training sets and produced with the training procedure under investigation.

The expected generalization error is thus given by

$$\mathrm{E}_{\mathcal{D}}\left[Err\left(\widehat{\mu}_{\mathcal{D}}\right)\right] = \mathrm{E}_{\mathcal{D}}\left[\mathrm{E}_{X}\left[Err\left(\widehat{\mu}_{\mathcal{D}}(\boldsymbol{X})\right)\right]\right], \tag{2.3.1}$$

which can also be expressed as

$$\mathrm{E}_{\mathcal{D}}\left[Err\left(\widehat{\mu}_{\mathcal{D}}\right)\right] = \mathrm{E}_{X}\left[\mathrm{E}_{\mathcal{D}}\left[Err\left(\widehat{\mu}_{\mathcal{D}}(\boldsymbol{X})\right)\right]\right]. \tag{2.3.2}$$

We can first determine the expected local error $\mathrm{E}_{\mathcal{D}}\left[Err\left(\widehat{\mu}_{\mathcal{D}}(\boldsymbol{X})\right)\right]$ in order to get the expected generalization error.

2.3.1 Squared Error Loss

When the loss function is the squared error loss, we know from Eq. (2.2.16) that the generalization error at $\boldsymbol{X} = \boldsymbol{x}$ writes

$$Err\left(\widehat{\mu}_{\mathcal{D}}(\boldsymbol{x})\right) = Err\left(\mu(\boldsymbol{x})\right) + \left(\mu(\boldsymbol{x}) - \widehat{\mu}_{\mathcal{D}}(\boldsymbol{x})\right)^2. \tag{2.3.3}$$

The true model μ is independent of the training set, so is the generalization error $Err\,(\mu(\boldsymbol{x}))$. The expected generalization error of $\widehat{\mu}$ at $\boldsymbol{X}=\boldsymbol{x}$ is then given by

$$\mathrm{E}_{\mathcal{D}}\left[Err\,(\widehat{\mu}_{\mathcal{D}}(\boldsymbol{x}))\right]=Err\,(\mu(\boldsymbol{x}))+\mathrm{E}_{\mathcal{D}}\left[(\mu(\boldsymbol{x})-\widehat{\mu}_{\mathcal{D}}(\boldsymbol{x}))^2\right].$$

The first term is the local generalization error of the true model while the second term is the expected estimation error at $\boldsymbol{X}=\boldsymbol{x}$, which can be re-expressed as

$$\begin{aligned}
&\mathrm{E}_{\mathcal{D}}\left[(\mu(\boldsymbol{x})-\widehat{\mu}_{\mathcal{D}}(\boldsymbol{x}))^2\right]\\
&=\mathrm{E}_{\mathcal{D}}\left[(\mu(\boldsymbol{x})-\mathrm{E}_{\mathcal{D}}\left[\widehat{\mu}_{\mathcal{D}}(\boldsymbol{x})\right]+\mathrm{E}_{\mathcal{D}}\left[\widehat{\mu}_{\mathcal{D}}(\boldsymbol{x})\right]-\widehat{\mu}_{\mathcal{D}}(\boldsymbol{x}))^2\right]\\
&=\mathrm{E}_{\mathcal{D}}\left[(\mu(\boldsymbol{x})-\mathrm{E}_{\mathcal{D}}\left[\widehat{\mu}_{\mathcal{D}}(\boldsymbol{x})\right])^2\right]+\mathrm{E}_{\mathcal{D}}\left[(\mathrm{E}_{\mathcal{D}}\left[\widehat{\mu}_{\mathcal{D}}(\boldsymbol{x})\right]-\widehat{\mu}_{\mathcal{D}}(\boldsymbol{x}))^2\right]\\
&\quad+2\mathrm{E}_{\mathcal{D}}\left[(\mu(\boldsymbol{x})-\mathrm{E}_{\mathcal{D}}\left[\widehat{\mu}_{\mathcal{D}}(\boldsymbol{x})\right])(\mathrm{E}_{\mathcal{D}}\left[\widehat{\mu}_{\mathcal{D}}(\boldsymbol{x})\right]-\widehat{\mu}_{\mathcal{D}}(\boldsymbol{x}))\right]\\
&=\mathrm{E}_{\mathcal{D}}\left[(\mu(\boldsymbol{x})-\mathrm{E}_{\mathcal{D}}\left[\widehat{\mu}_{\mathcal{D}}(\boldsymbol{x})\right])^2\right]+\mathrm{E}_{\mathcal{D}}\left[(\mathrm{E}_{\mathcal{D}}\left[\widehat{\mu}_{\mathcal{D}}(\boldsymbol{x})\right]-\widehat{\mu}_{\mathcal{D}}(\boldsymbol{x}))^2\right]\\
&=(\mu(\boldsymbol{x})-\mathrm{E}_{\mathcal{D}}\left[\widehat{\mu}_{\mathcal{D}}(\boldsymbol{x})\right])^2+\mathrm{E}_{\mathcal{D}}\left[(\mathrm{E}_{\mathcal{D}}\left[\widehat{\mu}_{\mathcal{D}}(\boldsymbol{x})\right]-\widehat{\mu}_{\mathcal{D}}(\boldsymbol{x}))^2\right]
\end{aligned}$$

since

$$\begin{aligned}
&\mathrm{E}_{\mathcal{D}}\left[(\mu(\boldsymbol{x})-\mathrm{E}_{\mathcal{D}}\left[\widehat{\mu}_{\mathcal{D}}(\boldsymbol{x})\right])(\mathrm{E}_{\mathcal{D}}\left[\widehat{\mu}_{\mathcal{D}}(\boldsymbol{x})\right]-\widehat{\mu}_{\mathcal{D}}(\boldsymbol{x}))\right]\\
&=(\mu(\boldsymbol{x})-\mathrm{E}_{\mathcal{D}}\left[\widehat{\mu}_{\mathcal{D}}(\boldsymbol{x})\right])\,\mathrm{E}_{\mathcal{D}}\left[(\mathrm{E}_{\mathcal{D}}\left[\widehat{\mu}_{\mathcal{D}}(\boldsymbol{x})\right]-\widehat{\mu}_{\mathcal{D}}(\boldsymbol{x}))\right]\\
&=(\mu(\boldsymbol{x})-\mathrm{E}_{\mathcal{D}}\left[\widehat{\mu}_{\mathcal{D}}(\boldsymbol{x})\right])(\mathrm{E}_{\mathcal{D}}\left[\widehat{\mu}_{\mathcal{D}}(\boldsymbol{x})\right]-\mathrm{E}_{\mathcal{D}}\left[\widehat{\mu}_{\mathcal{D}}(\boldsymbol{x})\right])\\
&=0.
\end{aligned}\tag{2.3.4}$$

Therefore, the expected generalization error at $\boldsymbol{X}=\boldsymbol{x}$ is given by

$$\begin{aligned}
\mathrm{E}_{\mathcal{D}}\left[Err\,(\widehat{\mu}_{\mathcal{D}}(\boldsymbol{x}))\right]=&\ Err\,(\mu(\boldsymbol{x}))+(\mu(\boldsymbol{x})-\mathrm{E}_{\mathcal{D}}\left[\widehat{\mu}_{\mathcal{D}}(\boldsymbol{x})\right])^2\\
&+\mathrm{E}_{\mathcal{D}}\left[(\mathrm{E}_{\mathcal{D}}\left[\widehat{\mu}_{\mathcal{D}}(\boldsymbol{x})\right]-\widehat{\mu}_{\mathcal{D}}(\boldsymbol{x}))^2\right].
\end{aligned}\tag{2.3.5}$$

This is the bias-variance decomposition of the expected generalization error.

The first term in (2.3.5) is the local generalization error of the true model, that is, the residual error. The residual error is independent of the training procedure and the training set, which provides in any case a lower bound for the expected generalization error. Notice that in practice, the computation of this lower bound is often unfeasible since the true model is usually unknown. The second term measures the discrepancy between the average estimate $\mathrm{E}_{\mathcal{D}}\left[\widehat{\mu}_{\mathcal{D}}(\boldsymbol{x})\right]$ and the value of the true model $\mu(\boldsymbol{x})$, and corresponds to the bias term. The third term measures the variability of the estimate $\widehat{\mu}_{\mathcal{D}}(\boldsymbol{x})$ over the models trained from all possible training sets, and corresponds to the variance term.

From (2.3.5), the expected generalization error writes

$$\begin{aligned}
\mathrm{E}_{\mathcal{D}}\left[Err\,(\widehat{\mu}_{\mathcal{D}})\right]=&\ Err\,(\mu)+\mathrm{E}_{\boldsymbol{X}}\left\{(\mu(\boldsymbol{X})-\mathrm{E}_{\mathcal{D}}\left[\widehat{\mu}_{\mathcal{D}}(\boldsymbol{X})\right])^2\right\}\\
&+\mathrm{E}_{\boldsymbol{X}}\left\{\mathrm{E}_{\mathcal{D}}\left[(\mathrm{E}_{\mathcal{D}}\left[\widehat{\mu}_{\mathcal{D}}(\boldsymbol{X})\right]-\widehat{\mu}_{\mathcal{D}}(\boldsymbol{X}))^2\right]\right\}.
\end{aligned}\tag{2.3.6}$$

2.3.2 Poisson Deviance Loss

In the case of the Poisson deviance loss, we know from Eq. (2.2.18) that the local generalization error writes

$$Err\left(\widehat{\mu}_{\mathcal{D}}(\boldsymbol{x})\right) = Err\left(\mu(\boldsymbol{x})\right) + \mathcal{E}^{\mathcal{P}}\left(\widehat{\mu}_{\mathcal{D}}(\boldsymbol{x})\right) \tag{2.3.7}$$

where

$$\mathcal{E}^{\mathcal{P}}\left(\widehat{\mu}_{\mathcal{D}}(\boldsymbol{x})\right) = 2\mu(\boldsymbol{x})\left(\frac{\widehat{\mu}_{\mathcal{D}}(\boldsymbol{x})}{\mu(\boldsymbol{x})} - 1 - \ln\left(\frac{\widehat{\mu}_{\mathcal{D}}(\boldsymbol{x})}{\mu(\boldsymbol{x})}\right)\right). \tag{2.3.8}$$

Because the true model μ is independent of the training set, the expected generalization error $\mathrm{E}_{\mathcal{D}}\left[Err\left(\widehat{\mu}_{\mathcal{D}}(\boldsymbol{x})\right)\right]$ can be expressed as

$$\mathrm{E}_{\mathcal{D}}\left[Err\left(\widehat{\mu}_{\mathcal{D}}(\boldsymbol{x})\right)\right] = Err\left(\mu(\boldsymbol{x})\right) + \mathrm{E}_{\mathcal{D}}\left[\mathcal{E}^{\mathcal{P}}\left(\widehat{\mu}_{\mathcal{D}}(\boldsymbol{x})\right)\right] \tag{2.3.9}$$

with

$$\mathrm{E}_{\mathcal{D}}\left[\mathcal{E}^{\mathcal{P}}\left(\widehat{\mu}_{\mathcal{D}}(\boldsymbol{x})\right)\right] = 2\mu(\boldsymbol{x})\left(\mathrm{E}_{\mathcal{D}}\left[\frac{\widehat{\mu}_{\mathcal{D}}(\boldsymbol{x})}{\mu(\boldsymbol{x})}\right] - 1 - \mathrm{E}_{\mathcal{D}}\left[\ln\left(\frac{\widehat{\mu}_{\mathcal{D}}(\boldsymbol{x})}{\mu(\boldsymbol{x})}\right)\right]\right). \tag{2.3.10}$$

Locally, the expected generalization error is equal to the generalization error of the true model plus the expected estimation error which can be attributed to the bias and the estimation fluctuation. Notice that the expected estimation error $\mathrm{E}_{\mathcal{D}}\left[\mathcal{E}^{\mathcal{P}}\left(\widehat{\mu}_{\mathcal{D}}(\boldsymbol{x})\right)\right]$ is positive since we have seen that the estimation error $\mathcal{E}^{\mathcal{P}}\left(\widehat{\mu}_{\mathcal{D}}(\boldsymbol{x})\right)$ is always positive. The generalization error of the true model is again a theoretical lower bound for the expected generalization error.

From (2.3.9) and (2.3.10), the expected generalization error writes

$$\mathrm{E}_{\mathcal{D}}\left[Err\left(\widehat{\mu}_{\mathcal{D}}\right)\right] = Err\left(\mu\right) + 2\mathrm{E}_{X}\left\{\mu(\boldsymbol{X})\left(\mathrm{E}_{\mathcal{D}}\left[\frac{\widehat{\mu}_{\mathcal{D}}(\boldsymbol{X})}{\mu(\boldsymbol{X})}\right] - 1 - \mathrm{E}_{\mathcal{D}}\left[\ln\left(\frac{\widehat{\mu}_{\mathcal{D}}(\boldsymbol{X})}{\mu(\boldsymbol{X})}\right)\right]\right)\right\}. \tag{2.3.11}$$

2.3.3 Gamma Deviance Loss

In the Gamma case, Eq. (2.2.20) tells us that the local generalization error is given by

$$Err\left(\widehat{\mu}_{\mathcal{D}}(\boldsymbol{x})\right) = Err\left(\mu(\boldsymbol{x})\right) + \mathcal{E}^{\mathcal{G}}\left(\widehat{\mu}_{\mathcal{D}}(\boldsymbol{x})\right), \tag{2.3.12}$$

where

$$\mathcal{E}^{\mathcal{G}}\left(\widehat{\mu}_{\mathcal{D}}(\boldsymbol{x})\right) = 2\left(\frac{\mu(\boldsymbol{x})}{\widehat{\mu}_{\mathcal{D}}(\boldsymbol{x})} - 1 - \ln\left(\frac{\mu(\boldsymbol{x})}{\widehat{\mu}_{\mathcal{D}}(\boldsymbol{x})}\right)\right). \tag{2.3.13}$$

Hence, the expected generalization error $\mathrm{E}_{\mathcal{D}}\left[Err\left(\widehat{\mu}_{\mathcal{D}}(\boldsymbol{x})\right)\right]$ can be expressed as

$$\mathrm{E}_{\mathcal{D}}\left[Err\left(\widehat{\mu}_{\mathcal{D}}(\boldsymbol{x})\right)\right] = Err\left(\mu(\boldsymbol{x})\right) + \mathrm{E}_{\mathcal{D}}\left[\mathcal{E}^{g}\left(\widehat{\mu}_{\mathcal{D}}(\boldsymbol{x})\right)\right] \tag{2.3.14}$$

with

$$\mathrm{E}_{\mathcal{D}}\left[\mathcal{E}^{g}\left(\widehat{\mu}_{\mathcal{D}}(\boldsymbol{x})\right)\right] = 2\left(\mathrm{E}_{\mathcal{D}}\left[\frac{\mu(\boldsymbol{x})}{\widehat{\mu}_{\mathcal{D}}(\boldsymbol{x})}\right] - 1 - \mathrm{E}_{\mathcal{D}}\left[\ln\left(\frac{\mu(\boldsymbol{x})}{\widehat{\mu}_{\mathcal{D}}(\boldsymbol{x})}\right)\right]\right). \tag{2.3.15}$$

Locally, the expected generalization error is equal to the generalization error of the true model plus the expected estimation error which can be attributed to the bias and the estimation fluctuation. Notice that the expected estimation error $\mathrm{E}_{\mathcal{D}}\left[\mathcal{E}^{g}\left(\widehat{\mu}_{\mathcal{D}}(\boldsymbol{x})\right)\right]$ is positive since we have seen that the estimation error $\mathcal{E}^{g}\left(\widehat{\mu}_{\mathcal{D}}(\boldsymbol{x})\right)$ is always positive. The generalization error of the true model is a theoretical lower bound for the expected generalization error.

From (2.3.14), the expected generalization error writes

$$\mathrm{E}_{\mathcal{D}}\left[Err\left(\widehat{\mu}_{\mathcal{D}}\right)\right] = Err\left(\mu\right) + 2\mathrm{E}_{\boldsymbol{X}}\left\{\mathrm{E}_{\mathcal{D}}\left[\frac{\mu(\boldsymbol{X})}{\widehat{\mu}_{\mathcal{D}}(\boldsymbol{X})}\right] - 1 - \mathrm{E}_{\mathcal{D}}\left[\ln\left(\frac{\mu(\boldsymbol{X})}{\widehat{\mu}_{\mathcal{D}}(\boldsymbol{X})}\right)\right]\right\}. \tag{2.3.16}$$

2.3.4 Bias and Variance

In order to minimise the expected generalization error, it might appear desirable to sacrifice a bit on the bias provided we can reduce to a large extend the variability of the prediction over the models trained from all possible training sets. The bias-variance decomposition of the expected generalization error is used for justifying the performances of ensemble learning techniques studied in Chap. 4.

2.4 (Expected) Generalization Error for Randomized Training Procedures

A training procedure which always produces the same model $\widehat{\mu}_{\mathcal{D}}$ for a given training set $\mathcal{D}$ (and fixed values for the tuning parameters) is said to be deterministic. This is the case for instance of regression trees studied in Chap. 3.

There also exist randomized training procedures that can produce different models for a fixed training set (and fixed values for the tuning parameters), such as random forests and boosting trees discussed in Chaps. 4 and 5. In order to account for the randomness of the training procedure, we introduce a random vector Θ which is assumed to fully capture the randomness of the training procedure. The model $\widehat{\mu}$ resulting from the randomized training procedure depends on the training set $\mathcal{D}$ and also on the random vector Θ, so that we use both notations $\widehat{\mu}$ and $\widehat{\mu}_{\mathcal{D},\Theta}$ for the model under consideration.

The generalization error $Err(\widehat{\mu}_{\mathcal{D},\Theta})$ is thus evaluated conditional on the training set $\mathcal{D}$ and the random vector Θ. The expected generalization error, which aims to assess the general accuracy of the training procedure, is now obtained by taking the average of the generalization error $Err(\widehat{\mu}_{\mathcal{D},\Theta})$ over $\mathcal{D}$ and Θ. Expression (2.3.1) becomes

$$\mathrm{E}_{\mathcal{D},\Theta}\left[Err\left(\widehat{\mu}_{\mathcal{D},\Theta}\right)\right] = \mathrm{E}_{\mathcal{D},\Theta}\left[\mathrm{E}_{X}\left[Err\left(\widehat{\mu}_{\mathcal{D},\Theta}(\boldsymbol{X})\right)\right]\right], \tag{2.4.1}$$

which can also be expressed as

$$\mathrm{E}_{\mathcal{D},\Theta}\left[Err\left(\widehat{\mu}_{\mathcal{D},\Theta}\right)\right] = \mathrm{E}_{X}\left[\mathrm{E}_{\mathcal{D},\Theta}\left[Err\left(\widehat{\mu}_{\mathcal{D},\Theta}(\boldsymbol{X})\right)\right]\right]. \tag{2.4.2}$$

Again, we can first determine the expected local error $\mathrm{E}_{\mathcal{D},\Theta}\left[Err\left(\widehat{\mu}_{\mathcal{D},\Theta}(\boldsymbol{X})\right)\right]$ in order to get the expected generalization error.

Taking into account the additional source of randomness in the training procedure, expressions (2.3.5), (2.3.9) and (2.3.14) become respectively

$$\begin{aligned}\mathrm{E}_{\mathcal{D},\Theta}\left[Err\left(\widehat{\mu}_{\mathcal{D},\Theta}(\boldsymbol{x})\right)\right] &= Err\left(\mu(\boldsymbol{x})\right) + \left(\mu(\boldsymbol{x}) - \mathrm{E}_{\mathcal{D},\Theta}\left[\widehat{\mu}_{\mathcal{D},\Theta}(\boldsymbol{x})\right]\right)^2 \\ &\quad + \mathrm{E}_{\mathcal{D},\Theta}\left[\left(\mathrm{E}_{\mathcal{D},\Theta}\left[\widehat{\mu}_{\mathcal{D},\Theta}(\boldsymbol{x})\right] - \widehat{\mu}_{\mathcal{D},\Theta}(\boldsymbol{x})\right)^2\right],\end{aligned} \tag{2.4.3}$$

$$\mathrm{E}_{\mathcal{D},\Theta}\left[Err\left(\widehat{\mu}_{\mathcal{D},\Theta}(\boldsymbol{x})\right)\right] = Err\left(\mu(\boldsymbol{x})\right) + \mathrm{E}_{\mathcal{D},\Theta}\left[\mathcal{E}^{\mathcal{P}}\left(\widehat{\mu}_{\mathcal{D},\Theta}(\boldsymbol{x})\right)\right] \tag{2.4.4}$$

with

$$\mathrm{E}_{\mathcal{D},\Theta}\left[\mathcal{E}^{\mathcal{P}}\left(\widehat{\mu}_{\mathcal{D},\Theta}(\boldsymbol{x})\right)\right] = 2\mu(\boldsymbol{x})\left(\mathrm{E}_{\mathcal{D},\Theta}\left[\frac{\widehat{\mu}_{\mathcal{D},\Theta}(\boldsymbol{x})}{\mu(\boldsymbol{x})}\right] - 1 - \mathrm{E}_{\mathcal{D},\Theta}\left[\ln\left(\frac{\widehat{\mu}_{\mathcal{D},\Theta}(\boldsymbol{x})}{\mu(\boldsymbol{x})}\right)\right]\right), \tag{2.4.5}$$

and

$$\mathrm{E}_{\mathcal{D},\Theta}\left[Err\left(\widehat{\mu}_{\mathcal{D},\Theta}(\boldsymbol{x})\right)\right] = Err\left(\mu(\boldsymbol{x})\right) + \mathrm{E}_{\mathcal{D},\Theta}\left[\mathcal{E}^{\mathcal{G}}\left(\widehat{\mu}_{\mathcal{D},\Theta}(\boldsymbol{x})\right)\right] \tag{2.4.6}$$

with

$$\mathrm{E}_{\mathcal{D},\Theta}\left[\mathcal{E}^{\mathcal{G}}\left(\widehat{\mu}_{\mathcal{D},\Theta}(\boldsymbol{x})\right)\right] = 2\left(\mathrm{E}_{\mathcal{D},\Theta}\left[\frac{\mu(\boldsymbol{x})}{\widehat{\mu}_{\mathcal{D},\Theta}(\boldsymbol{x})}\right] - 1 - \mathrm{E}_{\mathcal{D},\Theta}\left[\ln\left(\frac{\mu(\boldsymbol{x})}{\widehat{\mu}_{\mathcal{D},\Theta}(\boldsymbol{x})}\right)\right]\right). \tag{2.4.7}$$

2.5 Bibliographic Notes and Further Reading

This chapter is mainly inspired from Louppe (2014) and the book of Hastie et al. (2009). We also find inspiration from Wüthrich and Buser (2019) for the choice of the loss function in our ED family setting as well as for the decomposition of the generalization error in the Poisson case.

References

Hastie, T., Tibshirani, R., Friedman, J. (2009). The Elements of Statistical Learning. Data Mining, Inference, and Prediction. Second Edition. Springer Series in Statistics

Louppe, G. (2014). Understanding random forests: from theory to practice. arXiv:14077502

Wüthrich, M. V., Buser, C. (2019). Data analytics for non-life insurance pricing. Lecture notes

Chapter 3
Regression Trees

3.1 Introduction

In this chapter, we present the regression trees introduced by Breiman et al. (1984). Regression trees are at the core of this second volume. They are the building blocks of the ensemble techniques described in Chaps. 4 and 5. We closely follow the seminal book of Breiman et al. (1984). The presentation is also mainly inspired from Hastie et al. (2009) and Wüthrich and Buser (2019).

3.2 Binary Regression Trees

A regression tree partitions the feature space χ into disjoint subsets $\{\chi_t\}_{t\in\mathcal{T}}$, where $\mathcal{T}$ is a set of indexes. On each subset χ_t, the prediction $\widehat{c}_t$ of the response on that part of the feature space is assumed to be constant. The resulting predictions $\widehat{\mu}(\boldsymbol{x})$ can be written as

$$\widehat{\mu}(\boldsymbol{x}) = \sum_{t\in\mathcal{T}} \widehat{c}_t \mathrm{I}\left[\boldsymbol{x} \in \chi_t\right]. \tag{3.2.1}$$

Binary regression trees recursively partition the feature space by a sequence of binary splits until a stopping rule is applied, as illustrated in Fig. 3.1.

The node t_0 is called the root or root node of the tree and represents the feature space χ itself, also denoted χ_{t_0} to make explicit the link with the root node t_0. The feature space χ_{t_0} is first split into two disjoint subspaces χ_{t_1} and χ_{t_2} such that $\chi_{t_1} \cup \chi_{t_2} = \chi_{t_0}$. The subspaces χ_{t_1} and χ_{t_2} correspond to nodes t_1 and t_2, respectively. The node t_0 is said to be the parent of nodes t_1 and t_2 or equivalently nodes t_1 and t_2 are said to be the children of t_0. More precisely, t_1 is called the left child of t_0 while t_2 is called the right child.

Then, subspaces χ_{t_1} and χ_{t_2} are in turn split into two disjoint subspaces. For instance, subspace χ_{t_1} is partitioned into disjoint subsets χ_{t_3} and χ_{t_4} such that

M. Denuit et al., *Effective Statistical Learning Methods for Actuaries II*,
Springer Actuarial, https://doi.org/10.1007/978-3-030-57556-4_3

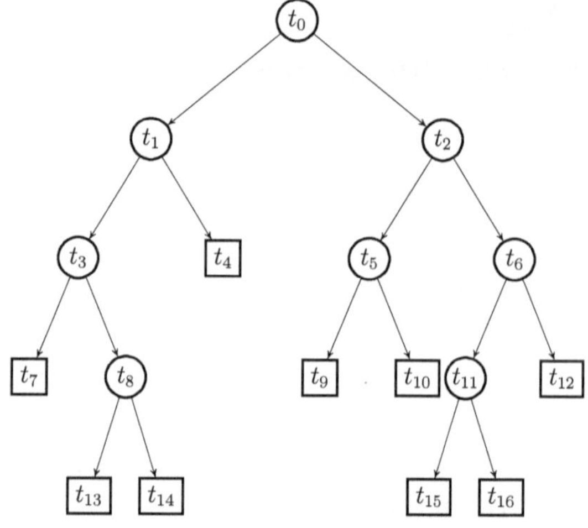

Fig. 3.1 Example of a binary regression tree. Circles indicate non-terminal nodes and rectangle boxes represent terminal nodes

$\chi_{t_3} \cup \chi_{t_4} = \chi_{t_1}$. The node t_1 is the parent of nodes t_3 and t_4 that correspond to subspaces χ_{t_3} and χ_{t_4}, while t_3 and t_4 are the left and right children of t_1, respectively. The node t_4 does not have children. Such a node is called terminal node or leaf of the tree, and is represented by a rectangle box. Non-terminal nodes are indicated by circles.

This process is continued until all nodes are designated terminals. One says that the tree stops growing. In Fig. 3.1, the terminal nodes are $t_4, t_7, t_9, t_{10}, t_{12}, t_{13}, t_{14}, t_{15}$ and t_{16}. The corresponding subspaces are disjoint and form together a partition of the feature space χ, namely $\chi_{t_4} \cup \chi_{t_7} \cup \chi_{t_9} \cup \chi_{t_{10}} \cup \chi_{t_{12}} \cup \chi_{t_{13}} \cup \chi_{t_{14}} \cup \chi_{t_{15}} \cup \chi_{t_{16}} = \chi$. The set of indexes $\mathcal{T}$ is then given by $\{t_4, t_7, t_9, t_{10}, t_{12}, t_{13}, t_{14}, t_{15}, t_{16}\}$.

In each terminal node $t \in \mathcal{T}$, the prediction of the response is denoted $\widehat{c}_t$, that is, $\widehat{\mu}(\boldsymbol{x}) = \widehat{c}_t$ for $\boldsymbol{x} \in \chi_t$.

Notice that each node t is split into a left child node t_L and a right child node t_R. In a more general case, we could consider multiway splits, resulting in more than two children nodes for t. Figure 3.2 represents a binary split with a multiway split resulting in four children nodes. However, the problem with multiway splits is that they partition the data too quickly, leaving insufficient observations at the next level down. Also, since multiway splits can be achieved by a series of binary splits, the latter are often preferred. Henceforth, we work with binary regression trees.

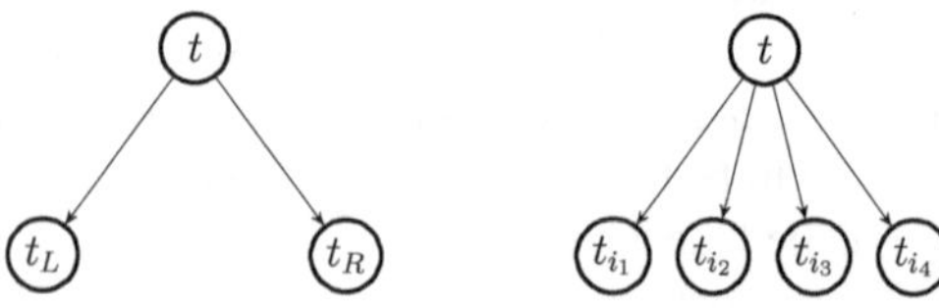

Fig. 3.2 Binary split versus multiway split with four children nodes

Several elements need to be discussed in order to determine $\widehat{\mu}(\boldsymbol{x})$:

1. The selection of the splits;
2. The decision to declare a node terminal;
3. The predictions in the terminal nodes.

Remark 3.2.1 Regression functions of the form (3.2.1) are piecewise constant functions. It is in general impossible to find the best piecewise constant regression function minimizing the generalization error. Regression trees are produced by following a greedy strategy.

3.2.1 Selection of the Splits

At each node t, the selection of the optimal split s_t requires to define a candidate set $\mathcal{S}_t$ of possible binary splits and a goodness of split criterion to pick the best one.

3.2.1.1 Candidate Set of Splits

Let us assume that node t is composed of observations with k distinct values for $\boldsymbol{x}$. The number of partitions of χ_t into two non-empty disjoint subsets χ_{t_L} and χ_{t_R} is then given by $2^{k-1} - 1$. Because of the exponential growth of the number of binary partitions with k, the strategy which consists in trying all partitions and taking the best one turns out to be unrealistic since often computationally intractable.

For this reason, the number of possible splits is restricted. Specifically, only standardized binary splits are usually considered. A standardized binary split is characterized as follows:

1. Depends on the value of only one single feature;
2. For an ordered feature x_j, only allows questions of the form $x_j \leq c$, where c is a constant;
3. For a categorical variable x_j, only allows questions of the form $x_j \in C$, where C is a subset of possible categories of the variable x_j.

An ordered feature x_j taking q different values $x_{j1}, \ldots, x_{jq}$ at node t generates $q - 1$ standardized binary splits at that node. The split questions are $x_j \leq c_i$, $i = 1, \ldots, q - 1$, where the constants c_i are taken halfway between consecutive values x_{ji} and $x_{j,i+1}$, that is $c_i = \frac{x_{j,i+1}+x_{ji}}{2}$.

For a categorical feature x_j with q different values at node t, the number of standardized binary splits generated at that node is $2^{q-1} - 1$. Let us notice that trivial standardized binary splits are excluded, meaning splits of χ_t that generate a subset (either χ_{t_L} or χ_{t_R}) with no observation. Therefore, at each node t, every possible standardized binary split is tested ant the best split is selected by means of a goodness of split criterion. In the feature space, working with standardized binary splits amounts to only consider splits that are perpendicular to the coordinate axes.

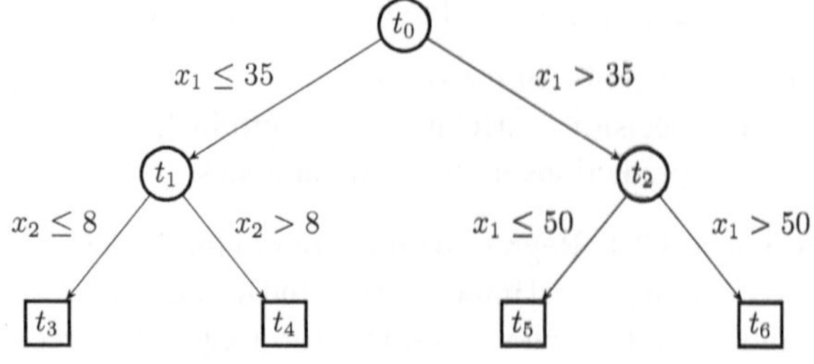

Fig. 3.3 Example of a tree with two features $18 \leq x_1 \leq 100$ and $0 \leq x_2 \leq 20$

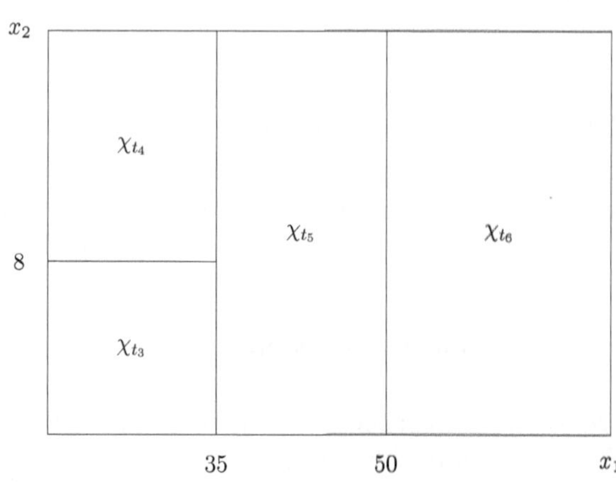

Fig. 3.4 Partition into rectangles of the feature space corresponding to the tree depicted in Fig. 3.3

Hence, the resulting regression trees recursively partition the feature space into hyper rectangles.

For example, let $p = 2$ with $18 \leq x_1 \leq 100$ and $0 \leq x_2 \leq 20$ and suppose that the resulting tree is the one described in Fig. 3.3. The feature space is then partitioned into the rectangles depicted in Fig. 3.4.

3.2.1.2 Goodness of Split

At a node t, the candidate set $\mathcal{S}_t$ is henceforth restricted to non-trivial standardized binary splits. In order to determine the best possible split $s_t \in \mathcal{S}_t$ for node t, a goodness of split criterion is needed.

In the ED family setting, the natural candidate is the deviance. That is, denoting by $D_{\chi_t}(\widehat{c}_t)$ the deviance on χ_t, the optimal split s_t solves

$$\min_{s \in \mathcal{S}_t} \left\{ D_{\chi_{t_L^{(s)}}}\left(\widehat{c}_{t_L^{(s)}}\right) + D_{\chi_{t_R^{(s)}}}\left(\widehat{c}_{t_R^{(s)}}\right) \right\}, \tag{3.2.2}$$

where $t_L^{(s)}$ and $t_R^{(s)}$ are the left and right children nodes of t resulting from split s and $\widehat{c}_{t_L^{(s)}}$ and $\widehat{c}_{t_R^{(s)}}$ are the corresponding predictions. The optimal split s_t then leads to children nodes $t_L^{(s_t)}$ and $t_R^{(s_t)}$ that we also denote t_L and t_R. Notice that solving (3.2.2) amounts to find $s \in \mathcal{S}_t$ that maximizes the decrease of the deviance at node t, namely

$$\max_{s \in \mathcal{S}_t} \left\{ D_{\chi_t}\left(\widehat{c}_t\right) - \left(D_{\chi_{t_L^{(s)}}}\left(\widehat{c}_{t_L^{(s)}}\right) + D_{\chi_{t_R^{(s)}}}\left(\widehat{c}_{t_R^{(s)}}\right) \right) \right\}. \tag{3.2.3}$$

A categorical feature x_j with q categories $x_{j1}, \ldots, x_{jq}$ at node t yields $2^{q-1} - 1$ possible splits. Hence, when q is large, computing (3.2.2) for every possible splits can be very time consuming. In order to speed up the procedure, each category x_{jl} can be replaced by the prediction $\widehat{c}_t(x_{jl})$ computed on $\chi_t \cap \{x_j = x_{jl}\}$. This procedure results in an ordered version of the categorical feature x_j, reducing the number of possible splits to $q - 1$. Notice that in the case the response records a number of claims, the number of observations in a given node can be too small to observe enough claims for some categories of the feature, especially in low frequency cases, so that the resulting predictions may not provide a relevant ordering of the categories. This potential issue is addressed in Sect. 3.2.2.2.

Remark 3.2.2 A tree starts from the whole feature space χ and is grown by iteratively dividing the subsets of χ into smaller subsets. This procedure consists in dividing each node t using the optimal split s_t that locally maximizes the decrease of the deviance. This greedy strategy could be improved by assessing the goodness of a split by also looking at those deeper in the tree. However, such an approach is more time consuming and does not seem to significantly improve the model performance.

3.2.2 *The Prediction in Each Terminal Node*

3.2.2.1 Maximum-Likelihood Estimator

For every terminal node $t \in \mathcal{T}$, we need to compute the prediction $\widehat{c}_t$. The deviance, equivalently denoted $D\left(\widehat{\mu}\right)$ and $D\left((\widehat{c}_t)_{t \in \mathcal{T}}\right)$ in the following, is given by

$$D\left(\widehat{\mu}\right) = \sum_{t \in \mathcal{T}} D_{\chi_t}\left(\widehat{c}_t\right).$$

Minimizing the deviance $D\left(\widehat{\mu}\right)$ with respect to $(\widehat{c}_t)_{t \in \mathcal{T}}$ is thus equivalent to minimizing each deviance $D_{\chi_t}\left(\widehat{c}_t\right)$ with respect to $\widehat{c}_t$. Therefore, the best estimate $\widehat{c}_t$ for the prediction is the maximum-likelihood estimate on χ_t.

3.2.2.2 Bayesian Estimator

Let the response Y be the number of claims. Given $\boldsymbol{X} = \boldsymbol{x}$, Y is assumed to be Poisson distributed with expected claim frequency $e\mu(\boldsymbol{x}) > 0$, where e is the exposure-to-risk. Let $\mathcal{D} = \{(y_i, \boldsymbol{x}_i);\ i \in \mathcal{I}\}$ be the observations available to train our model. For a node t, we denote by w_t the corresponding volume defined by

$$w_t = \sum_{i:\boldsymbol{x}_i \in \chi_t} e_i,$$

and by c_t the corresponding expected claim frequency defined by

$$c_t = \frac{1}{w_t} \sum_{i:\boldsymbol{x}_i \in \chi_t} e_i\, \mu(\boldsymbol{x}_i) > 0.$$

If the maximum-likelihood estimate $\widehat{c_t} = 0$, then we get a solution which makes no sense in practice. That is why we replace the maximum-likelihood estimate $\widehat{c_t}$ on χ_t by a Bayesian estimator $\widehat{c_t}^{Bayes}$.

In that goal, we assume that conditionally to $\Theta \sim \mathcal{G}am(\gamma, \gamma)$, the responses Y_i $(i \in \mathcal{I})$ are independent and distributed as

$$Y_i|\Theta \sim \mathcal{P}oi(e_i\mu(\boldsymbol{x}_i)\Theta).$$

The random variable Θ introduces uncertainty in the expected claim frequency. Given Θ, and denoting $Z_t = \sum_{i:\boldsymbol{x}_i \in \chi_t} Y_i$, we then have

$$Z_t|\Theta \sim \mathcal{P}oi(c_t w_t \Theta),$$

so that it comes

$$\begin{aligned} \mathrm{E}\left[c_t\Theta|Z_t\right] &= c_t\mathrm{E}\left[\Theta|Z_t\right] \\ &= \alpha_t\widehat{c_t} + (1-\alpha_t)c_t \end{aligned}$$

with maximum-likelihood estimate $\widehat{c_t}$ and credibility weight

$$\alpha_t = \frac{w_t\, c_t}{\gamma + w_t\, c_t}.$$

For more details about credibility theory and the latter calculations, we refer the interested reader to Bühlmann and Gisler (2005).

Therefore, given values for c_t and γ, we can compute $\mathrm{E}\left[c_t\Theta|Z_t\right]$, which is always positive, contrarily to $\widehat{c_t}$ that can be zero. However, c_t is obviously not known in practice. It could be replaced by $\widehat{c_t}$ or $\mathrm{E}\left[c_t\Theta|Z_t\right]$. The maximum-likelihood estimator $\widehat{c_t}$ does not solve our original problem. Turning to $\mathrm{E}\left[c_t\Theta|Z_t\right]$, we can compute it recursively as

$$\widehat{c_t}^{Bayes,k} = \widehat{\mathrm{E}}\left[c_t\Theta|Z_t\right] = \widehat{\alpha_t}\widehat{c_t} + (1-\widehat{\alpha_t})\widehat{c_t}^{Bayes,k-1},$$

with estimated credibility weights

$$\widehat{\alpha}_t = \frac{w_t \widehat{c_t}^{Bayes,k-1}}{\gamma + w_t \widehat{c_t}^{Bayes,k-1}}.$$

This recursive procedure can be initialized by

$$\widehat{c_t}^{Bayes,0} = \widehat{c}_0 = \frac{\sum_{i\in\mathcal{I}} Y_i}{\sum_{i\in\mathcal{I}} e_i},$$

which is the maximum-likelihood estimator for the expected claim frequency without making distinction between individuals. The remaining parameter γ, which enters the computation of the estimated credibility weights, still needs to be selected and is chosen externally.

Note that the R command `rpart` used in this chapter for the examples simplifies the recursive approach described above. It rather considers the estimator

$$\widehat{c_t}^{\texttt{rpart}} = \widehat{\alpha}_t^{\texttt{rpart}} \widehat{c}_t + (1 - \widehat{\alpha}_t^{\texttt{rpart}}) \widehat{c}_0$$

for $\mathrm{E}[c_t \Theta | Z_t]$, with estimated credibility weights

$$\widehat{\alpha}_t^{\texttt{rpart}} = \frac{w_t \widehat{c}_0}{\gamma + w_t \widehat{c}_0},$$

which is a special case of the Bühlmann–Straub model.

Remark 3.2.3 Considering the same Θ in each terminal node introduces dependence between the leaves. One way to remedy to this undesirable effect is to consider as many (independent) random effects as there are leaves in the tree. Notice that this latter approach requires the knowledge of the tree structure.

3.2.3 The Rule to Determine When a Node Is Terminal

A node t is inevitably terminal when χ_t can no longer be split. This is the case when the observations in node t share the same values for all the features, i.e. when $\boldsymbol{x}_i = \boldsymbol{x}_j$ for all $(y_i, \boldsymbol{x}_i)$ and $(y_j, \boldsymbol{x}_j)$ such that $\boldsymbol{x}_i, \boldsymbol{x}_j \in \chi_t$. In such a case, splitting node t would generate a subset with no observation. A node t is also necessarily terminal when it contains observations with the same value for the response, i.e. when $y_i = y_j$ for all $(y_i, \boldsymbol{x}_i)$ and $(y_j, \boldsymbol{x}_j)$ such that $\boldsymbol{x}_i, \boldsymbol{x}_j \in \chi_t$. In particular, this is the case when node t contains only one observation.

Those stopping criteria are inherent to the recursive partitioning procedure. Such inevitable terminal nodes lead to the biggest possible regression tree that we can

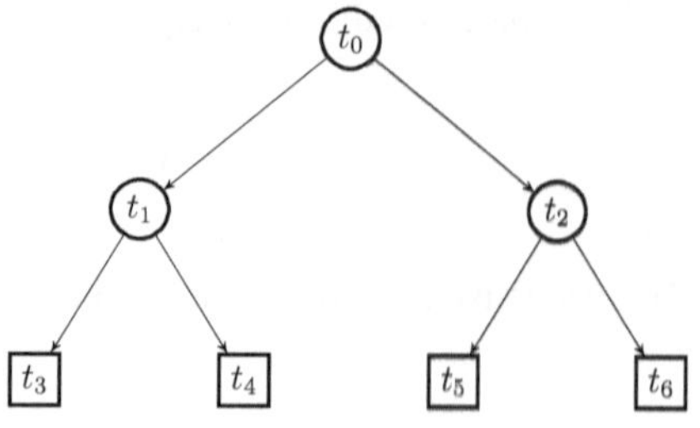

Fig. 3.5 Example of a tree with a maximal depth of two

grow on the training set. However such a tree is likely to capture noise in the training set and to cause overfitting.

In order to reduce the size of the tree and hence to prevent overfitting, these stopping criteria that are inherent to the recursive partitioning procedure are complemented with several rules. Three stopping rules that are commonly used can be formulated as follows:

- A node t is declared terminal when it contains less than a fixed number of observations.
- A node t is declared terminal if at least one of its children nodes t_L and t_R that results from the optimal split s_t contains less than a fixed number of observations.
- A node t is declared terminal when its depth is equal to a fixed maximal depth.

Notice that the depth of a node t is equal to d if it belongs to generation $d + 1$. For instance, in Fig. 3.5, t_0 has a depth of zero, t_1 and t_2 have a depth of one and terminal nodes t_3, t_4, t_5 and t_6 have a depth of two.

Another common stopping rule consists of setting a threshold and deciding that a node t is terminal if the decrease in deviance that we have by splitting node t with the optimal split s_t is less than this fixed threshold. Recall that splitting node t into children nodes t_L and t_R results in a decrease of the deviance given by

$$\Delta D_{\chi_t} = D_{\chi_t}\left(\widehat{c}_t\right) - \left(D_{\chi_{t_L}}\left(\widehat{c}_{t_L}\right) + D_{\chi_{t_R}}\left(\widehat{c}_{t_R}\right)\right). \tag{3.2.4}$$

It is interesting to notice that this deviance reduction ΔD_{χ_t} is always positive. Indeed, we have

$$\begin{aligned} D_{\chi_t}\left(\widehat{c}_t\right) &= D_{\chi_{t_L}}\left(\widehat{c}_t\right) + D_{\chi_{t_R}}\left(\widehat{c}_t\right) \\ &\geq D_{\chi_{t_L}}\left(\widehat{c}_{t_L}\right) + D_{\chi_{t_R}}\left(\widehat{c}_{t_R}\right) \end{aligned}$$

since the maximum-likelihood estimates $\widehat{c}_{t_L}$ and $\widehat{c}_{t_R}$ minimize the deviances $D_{\chi_{t_L}}\left(\widehat{c}_t\right)$ and $D_{\chi_{t_L}}\left(\widehat{c}_t\right)$, respectively. As a consequence, if for a partition $\{\chi_t\}_{t\in\mathcal{T}}$ of χ, we consider an additional standardized binary split s_t of χ_t for a given $t \in \mathcal{T}$, yielding the new set of indexes $\mathcal{T}' = \mathcal{T}\backslash\{t\} \cup \{t_L, t_R\}$ for the terminal nodes, we necessarily have

$$D\left((\widehat{c}_t)_{t\in\mathcal{T}}\right) \geq D\left((\widehat{c}_t)_{t\in\mathcal{T}'}\right).$$

So, an additional standardized binary split always decreases the deviance.

While the stopping rules presented above may give good results in practice, the strategy of stopping early the growing of the tree is in general unsatisfactory. For instance, the last rule provides a too large tree if the threshold is set too low. Increasing the threshold may lead to a too small tree, and a node t with a small deviance reduction ΔD_{χ_t} may have children nodes t_L and t_R with larger decreases in the deviance. Hence, by declaring t terminal, the good splits at nodes t_L and t_R would be never used.

That is why it is preferable to prune the tree instead of stopping the growing of the tree. Pruning a tree consists in fully developing the tree and then prune it upward until the optimal tree is found. This is discussed in Sect. 3.3.

3.2.4 Examples

The examples in this chapter are done with the `R` package `rpart`, which stands for `r`ecursive `part`itioning.

3.2.4.1 Simulated Dataset

We consider an example in car insurance. Four features $\boldsymbol{X} = (X_1, X_2, X_3, X_4)$ are supposed to be available, that are

- X_1 = Gender: policyholder's gender (female or male);
- X_2 = Age: policyholder's age (integer values from 18 to 65);
- X_3 = Split: whether the policyholder splits its annual premium or not (yes or no);
- X_4 = Sport: whether the policyholder's car is a sports car or not (yes or no).

The variables X_1, X_2, X_3 and X_4 are assumed to be independent and distributed as follows:

$$\begin{aligned}
&\mathrm{P}[X_1 = female] = \mathrm{P}[X_1 = male] = 0.5;\\
&\mathrm{P}[X_2 = 18] = \mathrm{P}[X_2 = 19] = \ldots = \mathrm{P}[X_2 = 65] = 1/48;\\
&\mathrm{P}[X_3 = yes] = \mathrm{P}[X_3 = no] = 0.5;\\
&\mathrm{P}[X_4 = yes] = \mathrm{P}[X_4 = no] = 0.5.
\end{aligned}$$

The values taken by a feature are thus equiprobable.

The response Y is supposed to be the number of claims. Given $\boldsymbol{X} = \boldsymbol{x}$, Y is assumed to be Poisson distributed with expected claim frequency given by

$$\begin{aligned}
\mu(\boldsymbol{x}) = 0.1 &\times (1 + 0.1\mathrm{I}[x_1 = male])\\
&\times (1 + 0.4\mathrm{I}[18 \leq x_2 < 30] + 0.2\mathrm{I}[30 \leq x_2 < 45])\\
&\times (1 + 0.15\mathrm{I}[x_4 = yes]).
\end{aligned}$$

Table 3.1 Ten first observations of the simulated dataset

	Y	X_1 (Gender)	X_2 (Age)	X_3 (Split)	X_4 (Sport)
1	0	Male	46	Yes	Yes
2	0	Male	57	No	No
3	0	Female	34	No	Yes
4	0	Female	27	Yes	No
5	0	Male	42	Yes	Yes
6	0	Female	27	No	Yes
7	0	Female	55	Yes	Yes
8	0	Female	23	No	Yes
9	0	Female	33	No	No
10	2	Male	36	No	Yes

Being a male increases the expected claim frequency by 10%, drivers between 18 and 29 (resp. 30 and 44) years old have expected claim frequencies 40% (resp. 20%) larger than policyholders older than 45 years old, splitting its premium does not influence the expected claim frequency while driving a sports car increases the expected claim frequency by 15%.

In this example, the true model $\mu(\boldsymbol{x})$ is known and we can simulate realizations of the random vector $(Y, \boldsymbol{X})$. Specifically, we generate $n = 500\,000$ independent realizations of $(Y, \boldsymbol{X})$, that is, we consider a learning set made of 500 000 observations $(y_1, \boldsymbol{x}_1), (y_2, \boldsymbol{x}_2), \ldots, (y_{500\,000}, \boldsymbol{x}_{500\,000})$. An observation represents a policy that has been observed during a whole year. In Table 3.1, we provide the ten first observations of the learning set. While the nine first policies made no claim over the past year, the tenth policyholder, who is a 36 years old man with a sports car and paying his premium annually, experienced two claims.

In this simulated dataset, the proportion of males is approximately 50%, so are the proportions of sports cars and policyholders splitting their premiums. For each age 18,19,...,65, there are between 10 188 and 10 739 policyholders.

We now aim to estimate the expected claim frequency $\mu(\boldsymbol{x})$. In that goal, we fit several trees on our simulated dataset with Poisson deviance as loss function. Here, we do not divide the learning set into a training set and a validation set, so that the whole learning set is used to train the models. The R command used is

```
> tree <- rpart(Y ~ Gender+Age+Split+Sport,
                data = dataset,
                method="poisson",
                control = rpart.control())
```

where `data` specifies the training set used to build the tree, `method` refers to the optimisation criterion applied at each split, here the Poisson deviance, and `control` enables to control the size of the tree.

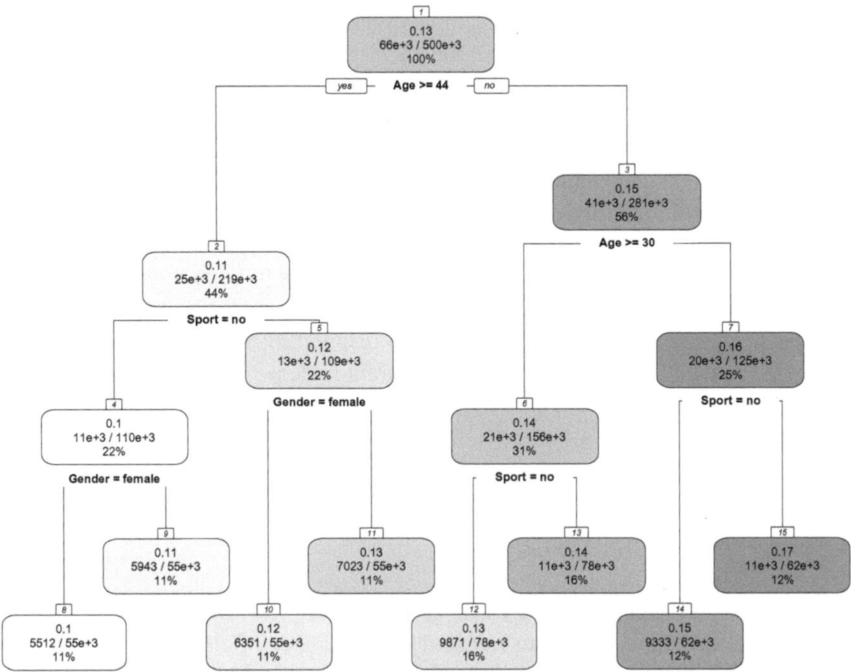

Fig. 3.6 Tree with a maximum depth equal to three as stopping rule

A first tree with a maximum depth of three as stopping rule is built. A node t is then terminal when its depth is equal to three, meaning that it belongs to the fourth generation of nodes. The resulting tree is depicted in Fig. 3.6 and presented in more details below:

```
> n= 500000

node), split, n, deviance, yval
      * denotes terminal node

 1) root 500000 279\,043.30 0.1317560
   2) Age>=44.5 218846 111779.70 0.1134548
     4) Sport=no 109659  53386.74 0.1044621
       8) Gender=female 54866  26084.89 0.1004673 *
       9) Gender=male 54793  27285.05 0.1084660 *
     5) Sport=yes 109187  58236.07 0.1224878
      10) Gender=female 54545  28257.29 0.1164381 *
      11) Gender=male 54642  29946.18 0.1285280 *
   3) Age< 44.5 281154 166261.40 0.1460015
     6) Age>=29.5 156219  88703.96 0.1355531
      12) Sport=no 77851  42515.21 0.1267940 *
      13) Sport=yes 78368  46100.83 0.1442541 *
     7) Age< 29.5 124935  77295.79 0.1590651
      14) Sport=no 62475  37323.97 0.1493856 *
      15) Sport=yes 62460  39898.18 0.1687435 *
```

We start with `n=500 000` observations in the training set. The nodes are numbered with the variable `node)`. The node `1)` is the root node, node `2)` corresponds to its left-child and node `3)` to its right-child, and so on, such that node `k)` corresponds to node t_{k-1} in our notations. For each node, the variable `split` specifies the split criterion applied, `n` the number of observations in the node, `deviance` the deviance at that node and `yval` the prediction (i.e. the estimate of the expected claim frequency) in that node. Also, * denotes terminal nodes, that are in this case nodes `8)` to `15)`.

In particular, terminal node `14)`, which corresponds to node t_{13}, is obtained by using feature x_4 (Sport) with answer no. It contains $n_{t_{13}} = 62\,475$ observations, the deviance is $D_{\chi_{t_{13}}}\left(\widehat{c}_{t_{13}}\right) = 37\,323.97$ and the corresponding estimated expected claim frequency is given by $\widehat{c}_{t_{13}} = 0.1493856$. Its parent node is t_6 and the decrease of the deviance resulting from the split of t_6 into nodes t_{13} and t_{14} is

$$
\begin{aligned}
\Delta D_{\chi_{t_6}} &= D_{\chi_{t_6}}\left(\widehat{c}_{t_6}\right) - \left(D_{\chi_{t_{13}}}\left(\widehat{c}_{t_{13}}\right) + D_{\chi_{t_{14}}}\left(\widehat{c}_{t_{14}}\right)\right) \\
&= 77295.79 - (37323.97 + 39898.18) \\
&= 73.64.
\end{aligned}
$$

In Fig. 3.6, each rectangle represents a node of the tree. In each node, one can see the estimated expected claim frequency, the number of claims, the number of observations as well as the proportion of the training set in the node. As an example, in node `3)` (or t_2), we observe the estimated expected claim frequency $\widehat{c}_{t_2} \approx 0.15$, the number of claims $\sum_{i:\boldsymbol{x}_i \in \chi_{t_2}} y_i \approx 41\,000$ and the number of observations $n_{t_2} \approx 281\,000$ which corresponds to approximately 56% of the training set. Below each non-terminal node t, we find the question that characterizes the split s_t. In case the answer is yes, one moves to the left-child node t_L while when the answer is no, one moves to the right-child node t_R. Each node is topped by a small rectangle box containing the number of the node which corresponds to the variable `node)`. Terminal nodes are nodes `8)` to `15)` and belong to the fourth generation of nodes as requested by the stopping rule. Finally, the darker the gray of a node, the higher the estimated expected claim frequency in that node.

The first split of the tree is defined by the question Age ≥ 44.5 (and not by Age ≥ 44 as it is suggested in Fig. 3.6). For feature Age, the set of possible questions is $x_2 \leq c_k$ with constants $c_k = \frac{18+19+2k-2}{2}$, $k = 1, \ldots, 47$. The best split for the root node is thus $x_2 \leq 44.5$. The feature Age is indeed the one that influences the most the expected claim frequency, up to a difference of 40% (resp. 20%) for policyholders with $18 \leq$ Age < 30 (resp. $30 \leq$ Age < 45) compared to those older than 45.

The left child node is node `2)` and comprised policyholders with Age ≥ 45. In that node, the feature Age does not influence the expected claim frequency, while features Sport and Gender can lead to expected claim frequencies that differ from 15% and 10%, respectively. The feature Sport is then naturally selected to define the best split at that node. The two resulting nodes are `4)` and `5)`, in which only the feature Gender still influences the expected claim frequency, hence the choice of the Gender to perform both splits leading to terminal nodes.

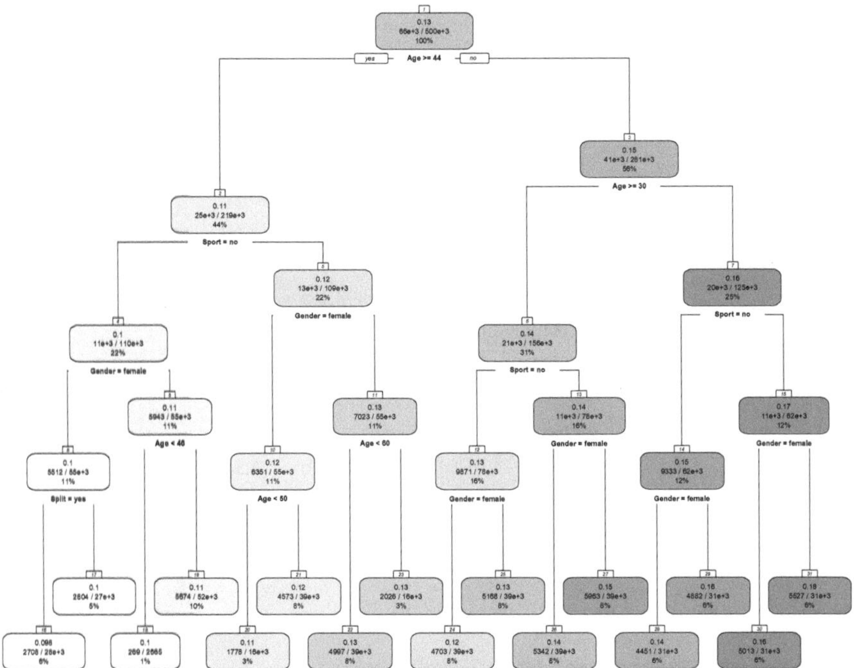

Fig. 3.7 Tree with a maximum depth equal to four as stopping rule

The right child node of the root is node `3)`. In that node, the feature Age is again the preferred feature as it can still produce a difference of $16.67\% = \frac{1.4}{1.2} - 1$ in expected claim frequencies. The children nodes `6)` and `7)` are then in turn split with the feature Sport since in both nodes, it yields to differences in expected claim frequencies of 15% while the feature Gender yields to differences of 10%. Notice that the feature Age is no longer relevant in these two nodes. The resulting nodes, namely `12)`, `13)`, `14)` and `15)`, are terminals since they all belongs to the fourth generation of nodes.

The tree that has been built with a maximum depth of three as stopping rule is not enough deep. Indeed, we notice that nodes `12)`, `13)`, `14)` and `15)` should still be split with the feature Gender since males have expected claim frequencies 10% higher than females. So, in this example, a maximum depth of three is too restrictive.

A tree with a maximum depth of four as stopping rule is then fitted. The resulting tree is showed in Fig. 3.7. The first four generations of nodes are obviously the same than in our previous tree depicted in Fig. 3.6. The only difference lies in the addition of a fifth generation of nodes, which enables to split nodes `12)`, `13)`, `14)` and `15)` with the feature Gender, as desired. However, while nodes `8)`, `9)`, `10)` and `11)` were terminal nodes in our previous tree, they now become non-terminal nodes as they do not belong to the fifth generation. Therefore, they are all split in order to meet our stopping rule. For instance, node `8)` is split with the feature Split, while we

Table 3.2 Decrease of the deviance ΔD_{χ_t} for each node t

Node `k)`	$\Delta D_{\chi_{t_{k-1}}}$
`1)`	279 043.30−(111779.70+166261.40) = 1002.20
`2)`	111779.70−(53386.74+58236.07) = 156.89
`3)`	166261.40−(88703.96+77295.79) = 261.65
`4)`	53386.74−(26084.89+27285.05) = 16.80
`5)`	58236.07−(28257.29+29946.18) = 32.60
`6)`	88703.96−(42515.21+46100.83) = 87.92
`7)`	77295.79−(37323.97+39898.18) = 73.64
`8)`	26084.89−(12962.76+13118.61) = 3.52
`9)`	27285.05−(1255.95+26027.61) = 1.49
`10)`	28257.29−(7964.89+20291.27) = 1.13
`11)`	29946.18−(21333.37+8612.48) = 0.33
`12)`	42515.21−(20701.11+21790.59) = 23.51
`13)`	46100.83−(22350.27+23718.03) = 32.53
`14)`	37323.97−(18211.52+19090.83) = 21.62
`15)`	39898.18−(19474.59+20396.94) = 26.65

know that this feature does not influence the expected claim frequency. Therefore, while we improve our model on the right hand side of the tree, i.e. node `3)` and its children, we start to overfit our dataset on the left hand side of the tree, i.e. node `2)` and subsequent.

In this example, specifying the stopping rule only with respect to the maximum depth does not lead to satisfying results. Instead, we could combine the rule of a maximum depth equal to four with a requirement of a minimum decrease of the deviance. In Table 3.2, we show the decrease of the deviance ΔD_{χ_t} observed at each node t. As expected, we notice that nodes `8)`, `9)`, `10)` and `11)` have the four smallest values for ΔD_{χ_t}.

If we select a threshold somewhere between 3.52 and 16.80 for the minimum decrease of the deviance allowed to split a node, which are the values of ΔD_{χ_t} at nodes `4)` and `8)`, respectively, we get the optimal tree presented in Fig. 3.8, in which nodes `16)` to `23)` now disappeared compared to our previous tree showed in Fig. 3.7. Hence, nodes `8)` to `11)` are now terminal nodes, as desired. Finally, we then get twelve terminal nodes.

In Table 3.3, we show the terminal nodes with their corresponding expected claim frequencies $\mu(\boldsymbol{x})$ as well as with their estimates $\widehat{\mu}(\boldsymbol{x})$.

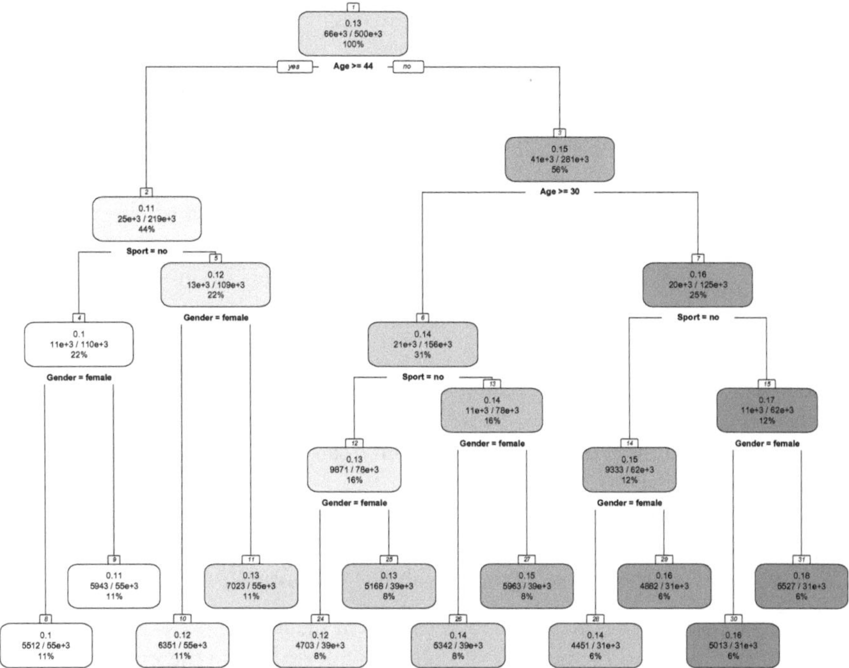

Fig. 3.8 Optimal tree

Table 3.3 Terminal nodes with their corresponding expected claim frequencies $\mu(\boldsymbol{x})$ and estimated expected claim frequencies $\widehat{\mu}(\boldsymbol{x})$

Node k)	$\boldsymbol{x}$			$\mu(\boldsymbol{x})$	$\widehat{\mu}(\boldsymbol{x})$
	x_1 (Gender)	x_2 (Age)	x_4 (Sport)		
Node 8)	Female	$x_2 \geq 44.5$	No	0.1000	0.1005
Node 9)	Male	$x_2 \geq 44.5$	No	0.1100	0.1085
Node 10)	Female	$x_2 \geq 44.5$	Yes	0.1150	0.1164
Node 11)	Male	$x_2 \geq 44.5$	Yes	0.1265	0.1285
Node 24)	Female	$29.5 \leq x_2 < 44.5$	No	0.1200	0.1206
Node 25)	Male	$29.5 \leq x_2 < 44.5$	No	0.1320	0.1330
Node 26)	Female	$29.5 \leq x_2 < 44.5$	Yes	0.1380	0.1365
Node 27)	Male	$29.5 \leq x_2 < 44.5$	Yes	0.1518	0.1520
Node 28)	Female	$x_2 < 29.5$	No	0.1400	0.1422
Node 29)	Male	$x_2 < 29.5$	No	0.1540	0.1566
Node 30)	Female	$x_2 < 29.5$	Yes	0.1610	0.1603
Node 31)	Male	$x_2 < 29.5$	Yes	0.1771	0.1772

3.2.4.2 Real Dataset

In this second example, we consider a motor third-party liability insurance portfolio of an insurance company operating in the EU that has been observed during one year. The portfolio comprises $n = 160\,944$ insurance policies. For each policy i, $i = 1, \ldots, n$, we have the numbers of claims y_i filed by the policyholder, the exposure-to-risk $e_i \leq 1$ expressed in year, which is the duration of observation for policy i, and eight features $\boldsymbol{x}_i = (x_{i1}, \ldots, x_{i8})$, namely

- x_{i1} = AgePh: policyholder's age;
- x_{i2} = AgeCar: age of the car;
- x_{i3} = Fuel: fuel of the car, with two categories (gas or diesel);
- x_{i4} = Split: splitting of the premium, with four categories (annually, semi-annually, quarterly or monthly);
- x_{i5} = Cover: extent of the coverage, with three categories (from compulsory third-party liability cover to comprehensive);
- x_{i6} = Gender: policyholder's gender, with two categories (female or male);
- x_{i7} = Use: use of the car, with two categories (private or professional);
- x_{i8} = PowerCat: the engine's power, with five categories.

Table 3.4 provides the then first observations of the dataset.

In Fig. 3.9, we give the number of policies with respect to the exposure-to-risk expressed in months. As we can see, the majority of the policies has been observed during the whole year of observation. Also, Fig. 3.10 displays the exposure-to-risk by category/value for each of the eight features. Finally, Table 3.5 shows the observed numbers of claims with their corresponding exposures-to-risk.

The observations are assumed to be independent and given $\boldsymbol{X} = \boldsymbol{x}$ and the exposure-to-risk e, the response Y is assumed to be Poisson distributed with expected claim frequency $e\mu(\boldsymbol{x})$. So, $\mu(\boldsymbol{x}_i)$ represents the expected annual claim frequency for policyholder i.

We fit a tree with Poisson deviance as loss function on the whole dataset, meaning that we do not isolate some observations to form a validation set. As stopping rule, we require a minimum number of observations in terminal nodes equal to 5000. The resulting tree is represented in Fig. 3.11. With such a stopping rule, we notice that terminal nodes do not necessarily have the same depth. For instance, terminal node `6)` is part of the third generation of nodes while terminal node `39)` is part of the sixth one.

Requiring at least 5000 observations within each terminal node is particularly relevant in this example as it guarantees a reasonable accuracy on the predictions $\widehat{c}_t$, $t \in \mathcal{T}$. Indeed, at portfolio level, the estimated expected annual claim frequency is $\widehat{\mu} = 13.9\%$ and the average exposure-to-risk is $e = 0.89$. One can then estimate a confidence interval of two standard deviations for $\widehat{\mu}$ given by

Table 3.4 Ten first observations of the simulated dataset, where ExpoR designates the exposure-to-risk

	Y	AgePh	AgeCar	Fuel	Split	Cover	Gender	Use	PowerCat	ExpoR
1	1	50	12	Gasoline	Monthly	TPL.Only	Male	Private	C2	1.000
2	0	64	3	Gasoline	Yearly	Limited.MD	Female	Private	C2	1.000
3	0	60	10	Diesel	Yearly	TPL.Only	Male	Private	C2	1.000
4	0	77	15	Gasoline	Yearly	TPL.Only	Male	Private	C2	1.000
5	1	28	7	Gasoline	Half-yearly	TPL.Only	Female	Private	C2	0.047
6	0	26	12	Gasoline	Quarterly	TPL.Only	Male	Private	C2	1.000
7	1	26	8	Gasoline	Half-yearly	Comprehensive	Male	Private	C2	1.000
8	0	58	14	Gasoline	Quarterly	TPL.Only	Female	Private	C2	0.403
9	0	59	6	Gasoline	Half-yearly	Limited.MD	Female	Private	C1	1.000
10	0	57	10	Gasoline	Half-yearly	Limited.MD	Female	Private	C1	1.000

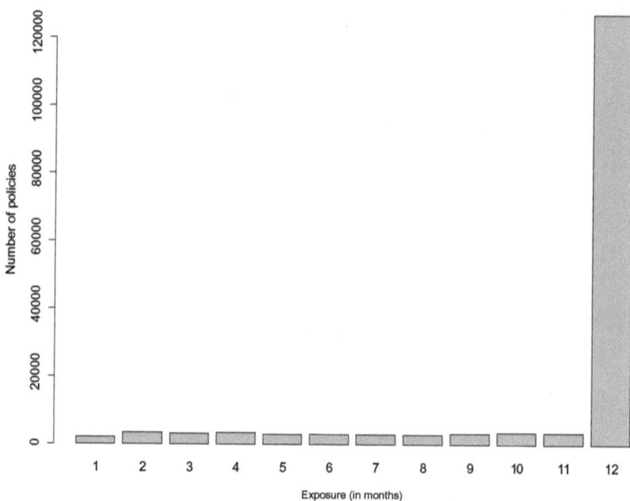

Fig. 3.9 Number of policies with respect to the exposure-to-risk expressed in months

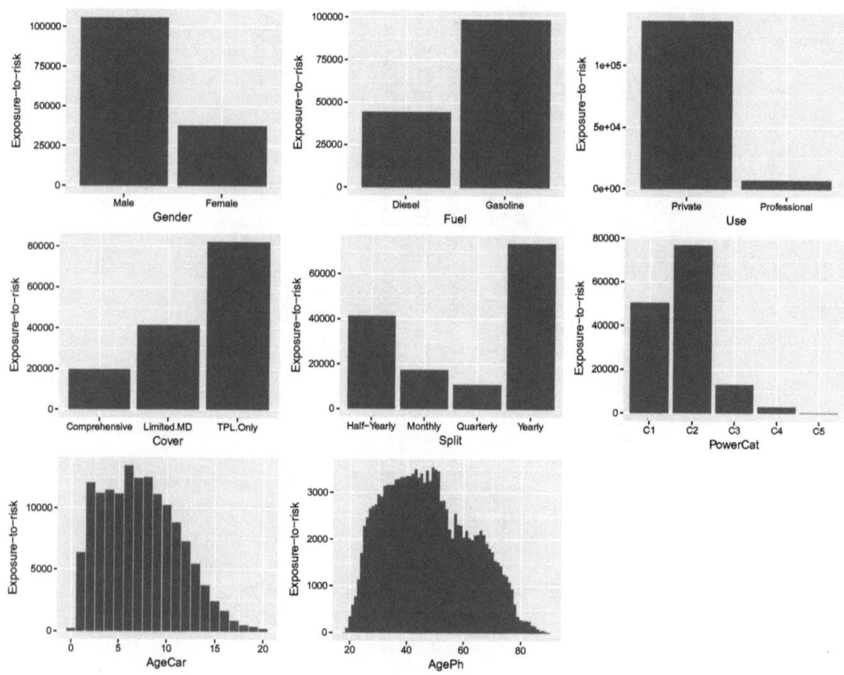

Fig. 3.10 Categories/values of the predictors and their corresponding exposures-to-risk

Table 3.5 Descriptive statistics for the number of claims

Number of claims	Exposure-to-risk
0	126 499.7
1	15 160.4
2	1424.9
3	145.4
4	14.3
5	1.4
≥6	0

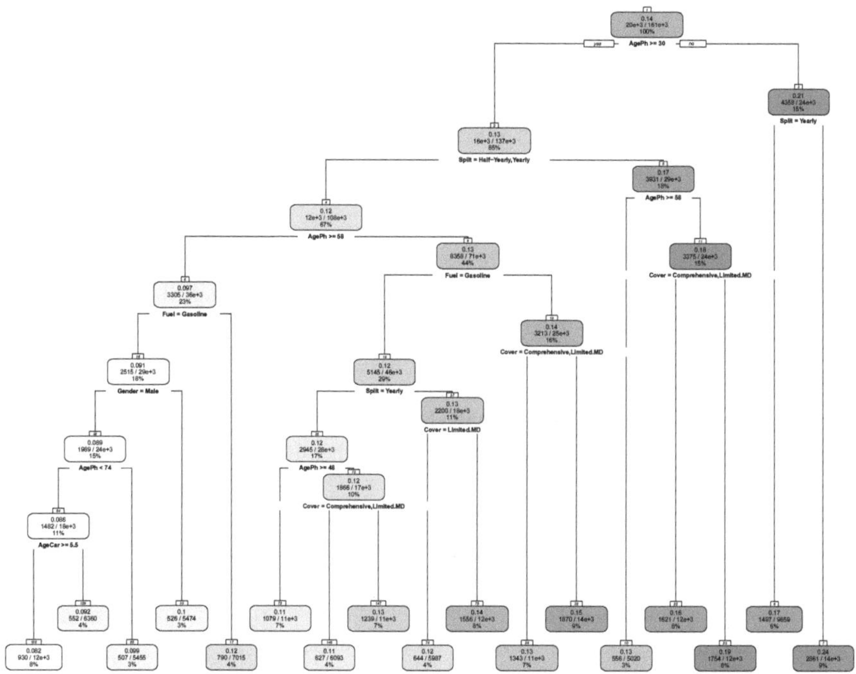

Fig. 3.11 Tree with at least 5000 observations in terminal nodes

$$
\begin{aligned}
&\left[\widehat{\mu}-2\sqrt{\frac{\widehat{\mu}}{e\times 5000}},\ \widehat{\mu}+2\sqrt{\frac{\widehat{\mu}}{e\times 5000}}\right]\\
&=\left[13.9\%-2\sqrt{\frac{13.9\%}{0.89\times 5000}},\ 13.9\%+2\sqrt{\frac{13.9\%}{0.89\times 5000}}\right]\\
&=[12.8\%,\ 15.0\%]\,.
\end{aligned}
$$

Roughly speaking, one sees that we obtain a precision of 1% in the average annual claim frequency $\widehat{\mu}$, which can be considered as satisfactory here. Notice that if we would have selected 1000 for the minimum number of observations in the terminal nodes, we would have get a precision of 2.5% in the average annual claim frequency $\widehat{\mu}$.

3.3 Right Sized Trees

We have seen several rules to declare a node t terminal. These rules have in common that they early stop the growth of the tree. Another way to find the right sized tree consists in fully developing the tree and then pruning it.

Henceforth, to ease the presentation, we denote a regression tree by T. A tree T is defined by a set of splits together with the order in which they are used and the predictions in the terminal nodes. When needed to make explicit the link with tree T, we use the notation $\mathcal{T}_{(T)}$ for the corresponding set of indexes $\mathcal{T}$ of the terminal nodes. In addition, we mean by $|T|$ the number of terminal nodes of tree T, that is $|\mathcal{T}_{(T)}|$.

In order to define the pruning process, one needs to specify the notion of a tree branch. A branch $T^{(t)}$ is the part of T that is composed of node t and all its descendants nodes. Pruning a branch $T^{(t)}$ of a tree T means deleting from T all descendants nodes of t. The resulting tree is denoted $T - T^{(t)}$. One says that $T - T^{(t)}$ is a pruned subtree of T. For instance, in Fig. 3.12, we represent a tree T, its branch $T^{(t_2)}$ and the subtree $T - T^{(t_2)}$ obtained from T by pruning the branch $T^{(t_2)}$. More generally, a tree T' that is obtained from T by successively pruning branches is called a pruned subtree of T, or simply a subtree of T, and is denoted $T' \preceq T$.

The first step of the pruning process is to grow the largest possible tree $T_{\max}$ by letting the splitting procedure continue until all terminal nodes contain either observations with identical values for the features (i.e. terminal nodes $t \in \mathcal{T}$ where $\boldsymbol{x}_i = \boldsymbol{x}_j$ for all $(y_i, \boldsymbol{x}_i)$ and $(y_j, \boldsymbol{x}_j)$ such that $\boldsymbol{x}_i, \boldsymbol{x}_j \in \chi_t$) or either observations with the same value for the response (i.e. terminal nodes $t \in \mathcal{T}$ where $y_i = y_j$ for all $(y_i, \boldsymbol{x}_i)$ and $(y_j, \boldsymbol{x}_j)$ such that $\boldsymbol{x}_i, \boldsymbol{x}_j \in \chi_t$).

Notice that the initial tree T_{init} can be smaller than $T_{\max}$. Indeed, let us assume that the pruning process starting with the largest tree $T_{\max}$ produces the subtree T_{prune}. Then, the pruning process will always lead to the same subtree T_{prune} if we start with any subtree T_{init} of $T_{\max}$ such that T_{prune} is a subtree of T_{init}.

We thus start the pruning process with a sufficiently large tree T_{init}. Then, the idea of pruning T_{init} consists in constructing a sequence of smaller and smaller trees

$$T_{\text{init}}, T_{|T_{\text{init}}|-1}, T_{|T_{\text{init}}|-2}, \ldots, T_1,$$

where T_k is a subtree of T_{init} with k terminal nodes, $k = 1, \ldots, |T_{\text{init}}| - 1$. In particular, T_1 is only composed of the root node t_0 of T_{init}.

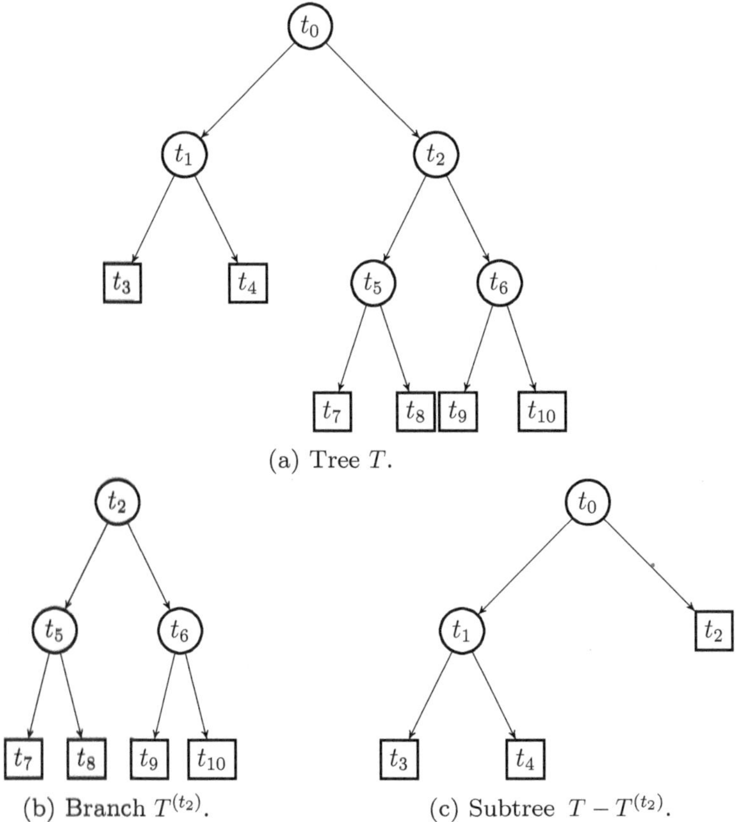

Fig. 3.12 Example of a branch $T^{(t_2)}$ for a tree T as well as the resulting subtree $T - T^{(t_2)}$ when pruning $T^{(t_2)}$

Let us denote by $\mathcal{C}(T_{\text{init}}, k)$ the class of all subtrees of T_{init} having k terminal nodes. An intuitive procedure to produce the sequence of trees $T_{\text{init}}, T_{|T_{\text{init}}|-1}, \ldots, T_1$ is to select, for every $k = 1, \ldots, |T_{\text{init}}| - 1$, the tree T_k which minimizes the deviance $D\left((\widehat{c}_t)_{t \in \mathcal{T}_{(T)}}\right)$ among all subtrees T of T_{init} with k terminal nodes, that is

$$D\left((\widehat{c}_t)_{t \in \mathcal{T}_{(T_k)}}\right) = \min_{T \in \mathcal{C}(T_{\text{init}}, k)} D\left((\widehat{c}_t)_{t \in \mathcal{T}_{(T)}}\right).$$

Thus, for every $k = 1, \ldots, |T_{\text{init}}| - 1$, T_k is the best subtree of T_{init} with k terminal nodes according to the deviance loss function.

It is natural to use the deviance in comparing subtrees with the same number of terminal nodes. However, the deviance is not helpful for comparing subtrees $T_{\text{init}}, T_{|T_{\text{init}}|-1}, T_{|T_{\text{init}}|-2}, \ldots, T_1$. Indeed, as noticed in Sect. 3.2.3, if T_k is a subtree of $T'_{k+1} \in \mathcal{C}(T_{\text{init}}, k+1)$, we necessarily have

$$D\left((\widehat{c}_t)_{t\in\mathcal{T}_{(T_k)}}\right) \geq D\left((\widehat{c}_t)_{t\in\mathcal{T}_{(T'_{k+1})}}\right)$$
$$\geq D\left((\widehat{c}_t)_{t\in\mathcal{T}_{(T_{k+1})}}\right).$$

Therefore, the selection of the best subtree T_{prune} among the sequence

$$T_{\text{init}}, T_{|T_{\text{init}}|-1}, T_{|T_{\text{init}}|-2}, \ldots, T_1$$

that is based on the deviance, or equivalently on the training sample estimate of the generalization error, will always lead to the largest tree T_{init}.

That is why the generalization errors of the pruned subtrees

$$T_{\text{init}}, T_{|T_{\text{init}}|-1}, T_{|T_{\text{init}}|-2}, \ldots, T_1$$

should be estimated on a validation set in order to determine T_{prune}. The choice of T_{prune} is also often done by cross-validation, as we will see in the following.

Such an intuitive procedure for constructing the sequence of trees

$$T_{\text{init}}, T_{|T_{\text{init}}|-1}, T_{|T_{\text{init}}|-2}, \ldots, T_1$$

has some drawbacks. One of them is to produce subtrees of T_{init} that are not nested, meaning that subtree T_k is not necessarily a subtree of T_{k+1}. Hence, a node t of T_{init} can reappear in tree T_k while it was cut off in tree T_{k+1}.

That is why the minimal cost-complexity pruning presented hereafter is usually preferred.

3.3.1 Minimal Cost-Complexity Pruning

We define the cost-complexity measure of a tree T as

$$R_\alpha(T) = D\left((\widehat{c}_t)_{t\in\mathcal{T}_{(T)}}\right) + \alpha|T|,$$

where the parameter α is a positive real number. The number of terminal nodes $|T|$ is called the complexity of the tree T. Thus, the cost-complexity measure $R_\alpha(T)$ is a combination of the deviance $D\left((\widehat{c}_t)_{t\in\mathcal{T}_{(T)}}\right)$ and a penalty for the complexity of the tree $\alpha|T|$. The parameter α can be interpreted as the increase in the penalty for having one more terminal node.

When we increase by one the number of terminal nodes of a tree T by splitting one of its terminal node t into two children nodes t_L and t_R, then we know that the deviance of the resulting tree T' is smaller than the deviance of the original tree T, that is

$$D\left((\widehat{c}_t)_{t\in\mathcal{T}_{(T')}}\right) \leq D\left((\widehat{c}_t)_{t\in\mathcal{T}_{(T)}}\right).$$

The deviance will always favor the more complex tree T' over T.

By introducing a penalty for the complexity of the tree, the cost-complexity measure may now prefer the original tree T over the most complex one T'. Indeed, the cost-complexity measure of T' can be written as

$$\begin{aligned} R_\alpha(T') &= D\left((\widehat{c}_t)_{t\in\mathcal{T}_{(T')}}\right) + \alpha|T'| \\ &= D\left((\widehat{c}_t)_{t\in\mathcal{T}_{(T')}}\right) + \alpha(|T|+1) \\ &= R_\alpha(T) + D\left((\widehat{c}_t)_{t\in\mathcal{T}_{(T')}}\right) - D\left((\widehat{c}_t)_{t\in\mathcal{T}_{(T)}}\right) + \alpha. \end{aligned}$$

Hence, depending on the value of α, the more complex tree T' may have a higher cost-complexity measure than T. We have $R_\alpha(T') \geq R_\alpha(T)$ if and only if

$$\alpha \geq D\left((\widehat{c}_t)_{t\in\mathcal{T}_{(T)}}\right) - D\left((\widehat{c}_t)_{t\in\mathcal{T}_{(T')}}\right). \tag{3.3.1}$$

In words, $R_\alpha(T') \geq R_\alpha(T)$ if and only if the deviance reduction that we get by producing tree T' is smaller than the increase in the penalty for having one more terminal node. In such cases, the deviance reduction is not sufficient to compensate the resulting increase in the penalty.

Therefore, using the cost-complexity measure for model selection may now lead to chose the simplest tree T over T'. If the value of α is such that condition (3.3.1) is fulfilled, then tree T will be preferred over T'. Otherwise, in case the value of α does not satisfy condition (3.3.1), the more complex tree T' will be preferred, meaning that the deviance reduction is higher than the corresponding increase in the penalty. In particular, this is the case when $\alpha = 0$, the cost-complexity measure then coinciding with the deviance.

Let T_{init} be the large tree that is to be pruned to the right sized tree T_{prune}. Of course, T_{init} may correspond to the largest possible tree $T_{\max}$. For a fixed value of α, we can now define $T(\alpha)$ as the subtree of T_{init} that minimizes the cost-complexity measure $R_\alpha(T)$, namely

$$R_\alpha(T(\alpha)) = \min_{T \preceq T_{\text{init}}} R_\alpha(T). \tag{3.3.2}$$

Hence, at this value of α, there is no subtree of T_{init} with lower cost-complexity measure than $T(\alpha)$.

When the penalty α per terminal node is small, the penalty $\alpha|T|$ for having a large tree is small as well so that $T(\alpha)$ will be large. For instance, if $\alpha = 0$, $R_\alpha(T)$ coincides with the deviance such that the largest tree T_{init} minimizes $R_\alpha(T)$. As the parameter α increases, the penalty for having a large tree increases and the subtrees $T(\alpha)$ will have less and less terminal nodes. Finally, for sufficiently large values of α, the subtree $T(\alpha)$ will consist of the root node only.

Because there can be more than one subtree of T_{init} minimizing $R_\alpha(T)$, we complement condition (3.3.2) with the following one:

$$\text{If } R_\alpha(T) = R_\alpha(T(\alpha)), \text{ then } T(\alpha) \preceq T. \tag{3.3.3}$$

This additional condition says that if there is a tie, namely more than one subtree of T_{init} minimizing $R_\alpha(T)$, then we select for $T(\alpha)$ the smallest tree, that is the one that is a subtree of all others satisfying (3.3.2). The resulting subtrees $T(\alpha)$ are called the smallest minimizing subtrees.

It is obvious that for every value of α there is at least one subtree of T_{init} that minimizes $R_\alpha(T)$ since there are only finitely many pruned subtrees of T_{init}. However, it is not clear whether the additional condition (3.3.3) can be met for every value of α. Indeed, this says that we cannot have two subtrees that minimize $R_\alpha(T)$ such that neither is a subtree of the other. This is guaranteed by the next proposition.

Proposition 3.3.1 *For every value of α, there exists a smallest minimizing subtree $T(\alpha)$.*

Proof We refer the interested reader to Sect. 10.2 (Theorem 10.7) in Breiman et al. (1984). □

Thus, for every value of $\alpha \geq 0$, there exists a unique subtree $T(\alpha)$ of T_{init} that minimizes $R_\alpha(T)$ and which satisfies $T(\alpha) \preceq T$ for all subtrees T minimizing $R_\alpha(T)$.

The large tree T_{init} has only a finite number of subtrees. Hence, even if α goes from zero to infinity in a continuous way, the set of the smallest minimizing subtrees $\{T(\alpha)\}_{\alpha\geq 0}$ only contains a finite number of subtrees of T_{init}.

Let $\alpha = 0$. We start from T_{init} and we find any pair of terminal nodes with a common parent node t such that the branch $T_{\text{init}}^{(t)}$ can be pruned without increasing the cost-complexity measure. We continue until we cannot longer find such pair in order to obtain a subtree of T_{init} with the same cost-complexity measure as T_{init} for $\alpha = 0$. We define $\alpha_0 = 0$ and we denote T_{α_0} the resulting subtree of T_{init}.

When α increases, it may become optimal to prune the branch $T_{\alpha_0}^{(t)}$ for a certain node t of T_{α_0}, meaning that the smaller tree $T_{\alpha_0} - T_{\alpha_0}^{(t)}$ becomes better than T_{α_0}. This will be the case once α is high enough to have

$$R_\alpha(T_{\alpha_0}) \geq R_\alpha\left(T_{\alpha_0} - T_{\alpha_0}^{(t)}\right).$$

The deviance of T_{α_0} can be written as

$$\begin{aligned}
D\left((\widehat{c}_s)_{s\in\mathcal{T}_{(T_{\alpha_0})}}\right) &= \sum_{s\in\mathcal{T}_{(T_{\alpha_0})}} D_{\chi_s}(\widehat{c}_s) \\
&= \sum_{s\in\mathcal{T}_{\left(T_{\alpha_0}-T_{\alpha_0}^{(t)}\right)}} D_{\chi_s}(\widehat{c}_s) + \sum_{s\in\mathcal{T}_{\left(T_{\alpha_0}^{(t)}\right)}} D_{\chi_s}(\widehat{c}_s) - D\left((\widehat{c}_s)_{s\in\{t\}}\right) \\
&= D\left((\widehat{c}_s)_{s\in\mathcal{T}_{\left(T_{\alpha_0}-T_{\alpha_0}^{(t)}\right)}}\right) + D\left((\widehat{c}_s)_{s\in\mathcal{T}_{\left(T_{\alpha_0}^{(t)}\right)}}\right) - D\left((\widehat{c}_s)_{s\in\{t\}}\right).
\end{aligned}$$

Furthermore, we have

$$|T_{\alpha_0}| = |T_{\alpha_0} - T_{\alpha_0}^{(t)}| + |T_{\alpha_0}^{(t)}| - 1.$$

The cost-complexity measure $R_\alpha(T_{\alpha_0})$ can then be rewritten as

$$\begin{aligned} R_\alpha(T_{\alpha_0}) = & R_\alpha\left(T_{\alpha_0} - T_{\alpha_0}^{(t)}\right) \\ & + D\left((\widehat{C}_s)_{s\in\widetilde{\mathcal{T}}_{\left(T_{\alpha_0}^{(t)}\right)}}\right) + \alpha|T_{\alpha_0}^{(t)}| \\ & - D\left((\widehat{C}_s)_{s\in\{t\}}\right) - \alpha. \end{aligned} \tag{3.3.4}$$

Thus, we have $R_\alpha(T_{\alpha_0}) \geq R_\alpha\left(T_{\alpha_0} - T_{\alpha_0}^{(t)}\right)$ if and only if

$$D\left((\widehat{C}_s)_{s\in\widetilde{\mathcal{T}}_{\left(T_{\alpha_0}^{(t)}\right)}}\right) + \alpha|T_{\alpha_0}^{(t)}| \geq D\left((\widehat{C}_s)_{s\in\{t\}}\right) + \alpha. \tag{3.3.5}$$

The left-hand side of (3.3.5) is the cost-complexity measure of the branch $T_{\alpha_0}^{(t)}$ while the right-hand side is the cost-complexity measure of the node t. Therefore, it becomes optimal to cut the branch $T_{\alpha_0}^{(t)}$ once its cost-complexity measure $R_\alpha\left(T_{\alpha_0}^{(t)}\right)$ becomes higher than the cost complexity measure $R_\alpha(t)$ of its root node t. This happens for values of α satisfying

$$\alpha \geq \frac{D\left((\widehat{C}_s)_{s\in\{t\}}\right) - D\left((\widehat{C}_s)_{s\in\widetilde{\mathcal{T}}_{\left(T_{\alpha_0}^{(t)}\right)}}\right)}{|T_{\alpha_0}^{(t)}| - 1}. \tag{3.3.6}$$

We denote by $\alpha_1^{(t)}$ the right-hand side of (3.3.6). For each non-terminal node t of T_{α_0} we can compute $\alpha_1^{(t)}$. For a tree T, let us denote by $\widetilde{\mathcal{T}}_{(T)}$ the set of its non-terminal nodes. Then, we define the weakest links of T_{α_0} as the non-terminal nodes t for which $\alpha_1^{(t)}$ is the smallest and we denote α_1 this minimum value, i.e.

$$\alpha_1 = \min_{t\in\widetilde{\mathcal{T}}_{(T_{\alpha_0})}} \alpha_1^{(t)}.$$

Cutting any branch of T_{α_0} is not optimal as long as $\alpha < \alpha_1$. However, once the parameter α reaches the value α_1, it becomes preferable to prune T_{α_0} at its weakest links. The resulting tree is then denoted T_{α_1}. Notice that it is appropriate to prune the tree T_{α_0} by exploring the nodes in top-down order. In such a way, we avoid cutting in a node t that will disappear later on when pruning T_{α_0}.

Now, we repeat the same process for T_{α_1}. Namely, for a non-terminal node t of T_{α_1}, it will be preferable to cut the branch $T_{\alpha_1}^{(t)}$ when

$$\alpha \geq \frac{D\left((\widehat{C}_s)_{s\in\{t\}}\right) - D\left((\widehat{C}_s)_{s\in\widetilde{\mathcal{T}}_{\left(T_{\alpha_1}^{(t)}\right)}}\right)}{|T_{\alpha_1}^{(t)}| - 1}. \tag{3.3.7}$$

We denote by $\alpha_2^{(t)}$ the right-hand side of (3.3.6) and we define

$$\alpha_2 = \min_{t \in \widetilde{\mathcal{T}}_{(T_{\alpha_1})}} \alpha_2^{(t)}.$$

The non-terminal nodes t of T_{α_1} for which $\alpha_2^{(t)} = \alpha_2$ are called the weakest links of T_{α_1} and it becomes better to cut in these nodes once α reaches the value α_2 in order to produce T_{α_2}.

Then we continue the same process for T_{α_2}, and so on until we reach the root node $\{t_0\}$. Finally, we come up with the sequence of trees

$$T_{\alpha_0}, T_{\alpha_1}, T_{\alpha_2}, \ldots, T_{\alpha_\kappa} = \{t_0\}.$$

In the sequence, we can obtain the next tree by pruning the current one.

The next proposition makes the link between this sequence of trees and the smallest minimizing subtrees $T(\alpha)$.

Proposition 3.3.2 *We have* $0 = \alpha_0 < \alpha_1 < \ldots < \alpha_\kappa < \alpha_{\kappa+1} = \infty$. *Furthermore, for all* $k = 0, 1, \ldots, \kappa$, *we have*

$$T(\alpha) = T_{\alpha_k} \quad \textit{for all } \alpha \in [\alpha_k, \alpha_{k+1}).$$

Proof We refer the interested reader to Sect. 10.2 in Breiman et al. (1984). □

This result is important since it gives the instructions to find the smallest minimizing subtrees $T(\alpha)$. It suffices to apply the recursive pruning steps described above. This recursive procedure is an efficient algorithm, any tree in the sequence being obtained by pruning the previous one. Hence, this algorithm only requires to consider a small fraction of the total possible subtrees of T_{init}.

The cost-complexity measure $R_\alpha(T)$ is given by

$$R_\alpha(T) = D\left((\widehat{c}_t)_{t \in \mathcal{T}_{(T)}}\right) + \alpha |T|,$$

where the parameter α is called the regularization parameter. The parameter α has the same unit as the deviance. It is often convenient to normalize the regularization parameter by the deviance of the root tree. We define the cost-complexity parameter as

$$\text{cp} = \frac{\alpha}{D\left((\widehat{c}_t)_{t \in \{t_0\}}\right)},$$

where $D\left((\widehat{c}_t)_{t \in \{t_0\}}\right)$ is the deviance of the root tree. The cost-complexity measure $R_\alpha(T)$ can then be rewritten as

$$\begin{aligned} R_\alpha(T) &= D\left((\widehat{c}_t)_{t \in \mathcal{T}_{(T)}}\right) + \alpha |T| \\ &= D\left((\widehat{c}_t)_{t \in \mathcal{T}_{(T)}}\right) + \text{cp}\, D\left((\widehat{c}_t)_{t \in \{t_0\}}\right) |T|. \end{aligned} \tag{3.3.8}$$

The sequence α_k, $k = 0, \ldots, \kappa$, defines the sequence cp_k, $k = 0, \ldots, \kappa$, for the cost-complexity parameters, with

$$\text{cp}_k = \frac{\alpha_k}{D\left((\widehat{c}_t)_{t\in\{t_0\}}\right)}.$$

3.3.1.1 Example

Consider the simulated dataset presented in Sect. 3.2.4.1. We use as T_{init} the tree depicted in Fig. 3.7, where a maximum depth of four has been used as stopping rule. The right sized tree T_{prune} is shown in Fig. 3.8 and is a subtree of T_{init}. The initial tree T_{init} is then large enough to start the pruning process.

Table 3.2 presents the decrease of the deviance ΔD_{χ_t} at each non-terminal node t of the initial tree T_{init}. The smallest decrease of the deviance is observed for node `11)`, also denoted t_{10}. Here, node t_{10} is the weakest node of T_{init}. It becomes optimal to cut the branch $T_{\text{init}}^{(t_{10})}$ for values of α satisfying

$$\begin{aligned}\alpha &\geq \frac{D\left((\widehat{c}_s)_{s\in\{t_{10}\}}\right) - D\left((\widehat{c}_s)_{s\in\mathcal{T}\left(T_{\text{init}}^{(t_{10})}\right)}\right)}{|T_{\text{init}}^{(t_{10})}| - 1}\\ &= \Delta D_{\chi_{t_{10}}}\\ &= 0.3254482.\end{aligned}$$

Therefore, we get $\alpha_1 = 0.3254482$ and hence, since $D\left((\widehat{c}_t)_{t\in\{t_0\}}\right) = 279\,043.30$,

$$\text{cp}_1 = \frac{\alpha_1}{D\left((\widehat{c}_t)_{t\in\{t_0\}}\right)} = \frac{0.3254482}{279\,043.30} = 1.1663\,10^{-6}.$$

Tree T_{α_1} is depicted in Fig. 3.13. Terminal nodes `22)` and `23)` of T_{init} disappeared in T_{α_1} and node `11)` becomes a terminal node.

The weakest node of T_{α_1} is node `10)` or t_9 with a decrease of the deviance $\Delta D_{\chi_{t_9}} = 1.13$. We cut the branch $T_{\alpha_1}^{(t_9)}$ once α satisfies

$$\begin{aligned}\alpha &\geq \frac{D\left((\widehat{c}_s)_{s\in\{t_9\}}\right) - D\left((\widehat{c}_s)_{s\in\mathcal{T}\left(T_{\alpha_1}^{(t_9)}\right)}\right)}{|T_{\alpha_1}^{(t_9)}| - 1}\\ &= \Delta D_{\chi_{t_9}}\\ &= 1.135009.\end{aligned}$$

We then have $\alpha_2 = 1.135009$ and

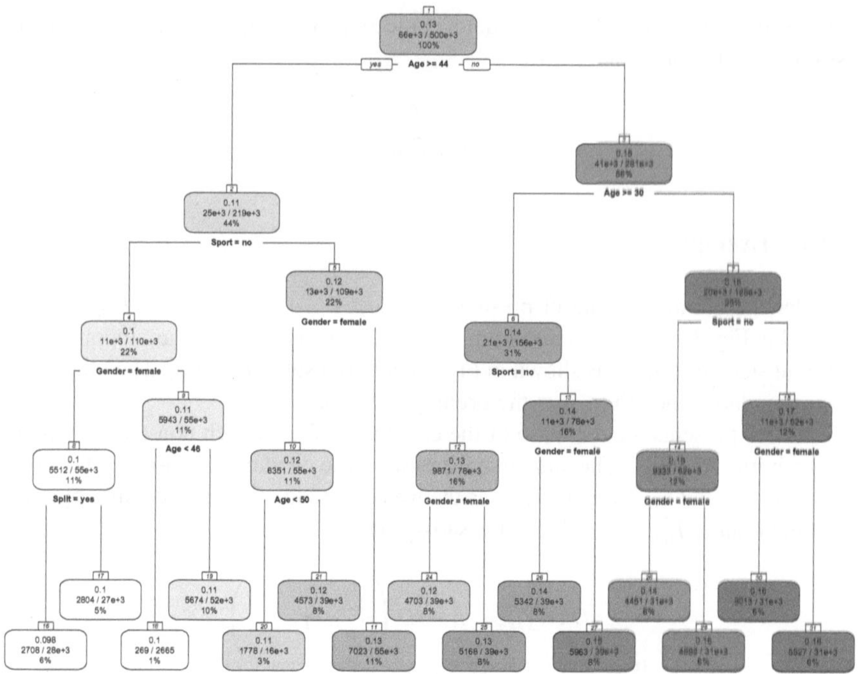

Fig. 3.13 Tree T_{α_1}

$$\text{cp}_2 = \frac{\alpha_2}{D\left((\widehat{c}_t)_{t \in \{t_0\}}\right)} = \frac{1.135009}{279\,043.30} = 4.0676\,10^{-6}.$$

Tree T_{α_2} is shown in Fig. 3.14 where node `10)` is now terminal.

In tree T_{α_2}, the smallest decrease of the deviance is at node `9)`. We have

$$\begin{aligned}
\alpha_3 &= \frac{D\left((\widehat{c}_s)_{s \in \{t_8\}}\right) - D\left((\widehat{c}_s)_{s \in \mathcal{T}_{\left(T_{\alpha_2}^{(t_8)}\right)}}\right)}{|T_{\alpha_2}^{(t_8)}| - 1} \\
&= \Delta D_{\chi_{t_8}} \\
&= 1.495393
\end{aligned}$$

and

$$\text{cp}_3 = \frac{\alpha_3}{D\left((\widehat{c}_t)_{t \in \{t_0\}}\right)} = \frac{1.495393}{279\,043.30} = 5.3590\,10^{-6}.$$

Tree T_{α_3} is shown in Fig. 3.15, node `9)` becomes a terminal node.

The smallest decrease of the deviance in T_{α_3} is at node `8)`. We have

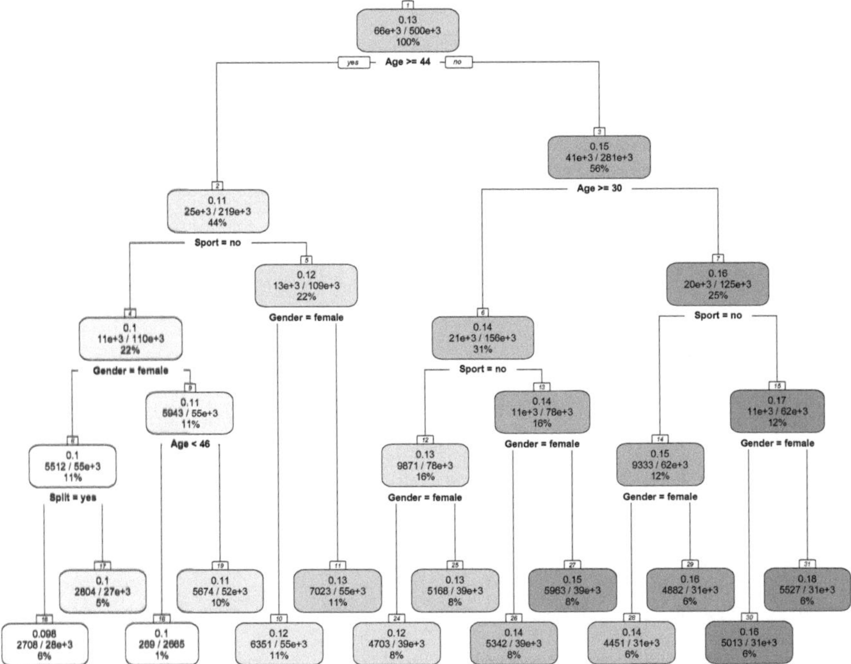

Fig. 3.14 Tree T_{α_2}

$$
\begin{aligned}
\alpha_4 &= \frac{D\left((\widehat{c}_s)_{s\in\{t_7\}}\right) - D\left((\widehat{c}_s)_{s\in\mathcal{T}\left(T_{\alpha_3}^{(t_7)}\right)}\right)}{|T_{\alpha_3}^{(t_7)}| - 1} \\
&= \Delta D_{\chi_{t_7}} \\
&= 3.515387
\end{aligned}
$$

and

$$
\mathrm{cp}_4 = \frac{\alpha_4}{D\left((\widehat{c}_t)_{t\in\{t_0\}}\right)} = \frac{3.515387}{279\,043.30} = 1.2598\,10^{-5}.
$$

Tree T_{α_4} is presented in Fig. 3.16, node 8) being now terminal.

Tree T_{α_4} corresponds to the right sized tree T_{prune} shown in Fig. 3.8. We then stop here the pruning process.

Of course, in situations of practical relevance, the right sized tree is usually not known. The pruning process is then stopped once we reach the root node t_0.

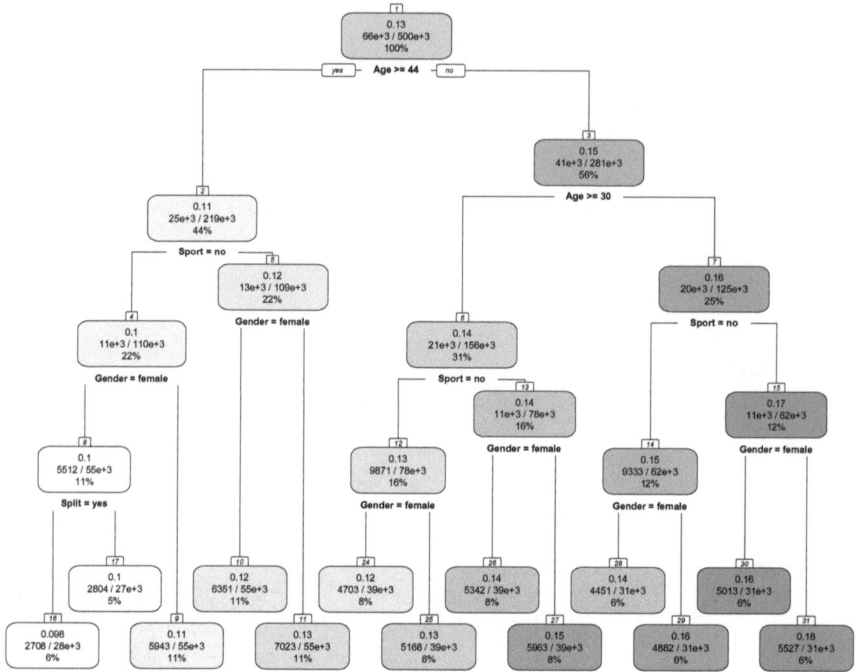

Fig. 3.15 Tree T_{α_3}

3.3.2 Choice of the Best Pruned Tree

Once the sequence of trees $T_{\alpha_0}, T_{\alpha_1}, T_{\alpha_2}, \ldots, T_{\alpha_\kappa} = \{t_0\}$ has been built, we need to select the best pruned tree. One way to proceed is to rely on cross-validation. We set

$$\tilde{\alpha}_k = \sqrt{\alpha_k \alpha_{k+1}},, \quad k = 0, 1, \ldots, \kappa,$$

where $\tilde{\alpha}_k$ is considered as a typical value for $[\alpha_k, \alpha_{k+1})$ and hence as the value corresponding to T_{α_k}. The parameter $\tilde{\alpha}_k$ corresponds to the geometric mean of α_k and α_{k+1}. Notice that $\tilde{\alpha}_0 = 0$ and $\tilde{\alpha}_\kappa = \infty$ since $\alpha_0 = 0$ and $\alpha_{\kappa+1} = \infty$.

In K−fold cross-validation, the training set $\mathcal{D}$ is partitioned into K subsets $\mathcal{D}_1, \mathcal{D}_2, \ldots, \mathcal{D}_K$ of roughly equal size and we label by $\mathcal{I}_j \subset \mathcal{I}$ the observations in $\mathcal{D}_j$ for all $j = 1, \ldots, K$. The jth training set is defined as $\mathcal{D} \backslash \mathcal{D}_j$, $j = 1, \ldots, K$, so that it contains the observations of the training set $\mathcal{D}$ that are not in $\mathcal{D}_j$. For each training set $\mathcal{D} \backslash \mathcal{D}_j$, we build the corresponding sequence of smallest minimizing subtrees $T^{(j)}(\tilde{\alpha}_0), T^{(j)}(\tilde{\alpha}_1), \ldots, T^{(j)}(\tilde{\alpha}_\kappa)$, starting from a sufficiently large tree $T^{(j)}_{\text{init}}$. Each tree $T^{(j)}(\tilde{\alpha}_k)$ of the sequence provides a partition $\{\chi_t\}_{t \in \mathcal{T}_{(T^{(j)}(\tilde{\alpha}_k))}}$ of the feature space χ and predictions $(\widehat{c}_t)_{t \in \mathcal{T}_{(T^{(j)}(\tilde{\alpha}_k))}}$ on that partition computed on the training set $\mathcal{D} \backslash \mathcal{D}_j$.

Since the observations in $\mathcal{D}_j$ have not been used to build the trees

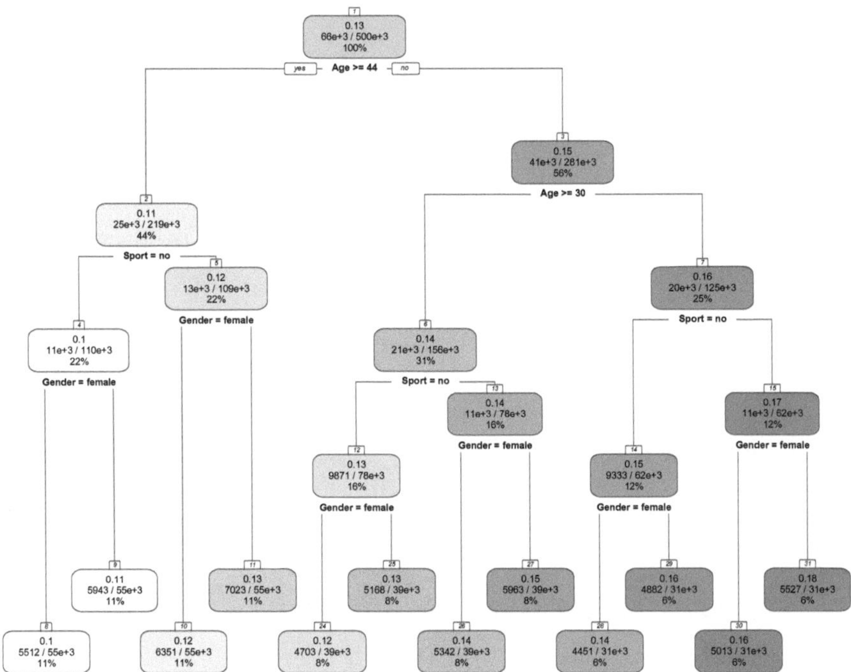

Fig. 3.16 Tree $T_{\alpha 4}$

$$T^{(j)}(\tilde{\alpha}_0), T^{(j)}(\tilde{\alpha}_1), \ldots, T^{(j)}(\tilde{\alpha}_\kappa),$$

$\mathcal{D}_j$ can play the role of validation set for those trees. If we denote by $\widehat{\mu}_{T^{(j)}(\tilde{\alpha}_k)}$ the model produced by tree $T^{(j)}(\tilde{\alpha}_k)$, that is

$$\widehat{\mu}_{T^{(j)}(\tilde{\alpha}_k)}(\boldsymbol{x}) = \sum_{t \in \mathcal{T}_{\left(T^{(j)}(\tilde{\alpha}_k)\right)}} \widehat{c}_t \mathrm{I}\left[\boldsymbol{x} \in \chi_t\right],$$

then an estimate of the generalization error of $\widehat{\mu}_{T^{(j)}(\tilde{\alpha}_k)}$ on $\mathcal{D}_j$ is given by

$$\widehat{Err}^{\text{val}}\left(\widehat{\mu}_{T^{(j)}(\tilde{\alpha}_k)}\right) = \frac{1}{|\mathcal{I}_j|} \sum_{i \in \mathcal{I}_j} L\left(y_i, \hat{\mu}_{T^{(j)}(\tilde{\alpha}_k)}(\boldsymbol{x}_i)\right).$$

So, the $K-$fold cross-validation estimate of the generalization error for the regularization parameter $\tilde{\alpha}_k$ is given by

$$\widehat{Err}^{CV}(\tilde{\alpha}_k) = \sum_{j=1}^{K} \frac{|\mathcal{I}_j|}{|\mathcal{I}|} \widehat{Err}^{val}\left(\widehat{\mu}_{T^{(j)}(\tilde{\alpha}_k)}\right)$$

$$= \frac{1}{|\mathcal{I}|} \sum_{j=1}^{K} \sum_{i \in \mathcal{I}_j} L\left(y_i, \widehat{\mu}_{T^{(j)}(\tilde{\alpha}_k)}(\boldsymbol{x}_i)\right).$$

According to the minimum cross-validation error principle, the right sized tree T_{prune} is then selected as the tree $T_{\alpha_{k^*}}$ of the sequence $T_{\alpha_0}, T_{\alpha_1}, T_{\alpha_2}, \ldots, T_{\alpha_\kappa}$ such that

$$\widehat{Err}^{CV}(\tilde{\alpha}_{k^*}) = \min_{k \in \{0,1,\ldots,\kappa\}} \widehat{Err}^{CV}(\tilde{\alpha}_k).$$

3.3.2.1 Example 1

Consider the simulated dataset of Sect. 3.2.4.1. Starting with the same initial tree T_{init} than in example of Sect. 3.3.1.1, we have

$$\alpha_1 = 0.3254482, \quad \alpha_2 = 1.135009, \quad \alpha_3 = 1.495393, \quad \alpha_4 = 3.515387, \quad \ldots$$

so that

$$\tilde{\alpha}_1 = 0.6077719, \quad \tilde{\alpha}_2 = 1.302799, \quad \tilde{\alpha}_3 = 2.29279, \quad \ldots$$

The values of cp_k and hence of α_k, $k = 1, \ldots, \kappa$, can be obtained from the R command `printcp()` so that we get

	CP	nsplit	rel error	xerror	xstd
1	3.5920e-03	0	1.00000	1.00000	0.0023034
2	9.3746e-04	1	0.99641	0.99642	0.0022958
3	5.6213e-04	2	0.99547	0.99549	0.0022950
4	3.1509e-04	3	0.99491	0.99493	0.0022945
5	2.6391e-04	4	0.99459	0.99470	0.0022944
6	1.1682e-04	5	0.99433	0.99437	0.0022934
7	1.1658e-04	6	0.99421	0.99436	0.0022936
8	9.5494e-05	7	0.99410	0.99424	0.0022936
9	8.4238e-05	8	0.99400	0.99413	0.0022938
10	7.7462e-05	9	0.99392	0.99406	0.0022936
11	6.0203e-05	10	0.99384	0.99396	0.0022936
12	1.2598e-05	11	0.99378	0.99386	0.0022932
13	5.3590e-06	12	0.99377	0.99392	0.0022936
14	4.0676e-06	13	0.99376	0.99393	0.0022936
15	1.1663e-06	14	0.99376	0.99395	0.0022937
16	0.0000e+00	15	0.99376	0.99395	0.0022937

The column `CP` contains the value of the complexity parameter $\text{cp}_k, k = 0, 1, \ldots, \kappa$. From bottom to top, we have $\text{cp}_0 = 0$, $\text{cp}_1 = 1.1663\,10^{-6}$, $\text{cp}_2 = 4.0676\,10^{-6}$, $\text{cp}_3 = 5.3590\,10^{-6}$, $\text{cp}_4 = 1.2598\,10^{-5}$ and so on up to $\text{cp}_{15} = 3.5920\,10^{-3}$ so that $\kappa = 15$. Of course, the values for cp_1 up to cp_4 coincide with those computed in example of Sect. 3.3.1.1. The value of α_k is obtained by multiplying cp_k by the deviance of the root tree $D\left((\widehat{c}_t)_{t\in\{t_0\}}\right) = 279\,043.30$.

The second column `nsplit` gives the number of splits of the corresponding tree T_{α_k}. One sees that $T_{\alpha_0} = T_{\text{init}}$ has 15 splits, T_{α_1} presents 14 splits and so on up to the root tree $T_{\alpha_{15}} = \{t_0\}$ which has 0 split.

The third column `rel error` provides the ratio between the deviance of T_{α_k} and the deviance of the root tree $\{t_0\}$, namely

$$\texttt{rel error}_k = \frac{D\left((\widehat{c}_t)_{t\in\mathcal{T}_{(T_{\alpha_k})}}\right)}{D\left((\widehat{c}_t)_{t\in\{t_0\}}\right)}.$$

This relative error is a training sample estimate. It starts from the smallest value $\texttt{rel error}_0 = 0.99376$ to increase up to $\texttt{rel error}_{15} = 1$ since the trees become smaller and smaller to reach the root tree. Obviously, choosing the optimal tree based on the relative error would always favor the largest tree.

Turning to the fourth column `xerror`, it provides the relative 10-fold cross-validation error estimate for parameter $\tilde{\alpha}_k$, namely

$$\texttt{xerror}_k = \frac{|\mathcal{I}|\,\widehat{Err}^{\text{CV}}(\tilde{\alpha}_k)}{D\left((\widehat{c}_t)_{t\in\{t_0\}}\right)},$$

where, in this example, $|\mathcal{I}| = n = 500\,000$. For instance, we have $\texttt{xerror}_0 = 0.99395$, $\texttt{xerror}_1 = 0.99395$ and so on up to $\texttt{xerror}_{15} = 1$. Hence, using the minimum cross-validation error principle, the right-sized tree turns out to be T_{α_4} with a relative 10-fold cross-validation error equal to 0.99386. The dataset being simulated, we know that T_{α_4} shown in Fig. 3.16 indeed coincides with the best possible tree.

The last column `xstd` will be commented in the next example.

3.3.2.2 Example 2

Consider the real dataset described in Sect. 3.2.4.2. The tree depicted in Fig. 3.11 with 17 terminal nodes, that has been obtained by requiring a minimum number of observations in terminal nodes equal to 5000, is used as T_{init}. We get the following results:

```
         CP nsplit rel error  xerror      xstd
1 9.3945e-03      0   1.00000 1.00003 0.0048524
2 4.0574e-03      1   0.99061 0.99115 0.0047885
3 2.1243e-03      2   0.98655 0.98719 0.0047662
4 1.4023e-03      3   0.98442 0.98524 0.0047565
5 5.1564e-04      4   0.98302 0.98376 0.0047441
6 4.5896e-04      5   0.98251 0.98344 0.0047427
7 4.1895e-04      6   0.98205 0.98312 0.0047437
```

```
8  2.3144e-04     7   0.98163 0.98254 0.0047440
9  1.9344e-04     8   0.98140 0.98244 0.0047428
10 1.6076e-04     9   0.98120 0.98233 0.0047405
11 1.5651e-04    10   0.98104 0.98225 0.0047400
12 1.4674e-04    11   0.98089 0.98225 0.0047400
13 1.0746e-04    12   0.98074 0.98213 0.0047390
14 8.6425e-05    13   0.98063 0.98200 0.0047385
15 8.6149e-05    14   0.98055 0.98189 0.0047374
16 5.3416e-05    15   0.98046 0.98197 0.0047391
17 0.0000e+00    16   0.98041 0.98204 0.0047396
```

The sequence of trees $T_{\alpha_0}, T_{\alpha_1}, \ldots, T_{\alpha_\kappa}$ is composed of 17 trees, namely $\kappa = 16$. The minimum 10-fold cross-validation error is equal to 0.98189 and corresponds to tree T_{α_2} with a complexity parameter $\text{cp}_2 = 8.6149\,10^{-5}$ and 14 splits. T_{α_2} is shown in Fig. 3.17. So, it is the tree with the minimum 10-fold cross-validation error when requiring at least 5000 observations in the terminal nodes. Notice that T_{α_2} with 15 terminal nodes is relatively big compared to T_{init} with 17 terminal nodes. The reason is that any terminal node of the initial tree must contain at least 5000 observations so that the size of T_{init} is limited.

For any $\tilde{\alpha}_k$, $k = 0, 1, \ldots, \kappa$, the standard deviation estimate

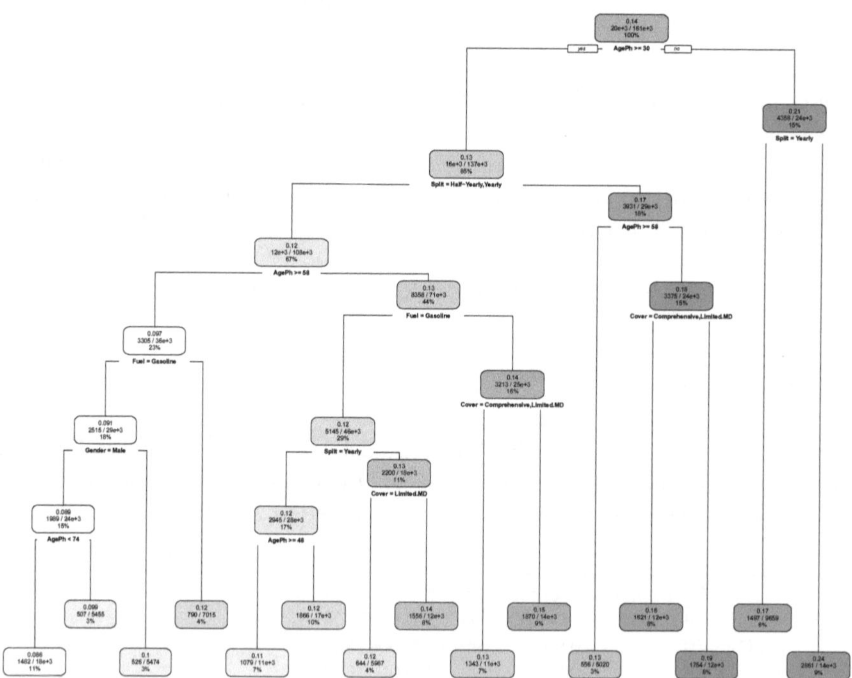

Fig. 3.17 Tree with minimum cross-validation error when requiring at least 5000 observations in the terminal nodes

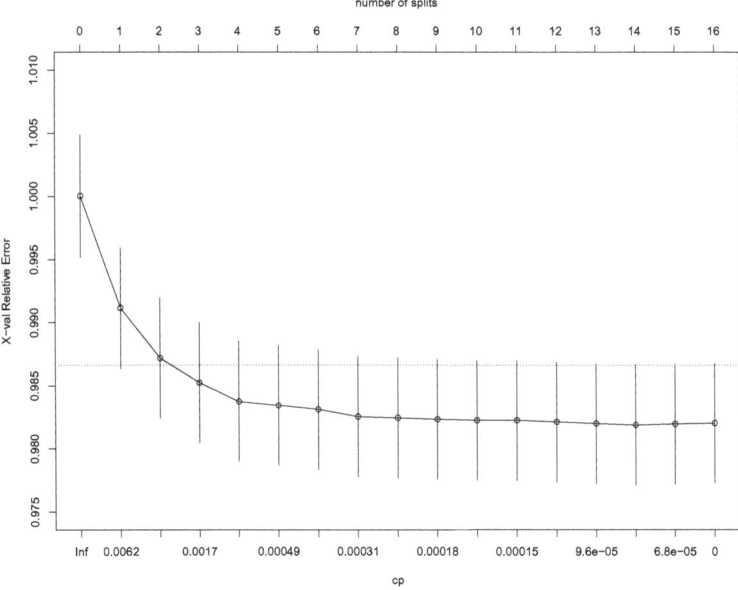

Fig. 3.18 Relative cross-validation error $\texttt{xerror}_k$ together with the relative standard error $\texttt{xstd}_k$

$$\widehat{Var}\left(\widehat{Err}^{CV}(\tilde{\alpha}_k)\right)^{1/2}$$

can be estimated empirically over the K estimates of the generalization error. The last column `xstd` provides an estimate of the relative standard error of the cross-validation error, namely

$$\texttt{xstd}_k = \frac{|\mathcal{I}|\,\widehat{Var}\left(\widehat{Err}^{CV}(\tilde{\alpha}_k)\right)^{1/2}}{D\left((\widehat{c}_t)_{t\in\{t_0\}}\right)},$$

where, in this example, $|\mathcal{I}| = n = 160\,944$. Figure 3.18 shows the relative cross-validation error $\texttt{xerror}_k$ together with the relative standard error $\texttt{xstd}_k$ for each tree T_{α_k}. From right to left, we start with T_{α_0} which corresponds to a complexity parameter equal to 0 and 16 splits to end with the root node tree $T_{\alpha_{16}}$ with 0 split.

As we can see, the tree of the sequence with only 3 splits, namely $T_{\alpha_{13}}$, is within one standard deviation (SD) of the tree T_{α_2} with the minimum cross-validation error. The 1-SD rule consists in selecting the smallest tree that is within one standard deviation of the tree that minimizes the cross-validation error. This rule recognizes that there is some uncertainty in the estimate of the cross-validation error and chooses the simplest tree whose accuracy is still judged acceptable. Hence, according to the 1-SD rule, the tree selected is $T_{\alpha_{k^{**}}}$ where k^{**} is the maximum $k \in \{k^*, k^*+1, \ldots, \kappa\}$ satisfying

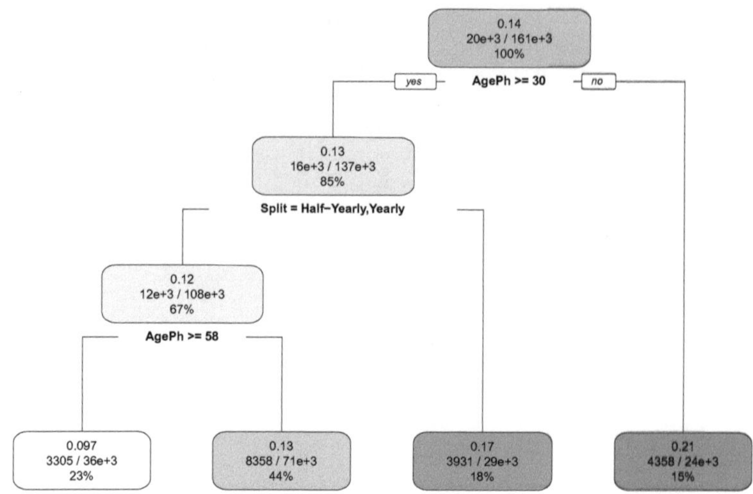

Fig. 3.19 Tree $T_{\alpha_{k^{**}}} = T_{\alpha 13}$

$$\widehat{Err}^{CV}(\tilde{\alpha}_k) \leq \widehat{Err}^{CV}(\tilde{\alpha}_{k^*}) + \widehat{Var}\left(\widehat{Err}^{CV}(\tilde{\alpha}_{k^*})\right)^{1/2}.$$

In our case, $T_{\alpha_{k^{**}}} = T_{\alpha_{13}}$, which is depicted in Fig. 3.19. Compared to tree $T_{\alpha_{k^*}} = T_{\alpha_2}$ which minimizes the cross-validation error, $T_{\alpha_{13}}$ is much more simpler. The number of terminal nodes decreases by 11, going from 15 in T_{α_2} to 4 in $T_{\alpha_{13}}$.

As a result, the minimum cross-validation principle selects T_{α_2} while the 1-SD rule chooses $T_{\alpha_{13}}$. Both trees can be compared on a validation set by computing their respective generalization errors.

3.3.2.3 Example 3

Consider once again the real dataset described in Sect. 3.2.4.2. This time, a stratified random split of the available dataset $\mathcal{L}$ is done to get a training set $\mathcal{D}$ and a validation set $\overline{\mathcal{D}}$. In this example, the training set $\mathcal{D}$ is assumed to be composed of 80% of the observations of the learning set $\mathcal{L}$ and the validation set $\overline{\mathcal{D}}$ of the 20% remaining observations.

The initial tree T_{init} is grown on the training set by requiring a minimum number of observations in terminal nodes equal to $4000 = 80\% \times 5000$. The resulting tree T_{init} is shown in Fig. 3.20 and only differs from the previous initial tree obtained on the whole dataset $\mathcal{L}$ at only one split in the bottom. Conducting the pruning process, we get the following results that are also illustrated in Fig. 3.21:

	CP	nsplit	rel error	xerror	xstd
1	9.1497e-03	0	1.00000	1.00001	0.0054547
2	3.7595e-03	1	0.99085	0.99160	0.0053964
3	2.2473e-03	2	0.98709	0.98853	0.0053797
4	1.5067e-03	3	0.98484	0.98607	0.0053688
5	6.0567e-04	4	0.98334	0.98417	0.0053489
6	4.6174e-04	5	0.98273	0.98362	0.0053447
7	4.5173e-04	6	0.98227	0.98337	0.0053502
8	2.1482e-04	7	0.98182	0.98301	0.0053520
9	2.0725e-04	9	0.98139	0.98304	0.0053529
10	1.6066e-04	10	0.98118	0.98297	0.0053534
11	1.5325e-04	11	0.98102	0.98293	0.0053548
12	1.1692e-04	12	0.98087	0.98294	0.0053558
13	9.0587e-05	13	0.98075	0.98302	0.0053580
14	8.4270e-05	15	0.98057	0.98292	0.0053562
15	0.0000e+00	16	0.98048	0.98285	0.0053555

The tree that minimizes the cross-validation error is the initial tree T_{α_0}, which means that requiring at least 4000 observations in the terminal nodes already prevents for overfitting. The 1-SD rule selects the tree $T_{\alpha_{11}}$ with only 3 splits, depicted in Fig. 3.22.

In order to compare both trees $T_{\alpha_{k^*}} = T_{\alpha_0}$ and $T_{\alpha_{k^{**}}} = T_{\alpha_{11}}$, we estimate their respective generalization errors on the validation set $\overline{\mathcal{D}}$. Remember that the validation

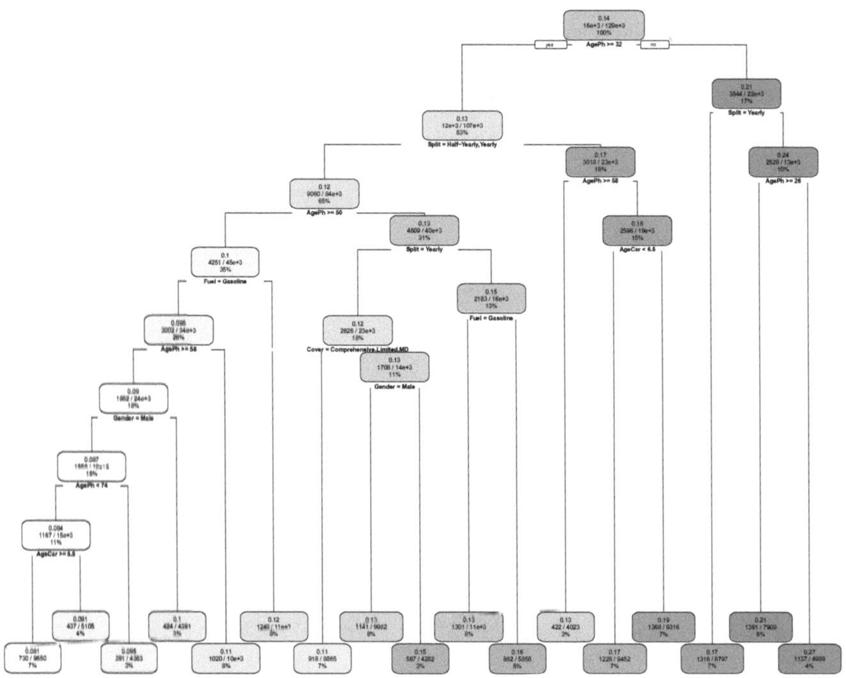

Fig. 3.20 Initial tree when requiring at least 4000 observations in the terminal nodes

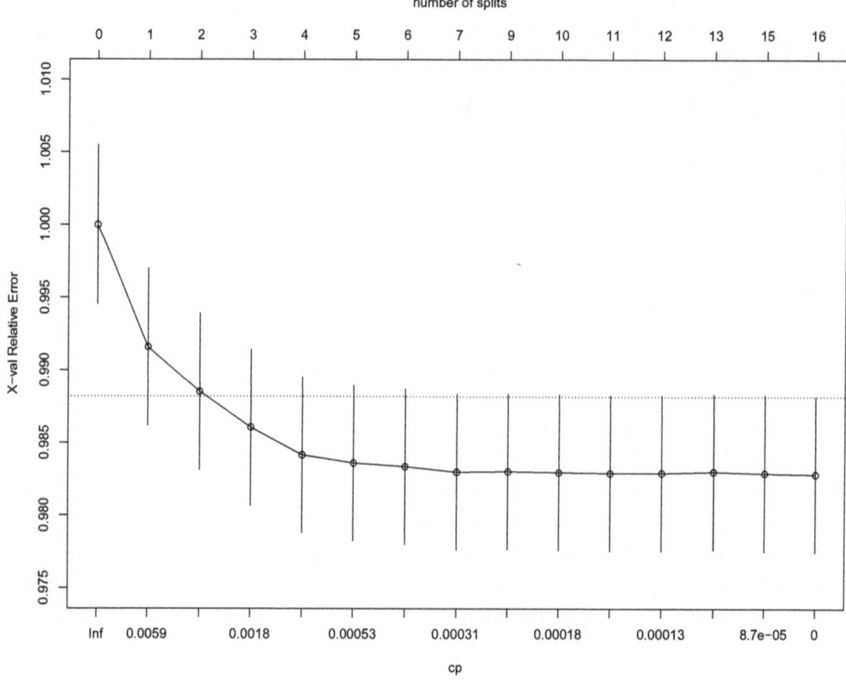

Fig. 3.21 Relative cross-validation error $\texttt{xerror}_k$ together with the relative standard error $\texttt{xstd}_k$

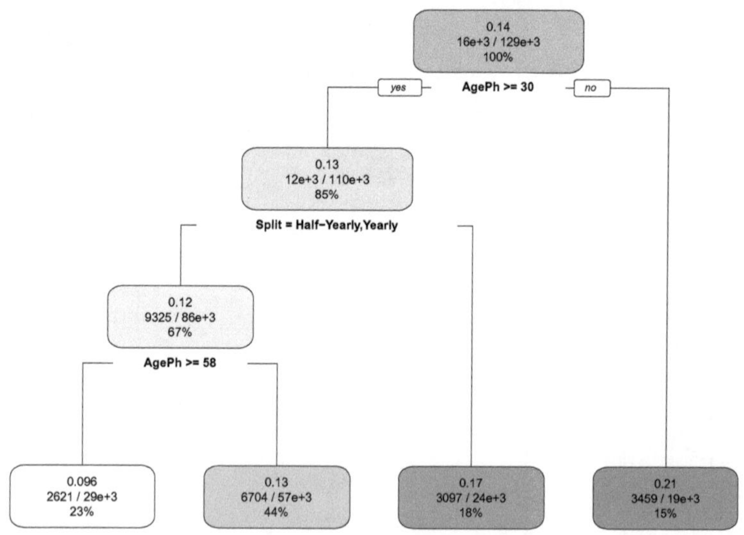

Fig. 3.22 Tree $T_{\alpha_{k^{**}}} = T_{\alpha_{11}}$

sample estimate of the generalization error for a tree T fitted on the training set is given by

$$\widehat{Err}^{\text{val}}\left(\widehat{\mu}_T\right) = \frac{1}{|\overline{\mathcal{I}}|} \sum_{i \in \overline{\mathcal{I}}} L\left(y_i, \widehat{\mu}_T(\boldsymbol{x}_i)e_i\right),$$

where $\widehat{\mu}_T$ corresponds to the model induced by tree T. We get

$$\widehat{Err}^{\text{val}}\left(\widehat{\mu}_{T_{\alpha_{k^*}}}\right) = 0.5452772$$

and

$$\widehat{Err}^{\text{val}}\left(\widehat{\mu}_{T_{\alpha_{k^{**}}}}\right) = 0.5464333.$$

The tree $T_{\alpha_{k^*}}$ that minimizes the cross-validation error is also the one that minimizes the generalization error estimated on the validation set. Hence, $T_{\alpha_{k^*}}$ has the best predictive accuracy compared to $T_{\alpha_{k^{**}}}$ and is thus judged as the best tree.

The difference between both generalization error estimates $\widehat{Err}^{\text{val}}\left(\widehat{\mu}_{T_{\alpha_{k^*}}}\right)$ and $\widehat{Err}^{\text{val}}\left(\widehat{\mu}_{T_{\alpha_{k^{**}}}}\right)$ is around 10^{-3}. Such a difference appears to be significant in this context. The validation sample estimate of the generalization error for the root node tree $\{t_0\}$ is given by

$$\widehat{Err}^{\text{val}}\left(\widehat{\mu}_{\{t_0\}}\right) = 0.54963.$$

The generalization error estimate only decreases by 0.0043528 from the root node tree $\{t_0\}$, that is the null model, to the optimal one $T_{\alpha_{k^*}}$.

The generalization error of a model $\widehat{\mu}$ can be decomposed as the sum of the generalization error of the true model μ and an estimation error that is positive. The generalization error of the true model μ is irreducible. Provided that the generalization error of the true model μ is large compared to the estimation error of the null model, a small decrease of the generalization error can actually mean a significant improvement.

3.4 Measure of Performance

The generalization error $Err\left(\widehat{\mu}\right)$ measures the performance of the model $\widehat{\mu}$. Specifically, its validation sample estimate $\widehat{Err}^{\text{val}}\left(\widehat{\mu}\right)$ enables to assess the predictive accuracy of $\widehat{\mu}$. However, as noticed in the previous section, there are some situations where the validation sample estimate only slightly reacts to a model change while such a small variation could actually reveal a significant improvement in terms of model accuracy.

Consider the simulated dataset of Sect. 3.2.4.1. Because the true model μ is known in this example, we can estimate the generalization error of the true model on the whole dataset, that is

$$\widehat{Err}(\mu) = 0.5546299. \tag{3.4.1}$$

Let $\widehat{\mu}_{\text{null}}$ be the model obtained by averaging the true expected claim frequencies over the observations. We get $\widehat{\mu}_{\text{null}} = 0.1312089$, such that its generalization error estimated on the whole dataset is

$$\widehat{Err}(\widehat{\mu}_{\text{null}}) = 0.5580889. \tag{3.4.2}$$

The difference between both error estimates $\widehat{Err}(\mu)$ and $\widehat{Err}(\widehat{\mu}_{\text{null}})$ is 0.003459. We observe that the improvement we get in terms of generalization error by using the true model μ instead of the null model $\widehat{\mu}_{\text{null}}$ is only of the order of 10^{-3}. A slight decrease of the generalization error can actually mean a real improvement in terms of model accuracy.

3.5 Relative Importance of Features

In insurance, the features are not all of equally importance for the response. Often, only a few of them have substantial influence on the response. Assessing the relative importances of the features to the response can thus be useful for the analyst.

For a tree T, the relative importance of feature x_j is the total reduction of deviance obtained by using this feature throughout the tree. The overall objective is to minimize the deviance. A feature that contributes a lot to this reduction will be more important than another one with a small or no deviance reduction. Specifically, denoting by $\widetilde{\mathcal{T}}_{(T)}(x_j)$ the set of all non-terminal nodes of tree T for which x_j was selected as the splitting feature, the relative importance of feature x_j is the sum of the deviance reductions ΔD_{χ_t} over the non-terminal nodes $t \in \widetilde{\mathcal{T}}_{(T)}(x_j)$, that is,

$$\mathcal{I}(x_j) = \sum_{t \in \widetilde{\mathcal{T}}_{(T)}(x_j)} \Delta D_{\chi_t}. \tag{3.5.1}$$

The features can be ordered with respect to their relative importances. The feature with the largest relative importance is the most important one, the feature with the second largest relative importance is the second most important one, and so on up to the least important feature. The most important features are those that appear higher in the tree or several times in the tree.

Note that the relative importances are relative measures, so that they can be normalized to improve their readability. It is customary to assign the largest a value of 100 and then scale the others accordingly. Another way is to scale the relative importances such that their sum equals 100, so that any relative importance can be interpreted as the percentage contribution to the overall model.

3.5.1 Example 1

As an illustration, let us compute the relative importances in the example of Sect. 3.2.4.1. Table 3.6 shows the deviance reduction ΔD_{χ_t} for any non-terminal node t of the optimal tree depicted in Fig. 3.8. We then get

$$
\begin{aligned}
\mathcal{I}(Age) &= 1002.20 + 261.65 = 1263.85 \\
\mathcal{I}(Sport) &= 156.89 + 87.92 + 73.64 = 318.45 \\
\mathcal{I}(Gender) &= 16.80 + 32.60 + 23.51 + 32.53 + 21.62 + 26.65 = 153.71 \\
\mathcal{I}(Split) &= 0.
\end{aligned}
$$

The sum of the relative importances is given by

$$\mathcal{I}(Age) + \mathcal{I}(Sport) + \mathcal{I}(Gender) + \mathcal{I}(Split) = 1736.01.$$

Normalizing the relative importances such that their sum equals 100, we get

Table 3.6 Decrease of the deviance ΔD_{χ_t} for any non-terminal node t of the optimal tree

Node k)	Splitting feature	$\Delta D_{\chi_{t_{k-1}}}$
1)	Age	279 043.30–(111779.70+166261.40) = 1002.20
2)	Sport	111779.70–(53386.74+58236.07) = 156.89
3)	Age	166261.40–(88703.96+77295.79) = 261.65
4)	Gender	53386.74–(26084.89+27285.05) = 16.80
5)	Gender	58236.07–(28257.29+29946.18) = 32.60
6)	Sport	88703.96–(42515.21+46100.83) = 87.92
7)	Sport	77295.79–(37323.97+39898.18) = 73.64
12)	Gender	42515.21–(20701.11+21790.59) = 23.51
13)	Gender	46100.83–(22350.27+23718.03) = 32.53
14)	Gender	37323.97–(18211.52+19090.83) = 21.62
15)	Gender	39898.18–(19474.59+20396.94) = 26.65

$$\mathcal{I}(Age) = 1263.85/17.3601 = 72.80$$
$$\mathcal{I}(Sport) = 318.45/17.3601 = 18.34$$
$$\mathcal{I}(Gender) = 153.71/17.3601 = 8.85$$
$$\mathcal{I}(Split) = 0.$$

The feature Age is the most important one, followed by Sport and Gender, which is in line with our expectation. Notice that the variable Split has an importance equals to 0 as it is not used in the tree.

3.5.2 Example 2

Consider the example of Sect. 3.2.4.2. The relative importances of features related to the tree depicted in Fig. 3.17 are shown in Fig. 3.23. One sees that the relative importance of features AgePh and Split represents approximately 90% of the total importance while the relative importance of the last five features (that are Cover, AgeCar, Gender, PowerCat and Use) is less than 5%.

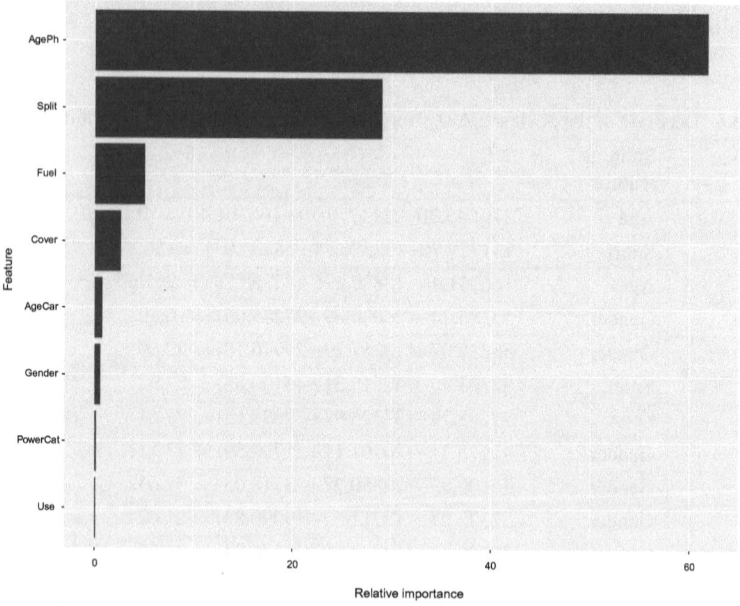

Fig. 3.23 Relative importances of the features related to the tree depicted in Fig. 3.17

3.5.3 *Effect of Correlated Features*

Most of the time, the features X_j are correlated in observational studies. When the correlation between some features becomes large, trees may run into trouble, as illustrated next. Such high correlations mean that the same information is encoded in several features.

Consider the example of Sect. 3.2.4.1 and assume that variables X_1 (Gender) and X_3 (Split) are correlated. In order to generate the announced correlation, we suppose that the distribution of females inside the portfolio differs according to the split of the premium. Specifically,

$$\begin{aligned} \mathrm{P}[X_1 = female|X_3 = yes] &= 1 - \mathrm{P}[X_1 = male|X_3 = yes] \\ &= \rho \end{aligned}$$

with $\rho \in [0.5, 1]$. Since $\mathrm{P}[X_1 = female] = 0.5$ and $\mathrm{P}[X_3 = yes] = 0.5$, we necessarily have

$$\begin{aligned} \mathrm{P}[X_1 = female|X_3 = no] &= 1 - \mathrm{P}[X_1 = male|X_3 = no] \\ &= 1 - \rho. \end{aligned}$$

The correlation between X_1 and X_3 increases with ρ. In particular, the case $\rho = 0.5$ corresponds to the independent case while both features X_1 and X_3 are perfectly correlated when $\rho = 1$.

We consider different values for ρ, from 0.5 to 1 by 0.1 steps. For any value of ρ, we generate a training set with 500 000 observations on which we build a tree minimizing the 10-fold cross validation error. The corresponding relative importances are depicted in Fig. 3.24. One sees that the importance of the variable Split increases with ρ, starting from 0 when $\rho = 0.5$ to completely replace the variable Gender when $\rho = 1$. Also, one observes that the more important the variable Split, the less the variable Gender.

The variable Split should not be used to explain the expected claim frequency, which is the case when Split and Gender are independent. However, introducing a correlation between Split and Gender leads to a transfer of importance from Gender to Split. As a result, the variable Split seems to be useful to explain the expected claim frequency while the variable Gender appears to be less important than it should be. This effect becomes even more pronounced as the correlation increases, the variable Split bringing more and more information about Gender. In the extreme case where Split and Gender are perfectly dependent, using Split instead of Gender always yields the same deviance reduction, so that both variables becomes fully equivalent. Note that in such a situation, the variable Split was automatically selected over Gender so that Gender has no importance in Fig. 3.24 for $\rho = 1$.

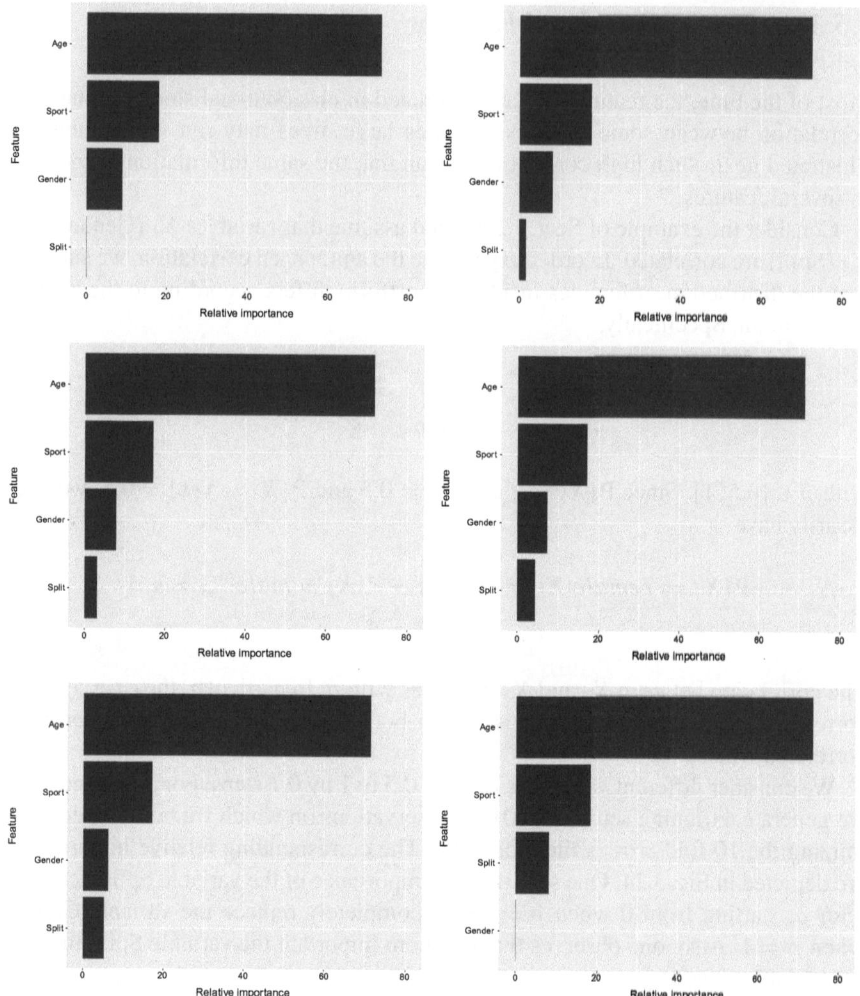

Fig. 3.24 Relative importances of the features for different values of ρ: $\rho = 0.5$ (top left), $\rho = 0.6$ (top right), $\rho = 0.7$ (middle left), $\rho = 0.8$ (middle right), $\rho = 0.9$ (bottom left) and $\rho = 1$ (bottom right)

For a set of features, here Split and Gender, we might have all the importances be small, yet we cannot delete them all. This little example shows that the relative importances could easily lead to wrong conclusions if they are not used carefully by the analyst.

3.6 Interactions

Interaction arises when the effect of a particular feature is reliant on the value of another. An example in motor insurance is given by driver's age and gender: often, young female drivers cause on average less claims compared to young male ones whereas this gender difference disappears (and sometimes even reverses) at older ages. Hence, the effect of age depends on gender. Regression trees automatically accounts for interactions.

Let us revisit the example of Sect. 3.2.4.1 for which we now assume the following expected annual claim frequencies:

$$\begin{aligned}\mu(\boldsymbol{x}) &= 0.1 \times (1 + 0.1\mathrm{I}[x_1 = male]) \\ &\quad \times (1 + (0.4 + 0.2\mathrm{I}[x_1 = male])\,\mathrm{I}[18 \leq x_2 < 30] + 0.2\mathrm{I}[30 \leq x_2 < 45]) \\ &\quad \times (1 + 0.15\mathrm{I}[x_4 = yes])\,. \end{aligned} \tag{3.6.1}$$

This time, the effect of the age on the expected claim frequency depends on the policyholder's gender. Young male drivers are indeed more risky than young female drivers in our example, young meaning here younger than 30 years old. Equivalently, the effect of the gender on the expected claim frequency depends on the policyholder's age. Indeed, a man and a woman both with $18 \leq \text{Age} < 30$ and with the same value for the feature Sport have expected claim frequencies that differ by $\frac{1.1 \times 1.6}{1.4} - 1 = 25.71\%$ while a man and a woman both with $\text{Age} \geq 30$ and with the same value for the feature Sport have expected claim frequencies that only differ by 10%. One says that features Gender and Age interact.

The optimal regression tree is shown in Fig. 3.25. By nature, the structure of a regression tree enables to account for potential interactions between features. In Fig. 3.25, the root node is split with the rule $\text{Age} \geq 30$. Hence, once the feature Gender is used for a split on the left hand side of the tree (node 2) and children), it only applies for policyholders with $\text{Age} \geq 30$, while when it appears on the right hand side of the tree (node 3) and children), it only applies for policyholders with $18 \leq \text{Age} < 30$. By construction, the impact of the feature Gender can then be different for categories $18 \leq \text{Age} < 30$ and $\text{Age} \geq 30$. The structure of a tree, which is a succession of binary splits, can thus easily reveal existing interactions between features.

Remark 3.6.1 Interaction has nothing to do with correlation. For instance, consider a motor insurance portfolio with the same age structure for males and females (so that the risk factors age and gender are mutually independent). If young male drivers are more dangerous compared to young female drivers whereas this ranking disappears or reverses at older ages, then age and gender interact despite being independent.

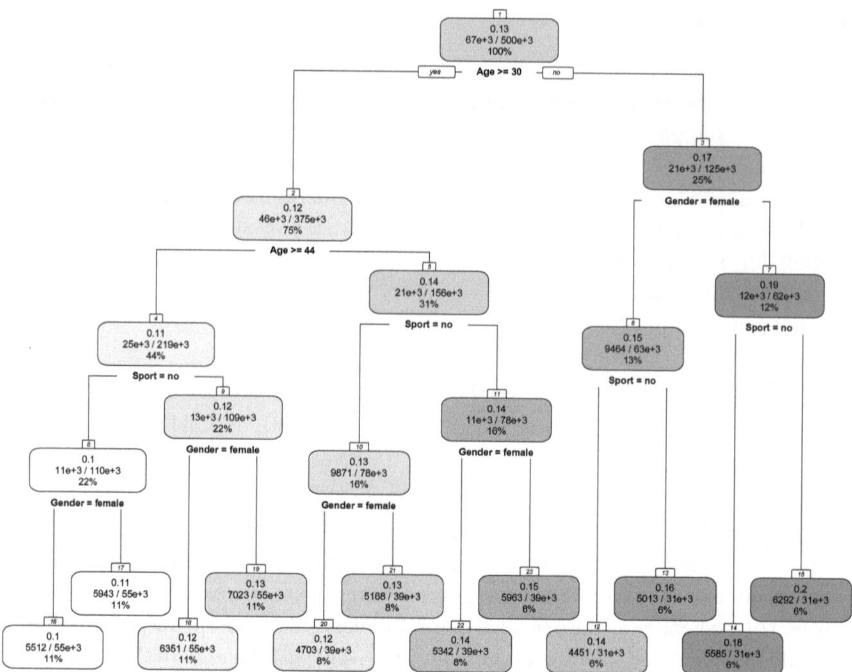

Fig. 3.25 Optimal tree

3.7 Limitations of Trees

3.7.1 *Model Instability*

One issue with trees is their high variance. There is a high variability of the prediction $\widehat{\mu}_{\mathcal{D}}(\boldsymbol{x})$ over the models trained from all possible training sets. The main reason is due to the hierarchical nature of the procedure. A change in one of the top splits is propagated to all the subsequent splits. A way to remedy that problem is to rely on bagging that averages many trees. This has the effect to reduce the variance. But the price to be paid for stabilizing the prediction is to deal with less comprehensive models so that we lose in terms of model interpretability. Instead of relying on one tree, Bagging works with many trees. Ensemble learning techniques such as Bagging will be studied in Chap. 4.

Consider the example of Sect. 3.2.4.1. The true expected claim frequency only takes twelve values, so that the complexity of the true model is small. We first generate 10 training sets with 5000 observations. On each training set, we fit a tree with only one split. Figure 3.26 shows the resulting trees. As we can see, even in this simple example, the feature used to make the split of the root node is not always the

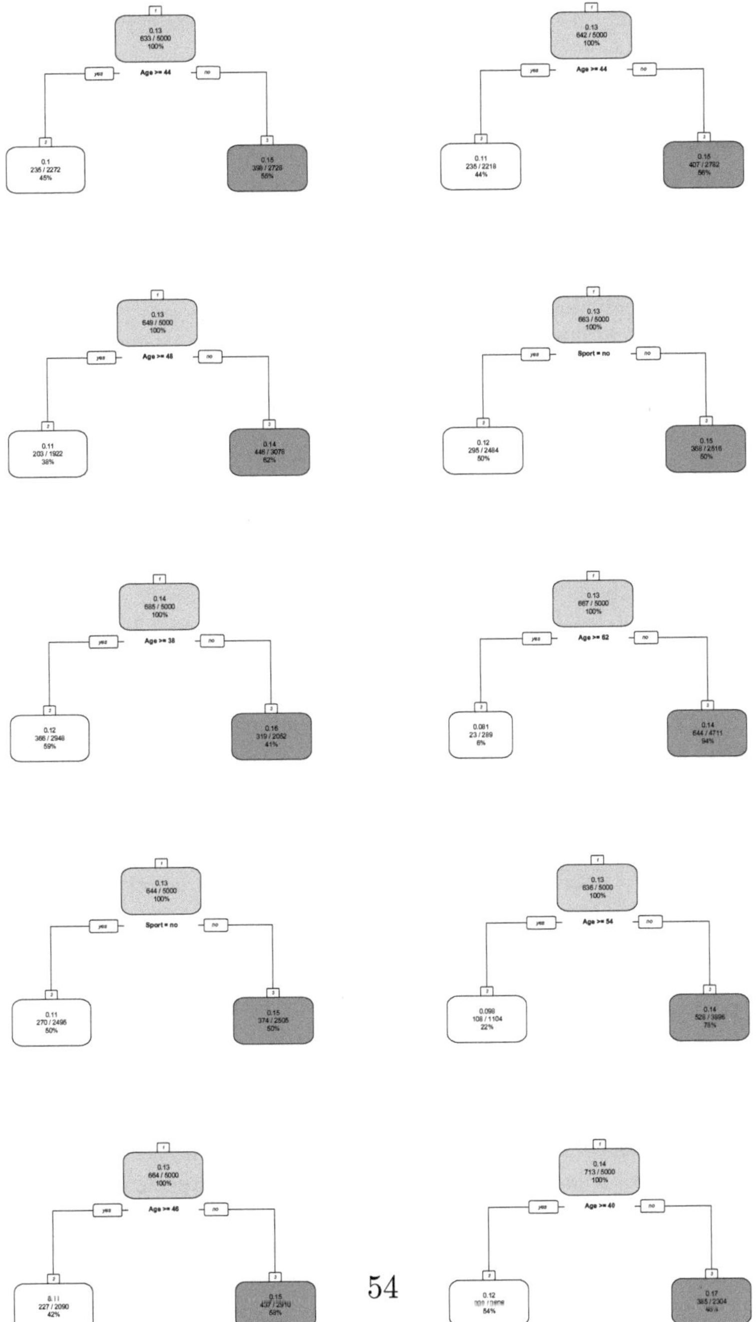

Fig. 3.26 Trees with only one split built on different training sets of size 5000

same. Furthermore, the value of the split when using the feature Age differs from one tree to another.

Increasing the size of the training set should decrease the variability of the trees with respect to the training set. In Figs. 3.27 and 3.28, we depict trees built on training sets with 50 000 and 500 000 observations, respectively. We notice that trees fitted on training sets with 50 000 observations always use the feature Age for the first split. Only the value of the split varies, but not too much. Finally, with training sets made of 500 000 observations, we observe that all the trees have the same structure, with $Age \geq 44$ as the split of the root node. In this case, the trees only differ on the corresponding predictions for $Age < 44$ and $Age \geq 44$.

The variability of the models with respect to the training set can be measured by the variance

$$\mathrm{E}_{\boldsymbol{X}}\left\{\mathrm{E}_{\mathcal{D}}\left[\left(\mathrm{E}_{\mathcal{D}}\left[\widehat{\mu}_{\mathcal{D}}(\boldsymbol{X})\right]-\widehat{\mu}_{\mathcal{D}}(\boldsymbol{X})\right)^2\right]\right\}, \tag{3.7.1}$$

which has been introduced in Chap. 2. In Table 3.7, we calculate the variance (3.7.1) by Monte-Carlo simulation for training sets of sizes 5000, 50 000 and 500 000. As expected, the variance decreases as the number of observations in the training set increases.

Even in this simple example, where the true model only takes twelve values and where we only have three important features (one of which being continuous), training sets with 50 000 observations can still lead to trees with different splits for the root node. Larger trees can even be more impacted due to the hierarchical nature of the procedure. A change in the split for the root node can be propagated to the subsequent nodes, which may lead to trees with very different structures.

In practice, the complexity of the true model is usually much larger than in this example. Typically, more features influence the true expected claim frequency and the impact of a continuous feature such as the age is often more complicated than being summarized with three categories. In this setting, let us assume that the true expected claim frequency is actually

$$\begin{aligned}\mu(\boldsymbol{x}) &= 0.1 \times (1+0.1\mathrm{I}\left[x_1 = male\right]) \\ &\quad \times \left(1+\frac{1}{\sqrt{x_2-17}}\right) \\ &\quad \times (1+0.15\mathrm{I}\left[x_4 = yes\right]).\end{aligned} \tag{3.7.2}$$

This time, the impact of the age on the expected claim frequency is more complex than in the previous example, and is depicted in Fig. 3.29. The expected claim frequency smoothly decreases with the age, young drivers being more risky. Even with training sets made of 500 000 observations, the resulting trees can still have different splits for the root node, as shown in Fig. 3.30. We get the following variance

$$\mathrm{E}_{\boldsymbol{X}}\left\{\mathrm{E}_{\mathcal{D}}\left[\left(\mathrm{E}_{\mathcal{D}}\left[\widehat{\mu}_{\mathcal{D}}(\boldsymbol{X})\right]-\widehat{\mu}_{\mathcal{D}}(\boldsymbol{X})\right)^2\right]\right\} = 0.03374507\,10^{-3}$$

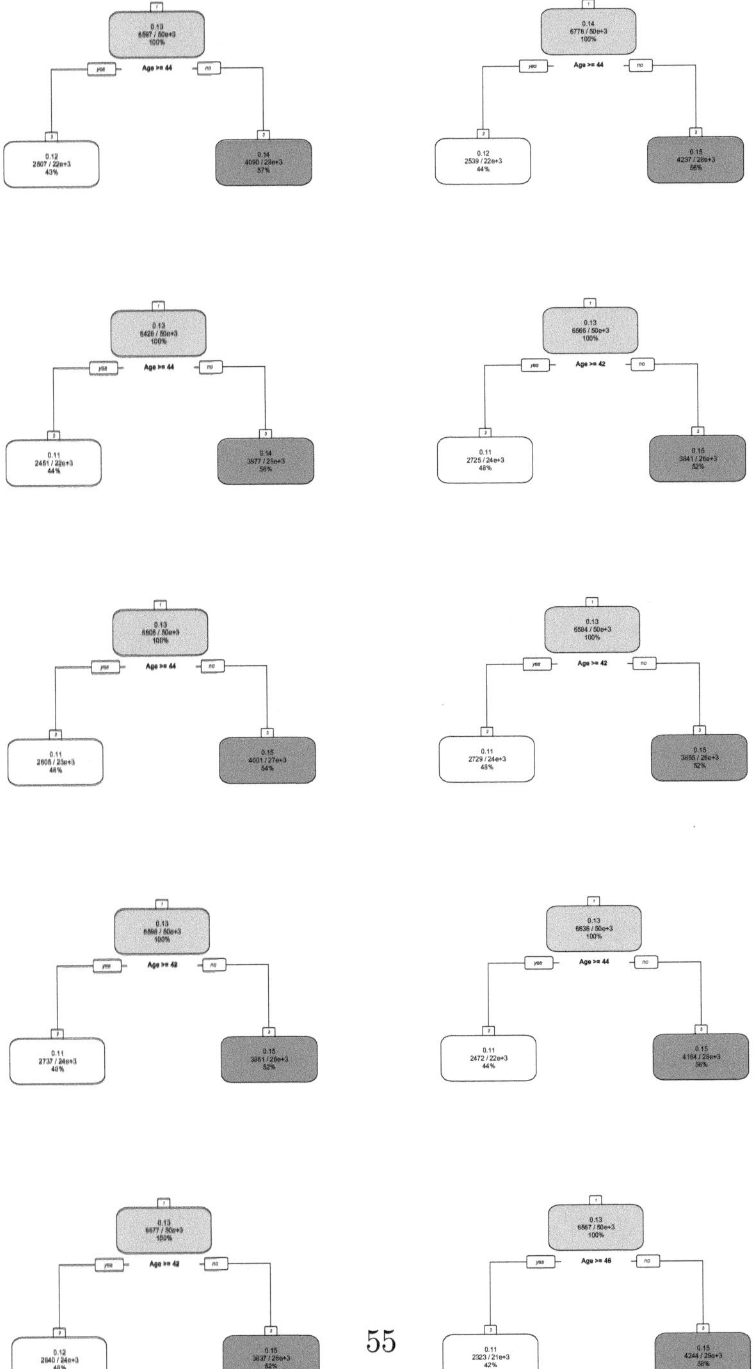

Fig. 3.27 Trees with only one split built on different training sets of size 50 000

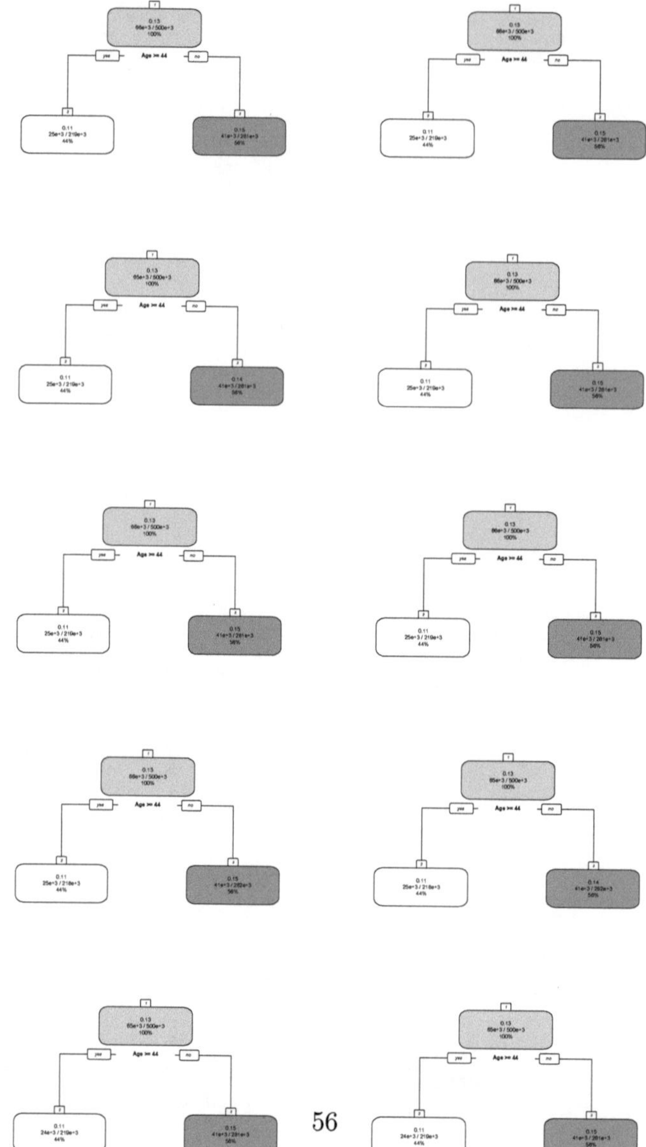

Fig. 3.28 Trees with only one split built on different training sets of size 500 000

Table 3.7 Estimation of (3.7.1) by Monte-Carlo simulation for training sets of sizes 5000, 50 000 and 500 000

Size of the training sets	$\mathrm{E}_{X}\left\{\mathrm{E}_{\mathcal{D}}\left[(\mathrm{E}_{\mathcal{D}}\left[\widehat{\mu}_{\mathcal{D}}(X)\right] - \widehat{\mu}_{\mathcal{D}}(X))^2\right]\right\}$
5000	$0.206928850\ 10^{-3}$
50 000	$0.061018808\ 10^{-3}$
500 000	$0.002361467\ 10^{-3}$

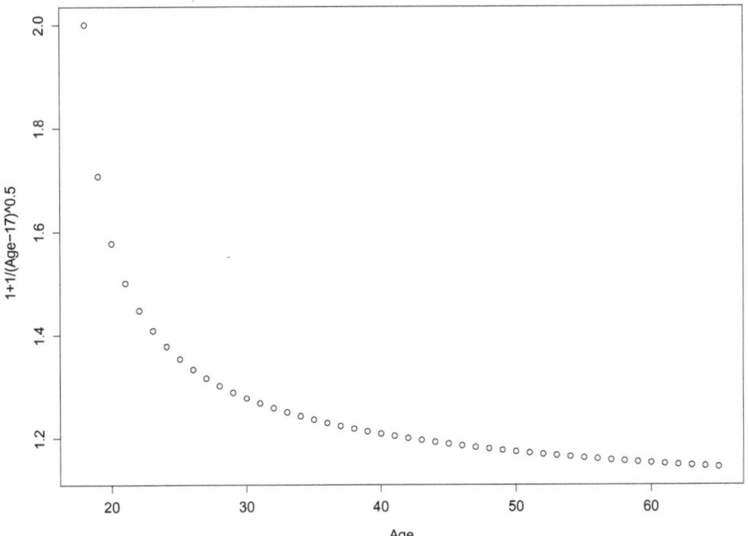

Fig. 3.29 Impact of the age on the expected claim frequency

by Monte-Carlo simulation, which is, as expected, larger than the variance in Table 3.7 for training sets with 500 000 observations.

In insurance, the complexity of the true model can be large so that training sets of reasonable size for an insurance portfolio often lead to unstable estimators $\widehat{\mu}_{\mathcal{D}}$ with respect to $\mathcal{D}$. In addition, correlated features can still increase model instability.

3.7.2 *Lack of Smoothness*

When the true model $\mu(\boldsymbol{x})$ is smooth, trees are unlikely to capture all the nuances of $\mu(\boldsymbol{x})$. One says that regression trees suffer from a lack of smoothness. Ensemble techniques described in Chaps. 4 and 5 will enable to address this issue.

As an illustration, we consider the example of Sect. 3.2.4.1 with expected claim frequencies given by (3.7.2). The expected claim frequency $\mu(\boldsymbol{x})$ smoothly decreases with the feature Age. We simulate a training set with 500 000 observations and we

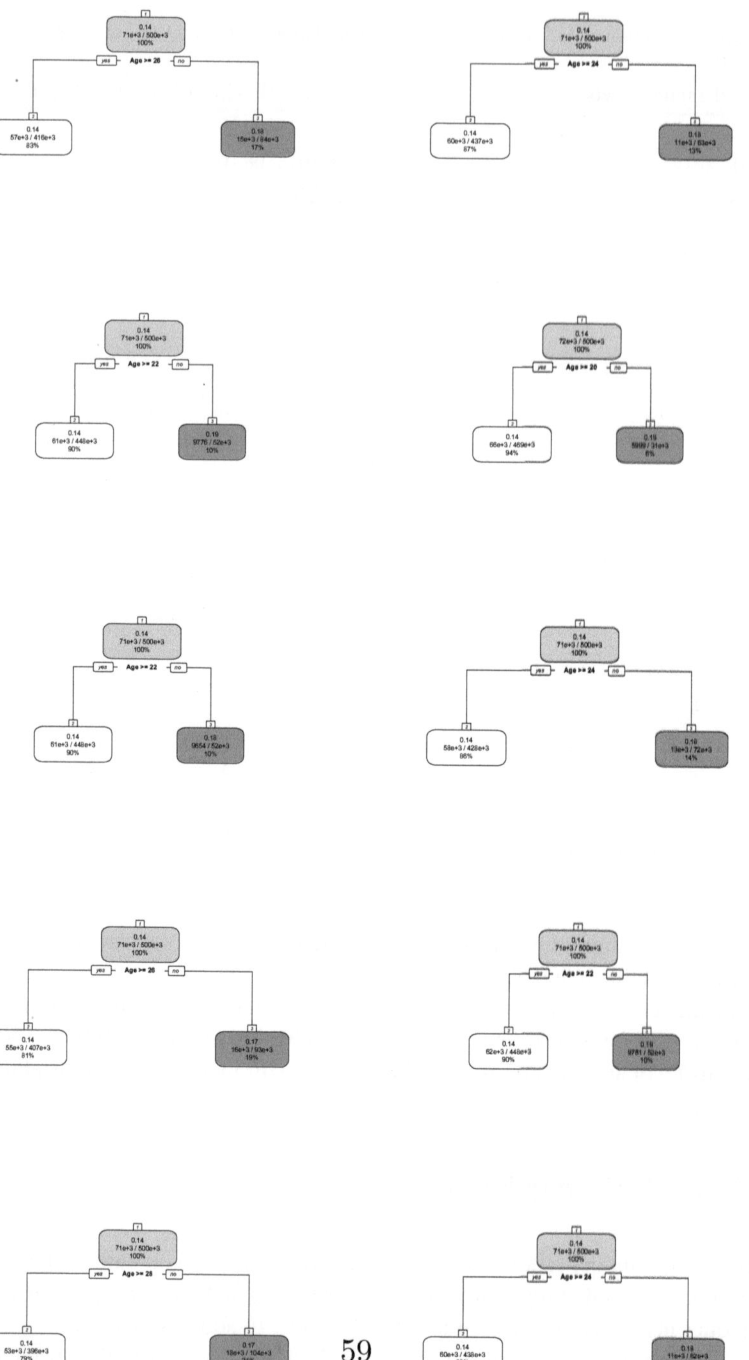

Fig. 3.30 Trees with only one split built on different training sets of size 500 000 for the true expected claim frequency given by (3.7.2)

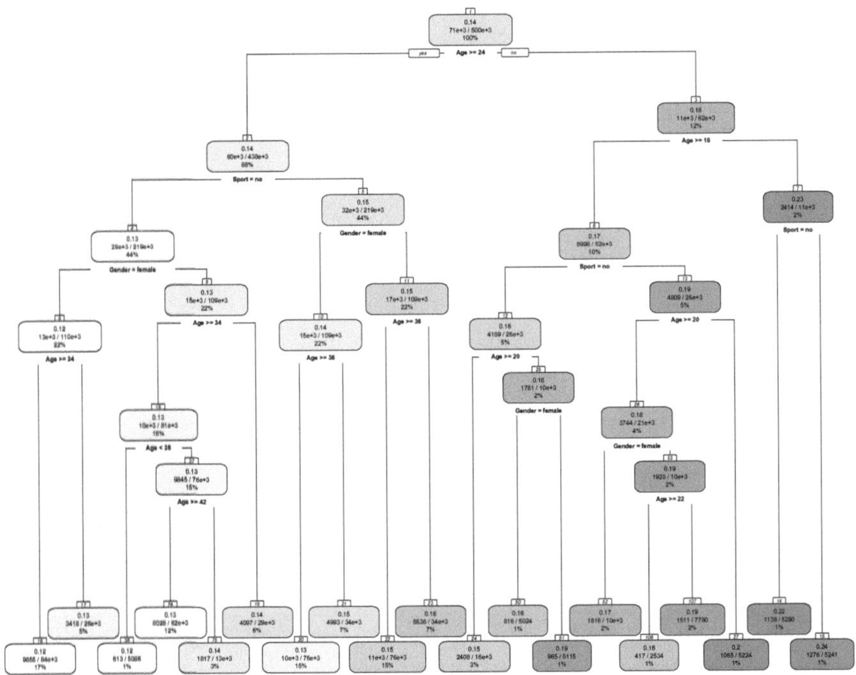

Fig. 3.31 Tree minimizing the 10-fold cross validation error

fit a tree which minimizes the 10-fold cross validation error, depicted in Fig. 3.31. Notice that the feature Split is not used by this tree, as desired.

The corresponding model $\widehat{\mu}$ is not satisfactory. Figure 3.32 illustrates both $\widehat{\mu}$ and μ as functions of policyholder's age for fixed values of variables Gender and Sport. One sees that $\widehat{\mu}$ does not reproduce the smooth decreasing behavior of the true model μ with respect to the age.

Typically, small trees suffer from a lack of smoothness because of their limited number of terminal nodes whereas large trees, that could overcome this limitation because of their larger number of leaves, tend to overfit the data. Trees with the highest predictive accuracy cannot be too large so that their limited possible predicted outcomes prevent obtaining models with smooth behaviors with respect to some features.

3.8 Bibliographic Notes and Further Reading

The decision trees are first due to Morgan and Sonquist (1963), who suggested a method called automatic interaction detector (AID) in the context of survey data. Several improvements were then proposed by Sonquist (1970), Messenger and

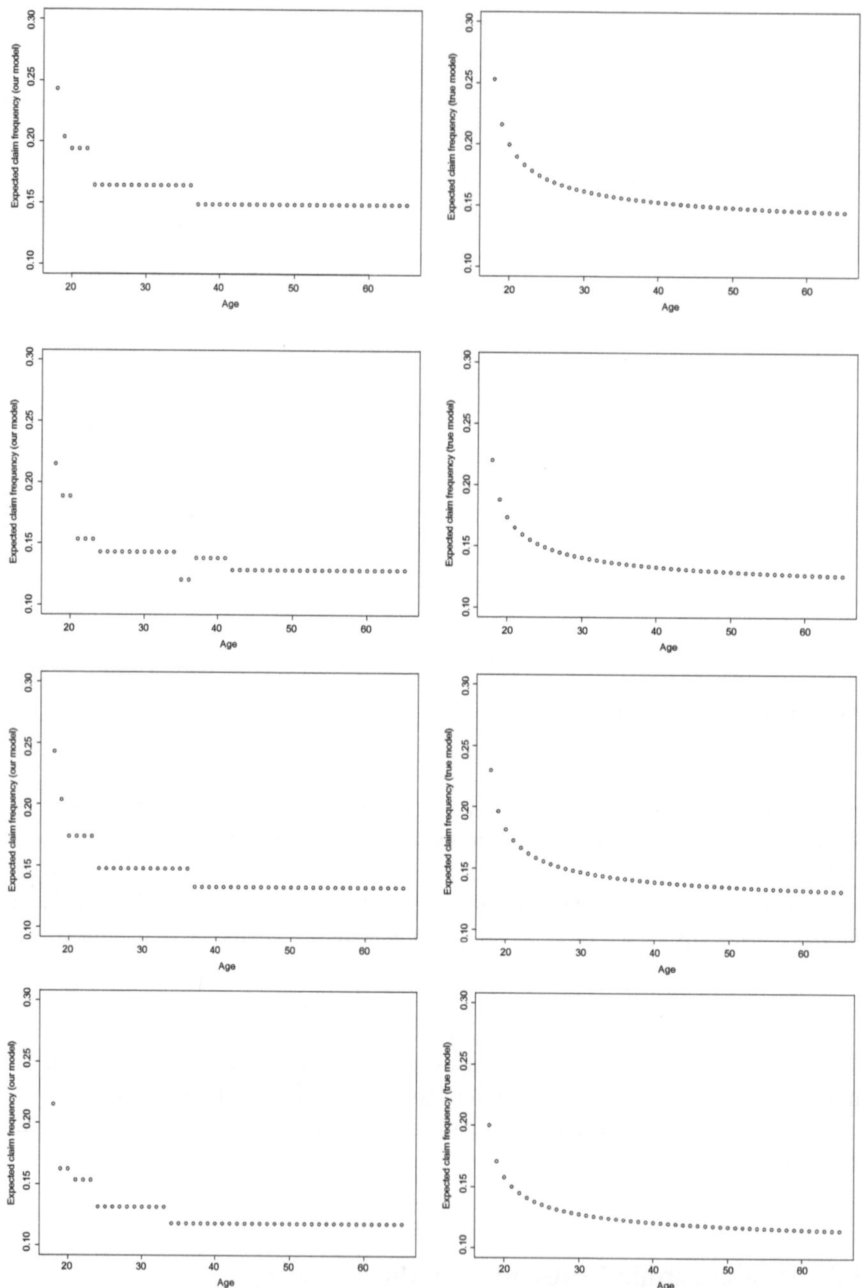

Fig. 3.32 Fitted model $\widehat{\mu}$ (on the left) and true model μ (on the right) as functions of policyholder's age. From top to bottom: (Gender = male, Sport = yes), (Gender = male, Sport = no), (Gender = female, Sport = yes) and (Gender = female, Sport = no)

Mandell (1972), Gillo (1972) and Sonquist et al. (1974). The most important contributors to modern methodological principles of decision trees are however Breiman (1978a, b), Friedman (1977, 1979) and Quinlan (1979, 1986) who proposed very similar algorithms for the construction of decision trees. The seminal work of Breiman et al. (1984), complemented by the work of Quinlan (1993), enabled to create a simple and consistent methodological framework for decision trees, the classification and regression tree (CART) techniques, which facilitated the diffusion of tree-based models towards a large audience. Our presentation is mainly inspired from Breiman et al. (1984), Hastie et al. (2009) and Wüthrich and Buser (2019), the latter reference adapting tree based methods to model claim frequencies. Also, Louppe (2014) made a good overview of the literature from which this section is greatly inspired.

References

Breiman L (1978a) Parsimonious binary classification trees. Preliminary report. Technology Service Corporation, Santa Monica, Calif

Breiman L (1978b) Description of chlorine tree development and use. Technical report, Technology Service Corporation, Santa Monica, CA

Breiman L, Friedman JH, Olshen RA, Stone CJ (1984) Classification and regression trees. Wadsworth statistics/probability series

Bühlmann H, Gisler A (2005) A course in credibility theory and its applications. Springer, Berlin

Denuit M, Hainaut D, Trufin J (2019) Effective statistical learning methods for actuaries I: GLMs and extensions. Springer actuarial lecture notes

Feelders A (2019) Classification trees. Lecture notes

Friedman JH (1977) A recursive partitioning decision rule for nonparametric classification. IEEE Trans Comput 100(4):404–408

Friedman JH (1979) A tree-structured approach to nonparametric multiple regression. Smoothing techniques for curve estimation. Springer, Berlin, pp 5–22

Geurts P (2002) Contributions to decision tree induction: bias/variance tradeoff and time series classification. PhD thesis

Gillo M (1972) Maid: a honeywell 600 program for an automatised survey analysis. Behav Sci 17:251–252

Hastie T, Tibshirani R, Friedman J (2009) The elements of statistical learning. Data mining, inference, and prediction, 2nd edn. Springer series in statistics

Louppe G (2014) Understanding random forests: from theory to practice. arXiv:14077502

Messenger R, Mandell L (1972) A modal search technique for predictive nominal scale multivariate analysis. J Am Stat Assoc 67(340):768–772

Morgan JN, Sonquist JA (1963) Problems in the analysis of survey data, and a proposal. J Am Stat Assoc 58(302):415–434

Quinlan JR (1979) Discovering rules by induction from large collections of examples. Expert systems in the micro electronic age. Edinburgh University Press, Edinburgh

Quinlan JR (1986) Induction of decision trees. Mach Learn 1:81–106

Quinlan JR (1993) C4.5: programs for machine learning, vol 1. Morgan Kaufmann, Burlington

Sonquist JA (1970) Multivariate model building: the validation of a search strategy. Survey Research Center, University of Michigan

Sonquist JA, Baker EL, Morgan JN (1974) Searching for structure: an approach to analysis of substantial bodies of micro-data and documentation for a computer program. Survey Research Center, University of Michigan Ann Arbor, MI

Wüthrich MV, Buser C (2019) Data analytics for non-life insurance pricing. Lecture notes

Chapter 4
Bagging Trees and Random Forests

4.1 Introduction

Two ensemble methods are considered in this chapter, namely bagging trees and random forests. One issue with regression trees is their high variance. There is a high variability of the prediction $\widehat{\mu}_{\mathcal{D}}(\boldsymbol{x})$ over the trees trained from all possible training sets $\mathcal{D}$. Bagging trees and random forests aim to reduce the variance without too much altering bias.

Ensemble methods are relevant tools to reduce the expected generalization error of a model by driving down the variance of the model without increasing too much the bias. The principle of ensemble methods (based on randomization) consists in introducing random perturbations into the training procedure in order to get different models from a single training set $\mathcal{D}$ and combining them to obtain the estimate of the ensemble.

Let us start with the average prediction $\mathrm{E}_{\mathcal{D}}\left[\widehat{\mu}_{\mathcal{D}}(\boldsymbol{x})\right]$. It has the same bias as $\widehat{\mu}_{\mathcal{D}}(\boldsymbol{x})$ since

$$\mathrm{E}_{\mathcal{D}}\left[\widehat{\mu}_{\mathcal{D}}(\boldsymbol{x})\right] = \mathrm{E}_{\mathcal{D}}\left[\mathrm{E}_{\mathcal{D}}\left[\widehat{\mu}_{\mathcal{D}}(\boldsymbol{x})\right]\right], \tag{4.1.1}$$

and zero variance, that is,

$$\mathrm{Var}_{\mathcal{D}}\left[\mathrm{E}_{\mathcal{D}}\left[\widehat{\mu}_{\mathcal{D}}(\boldsymbol{x})\right]\right] = 0. \tag{4.1.2}$$

Hence, finding a training procedure that produces a good approximation of the average model in order to stabilize model predictions seems to be a good strategy.

If we assume that we can draw as many training sets as we want, so that we have B training sets $\mathcal{D}^1, \mathcal{D}^2, \ldots, \mathcal{D}^B$ available, then an approximation of the average model can be obtained by averaging the regression trees built on these training sets, that is,

$$\widehat{\mathrm{E}}_{\mathcal{D}}\left[\widehat{\mu}_{\mathcal{D}}(\boldsymbol{x})\right] = \frac{1}{B}\sum_{b=1}^{B}\widehat{\mu}_{\mathcal{D}^b}(\boldsymbol{x}). \tag{4.1.3}$$

M. Denuit et al., *Effective Statistical Learning Methods for Actuaries II*, Springer Actuarial, https://doi.org/10.1007/978-3-030-57556-4_4

In such a case, the average of the estimate (4.1.3) with respect to the training sets $\mathcal{D}^1, \ldots, \mathcal{D}^B$ is the average prediction $\mathrm{E}_{\mathcal{D}}\left[\widehat{\mu}_{\mathcal{D}}(\boldsymbol{x})\right]$, that is,

$$\begin{aligned} \mathrm{E}_{\mathcal{D}^1,\ldots,\mathcal{D}^B}\left[\frac{1}{B}\sum_{b=1}^{B}\widehat{\mu}_{\mathcal{D}^b}(\boldsymbol{x})\right] &= \frac{1}{B}\sum_{b=1}^{B}\mathrm{E}_{\mathcal{D}^b}\left[\widehat{\mu}_{\mathcal{D}^b}(\boldsymbol{x})\right] \\ &= \mathrm{E}_{\mathcal{D}}\left[\widehat{\mu}_{\mathcal{D}}(\boldsymbol{x})\right], \end{aligned} \tag{4.1.4}$$

while the variance of (4.1.3) with respect to $\mathcal{D}^1, \ldots, \mathcal{D}^B$ is given by

$$\begin{aligned} \mathrm{Var}_{\mathcal{D}^1,\ldots,\mathcal{D}^B}\left[\frac{1}{B}\sum_{b=1}^{B}\widehat{\mu}_{\mathcal{D}^b}(\boldsymbol{x})\right] &= \frac{1}{B^2}\mathrm{Var}_{\mathcal{D}^1,\ldots,\mathcal{D}^B}\left[\sum_{b=1}^{B}\widehat{\mu}_{\mathcal{D}^b}(\boldsymbol{x})\right] \\ &= \frac{1}{B^2}\sum_{b=1}^{B}\mathrm{Var}_{\mathcal{D}^b}\left[\widehat{\mu}_{\mathcal{D}^b}(\boldsymbol{x})\right] \\ &= \frac{\mathrm{Var}_{\mathcal{D}}\left[\widehat{\mu}_{\mathcal{D}}(\boldsymbol{x})\right]}{B} \end{aligned} \tag{4.1.5}$$

since predictions $\widehat{\mu}_{\mathcal{D}^1}(\boldsymbol{x}), \ldots, \widehat{\mu}_{\mathcal{D}^B}(\boldsymbol{x})$ are independent and identically distributed. So, averaging over B estimates fitted on different training sets leaves the bias unchanged compared to each individual estimate while it divides the variance by B. The estimate (4.1.3) is then less variable than each individual one.

In practice, the probability distribution from which the observations of the training set are drawn is usually not known so that there is only one training set available. In this context, the bootstrap approach, used both in bagging trees and random forests, appears to be particularly useful.

4.2 Bootstrap

Suppose we have independent random variables $Y_1, Y_2, \ldots, Y_n$ with common distribution function F that is unknown and that we are interested in using them to estimate some quantity $\theta(F)$ associated with F. An estimator

$$\widehat{\theta} = g(Y_1, Y_2, \ldots, Y_n)$$

is available for $\theta(F)$. The distributional properties of $\widehat{\theta}$ in terms of the variables $Y_1, Y_2, \ldots, Y_n$ cannot be determined since the distribution function F is not known. The idea of bootstrap is to estimate F.

The empirical counterpart to F is defined as

$$\widehat{F}_n(x) = \frac{\#\{Y_i \text{ such that } Y_i \leq x\}}{n} = \frac{1}{n}\sum_{i=1}^{n}\mathrm{I}[Y_i \leq x].$$

Thus, the empirical distribution function $\widehat{F}_n$ puts an equal probability $\frac{1}{n}$ on each of the observed data points $Y_1, \ldots, Y_n$. The idea behind the non-parametric bootstrap is to simulate sets of independent random variables

$$Y_1^{(*b)}, Y_2^{(*b)}, \ldots, Y_n^{(*b)}$$

obeying the distribution function $\widehat{F}_n, b = 1, 2, \ldots, B$. This can be done by simulating $U_i \sim \mathcal{U}ni(0, 1)$ and setting

$$Y_i^{(*b)} = y_I \text{ with } I = [nU_i] + 1.$$

Then, for each $b = 1, \ldots, B$, we calculate

$$\widehat{\theta}^{(*b)} = g(Y_1^{(*b)}, Y_2^{(*b)}, \ldots, Y_n^{(*b)}),$$

so that the corresponding bootstrap distribution of $\widehat{\theta}$ is given by

$$F^*_{\widehat{\theta}}(x) = \frac{1}{B} \sum_{b=1}^{B} \mathrm{I}\left[\widehat{\theta}^{(*b)} \leq x\right].$$

4.3 Bagging Trees

Bagging is one of the first ensemble methods proposed in the literature. Consider a model fitted to our training set $\mathcal{D}$, obtaining the prediction $\widehat{\mu}_{\mathcal{D}}(\boldsymbol{x})$ at point $\boldsymbol{x}$. Bootstrap aggregation or bagging averages this prediction over a set of bootstrap samples in order to reduce its variance.

The probability distribution of the random vector $(Y, \boldsymbol{X})$ is usually not known. This latter distribution is then approximated by its empirical version which puts an equal probability $\frac{1}{|\mathcal{I}|}$ on each of the observations $\{(y_i, \boldsymbol{x}_i); i \in \mathcal{I}\}$ of the training set $\mathcal{D}$. Hence, instead of simulating B training sets $\mathcal{D}^1, \mathcal{D}^2, \ldots, \mathcal{D}^B$ from the probability distribution of $(Y, \boldsymbol{X})$, which is not possible in practice, the idea of bagging is rather to simulate B bootstrap samples $\mathcal{D}^{*1}, \mathcal{D}^{*2}, \ldots, \mathcal{D}^{*B}$ of the training set $\mathcal{D}$ from its empirical counterpart. Specifically, a bootstrap sample of $\mathcal{D}$ is obtained by simulating independently $|\mathcal{I}|$ observations from the empirical distribution of $(Y, \boldsymbol{X})$ defined above. A bootstrap sample is thus a random sample of $\mathcal{D}$ taken with replacement which has the same size as $\mathcal{D}$. Notice that, on average, 63.2% of the observations of the training set are represented at least once in a bootstrap sample. Indeed,

$$1 - \left(\frac{|\mathcal{I}| - 1}{|\mathcal{I}|}\right)^{|\mathcal{I}|}, \tag{4.3.1}$$

Table 4.1 Probability in (4.3.1) with respect to $|\mathcal{I}|$

$\|\mathcal{I}\|$	$1-\left(\frac{\|\mathcal{I}\|-1}{\|\mathcal{I}\|}\right)^{\|\mathcal{I}\|}$
10	0.651322
100	0.633968
1000	0.632305
10 000	0.632139
100 000	0.632122

which is computed in Table 4.1 for different values of $|\mathcal{I}|$, is the probability that a given observation of the training set is represented at least once. One can see that (4.3.1) quickly reaches the value of 63.2%.

Let $\mathcal{D}^{*1}, \mathcal{D}^{*2}, \ldots, \mathcal{D}^{*B}$ be B bootstrap samples of the training set $\mathcal{D}$. For each $\mathcal{D}^{*b}$, $b = 1, \ldots, B$, we fit our model, giving prediction $\widehat{\mu}_{\mathcal{D},\Theta_b}(\boldsymbol{x}) = \widehat{\mu}_{\mathcal{D}^{*b}}(\boldsymbol{x})$. The bagging prediction is then defined by

$$\widehat{\mu}^{\text{bag}}_{\mathcal{D},\Theta}(\boldsymbol{x}) = \frac{1}{B}\sum_{b=1}^{B} \widehat{\mu}_{\mathcal{D},\Theta_b}(\boldsymbol{x}), \tag{4.3.2}$$

where $\Theta = (\Theta_1, \ldots, \Theta_B)$. Random vectors $\Theta_1, \ldots, \Theta_B$ fully capture the randomness of the training procedure. For bagging, $\Theta_1, \ldots, \Theta_B$ are independent and identically distributed so that Θ_b is a vector of $|\mathcal{I}|$ integers randomly and uniformly drawn in $\mathcal{I}$. Each component of Θ_b indexes one observation of the training set selected in $\mathcal{D}^{*b}$. In this book, bagging is applied to unpruned regression trees. This provides the following algorithm:

Algorithm 4.1: Bagging trees

For $b = 1$ to B **do**

1. Generate a bootstrap sample $\mathcal{D}^{*b}$ of $\mathcal{D}$.
2. Fit an unpruned tree on $\mathcal{D}^{*b}$, which gives prediction $\widehat{\mu}_{\mathcal{D},\Theta_b}(\boldsymbol{x})$.

End for

Output: $\widehat{\mu}^{\text{bag}}_{\mathcal{D},\Theta}(\boldsymbol{x}) = \frac{1}{B}\sum_{b=1}^{B} \widehat{\mu}_{\mathcal{D},\Theta_b}(\boldsymbol{x})$.

As mentioned previously, two main drawbacks of regression trees are that they produce piece-wise constant estimates and that they are rather unstable under a small change in the observations of the training set. The construction of an ensemble of trees produces more stable and smoothed estimates under averaging.

4.3.1 Bias

For bagging, the bias is the same as the one of the individual sampled models. Indeed,

$$\begin{aligned}
Bias(\boldsymbol{x}) &= \mu(\boldsymbol{x}) - \mathrm{E}_{\mathcal{D},\Theta}\left[\widehat{\mu}^{\text{bag}}_{\mathcal{D},\Theta}(\boldsymbol{x})\right] \\
&= \mu(\boldsymbol{x}) - \mathrm{E}_{\mathcal{D},\Theta_1,\ldots,\Theta_B}\left[\frac{1}{B}\sum_{b=1}^{B}\widehat{\mu}_{\mathcal{D},\Theta_b}(\boldsymbol{x})\right] \\
&= \mu(\boldsymbol{x}) - \frac{1}{B}\sum_{b=1}^{B}\mathrm{E}_{\mathcal{D},\Theta_b}\left[\widehat{\mu}_{\mathcal{D},\Theta_b}(\boldsymbol{x})\right] \\
&= \mu(\boldsymbol{x}) - \mathrm{E}_{\mathcal{D},\Theta_b}\left[\widehat{\mu}_{\mathcal{D},\Theta_b}(\boldsymbol{x})\right] \qquad (4.3.3)
\end{aligned}$$

since predictions $\widehat{\mu}_{\mathcal{D},\Theta_1}(\boldsymbol{x}), \ldots, \widehat{\mu}_{\mathcal{D},\Theta_B}(\boldsymbol{x})$ are identically distributed.

However, the bias of $\widehat{\mu}_{\mathcal{D},\Theta_b}(\boldsymbol{x})$ is typically greater in absolute terms than the bias of $\widehat{\mu}_{\mathcal{D}}(\boldsymbol{x})$ fitted on $\mathcal{D}$ since the reduced sample $\mathcal{D}^{*b}$ imposes restrictions. The improvements in the estimation obtained by bagging will be a consequence of variance reduction.

Notice that trees are ideal candidates for bagging. They can handle complex interaction structures in the data and they have relatively low bias if grown sufficiently deep. Because they are noisy, they will greatly benefit from the averaging.

4.3.2 Variance

The variance of $\widehat{\mu}^{\text{bag}}_{\mathcal{D},\Theta}(\boldsymbol{x})$ can be written as

$$\begin{aligned}
\mathrm{Var}_{\mathcal{D},\Theta}\left[\widehat{\mu}^{\text{bag}}_{\mathcal{D},\Theta}(\boldsymbol{x})\right] &= \mathrm{Var}_{\mathcal{D},\Theta_1,\ldots,\Theta_B}\left[\frac{1}{B}\sum_{b=1}^{B}\widehat{\mu}_{\mathcal{D},\Theta_b}(\boldsymbol{x})\right] \\
&= \frac{1}{B^2}\mathrm{Var}_{\mathcal{D},\Theta_1,\ldots,\Theta_B}\left[\sum_{b=1}^{B}\widehat{\mu}_{\mathcal{D},\Theta_b}(\boldsymbol{x})\right] \\
&= \frac{1}{B^2}\left\{\mathrm{Var}_{\mathcal{D}}\left[\mathrm{E}_{\Theta_1,\ldots,\Theta_B}\left[\sum_{b=1}^{B}\widehat{\mu}_{\mathcal{D},\Theta_b}(\boldsymbol{x})\middle|\mathcal{D}\right]\right]\right. \\
&\qquad \left. +\mathrm{E}_{\mathcal{D}}\left[\mathrm{Var}_{\Theta_1,\ldots,\Theta_B}\left[\sum_{b=1}^{B}\widehat{\mu}_{\mathcal{D},\Theta_b}(\boldsymbol{x})\middle|\mathcal{D}\right]\right]\right\} \\
&= \mathrm{Var}_{\mathcal{D}}\left[\mathrm{E}_{\Theta_b}\left[\widehat{\mu}_{\mathcal{D},\Theta_b}(\boldsymbol{x})\middle|\mathcal{D}\right]\right] + \frac{1}{B}\mathrm{E}_{\mathcal{D}}\left[\mathrm{Var}_{\Theta_b}\left[\widehat{\mu}_{\mathcal{D},\Theta_b}(\boldsymbol{x})\middle|\mathcal{D}\right]\right]
\end{aligned}$$

(4.3.4)

since conditionally to $\mathcal{D}$, predictions $\widehat{\mu}_{\mathcal{D},\Theta_1}(\boldsymbol{x}), \ldots, \widehat{\mu}_{\mathcal{D},\Theta_B}(\boldsymbol{x})$ are independent and identically distributed. The second term is the within-$\mathcal{D}$ variance, a result of the randomization due to the bootstrap sampling. The first term is the sampling variance of the bagging ensemble, a result of the sampling variability of $\mathcal{D}$ itself. As the number of aggregated estimates gets arbitrarily large, i.e. as $B \to \infty$, the variance of $\widehat{\mu}^{\text{bag}}_{\mathcal{D},\Theta}(\boldsymbol{x})$ reduces to $\text{Var}_{\mathcal{D}}\left[\text{E}_{\Theta_b}\left[\widehat{\mu}_{\mathcal{D},\Theta_b}(\boldsymbol{x})\middle|\mathcal{D}\right]\right]$.

From (4.3.4) and

$$\text{Var}_{\mathcal{D},\Theta_b}\left[\widehat{\mu}_{\mathcal{D},\Theta_b}(\boldsymbol{x})\right] = \text{Var}_{\mathcal{D}}\left[\text{E}_{\Theta_b}\left[\widehat{\mu}_{\mathcal{D},\Theta_b}(\boldsymbol{x})\middle|\mathcal{D}\right]\right] + \text{E}_{\mathcal{D}}\left[\text{Var}_{\Theta_b}\left[\widehat{\mu}_{\mathcal{D},\Theta_b}(\boldsymbol{x})\middle|\mathcal{D}\right]\right], \tag{4.3.5}$$

we see that

$$\text{Var}_{\mathcal{D},\Theta}\left[\widehat{\mu}^{\text{bag}}_{\mathcal{D},\Theta}(\boldsymbol{x})\right] \leq \text{Var}_{\mathcal{D},\Theta_b}\left[\widehat{\mu}_{\mathcal{D},\Theta_b}(\boldsymbol{x})\right]. \tag{4.3.6}$$

The variance of the bagging prediction $\widehat{\mu}^{\text{bag}}_{\mathcal{D},\Theta}(\boldsymbol{x})$ is smaller than the variance of an individual prediction $\widehat{\mu}_{\mathcal{D},\Theta_b}(\boldsymbol{x})$. Actually, we learn from (4.3.4) and (4.3.5) that the variance reduction is given by

$$\text{Var}_{\mathcal{D},\Theta_b}\left[\widehat{\mu}_{\mathcal{D},\Theta_b}(\boldsymbol{x})\right] - \text{Var}_{\mathcal{D},\Theta}\left[\widehat{\mu}^{\text{bag}}_{\mathcal{D},\Theta}(\boldsymbol{x})\right] = \frac{B-1}{B}\text{E}_{\mathcal{D}}\left[\text{Var}_{\Theta_b}\left[\widehat{\mu}_{\mathcal{D},\Theta_b}(\boldsymbol{x})\middle|\mathcal{D}\right]\right], \tag{4.3.7}$$

which increases as B increases and tends to $\text{E}_{\mathcal{D}}\left[\text{Var}_{\Theta_b}\left[\widehat{\mu}_{\mathcal{D},\Theta_b}(\boldsymbol{x})\middle|\mathcal{D}\right]\right]$ when $B \to \infty$.

Let us introduce the correlation coefficient $\rho(\boldsymbol{x})$ between any pair of predictions used in the averaging which are built on the same training set but fitted on two different bootstrap samples. Using the definition of the Pearson's correlation coefficient, we get

$$\begin{aligned}\rho(\boldsymbol{x}) &= \frac{\text{Cov}_{\mathcal{D},\Theta_b,\Theta_{b'}}\left[\widehat{\mu}_{\mathcal{D},\Theta_b}(\boldsymbol{x}), \widehat{\mu}_{\mathcal{D},\Theta_{b'}}(\boldsymbol{x})\right]}{\sqrt{\text{Var}_{\mathcal{D},\Theta_b}\left[\widehat{\mu}_{\mathcal{D},\Theta_b}(\boldsymbol{x})\right]}\sqrt{\text{Var}_{\mathcal{D},\Theta_{b'}}\left[\widehat{\mu}_{\mathcal{D},\Theta_{b'}}(\boldsymbol{x})\right]}} \\ &= \frac{\text{Cov}_{\mathcal{D},\Theta_b,\Theta_{b'}}\left[\widehat{\mu}_{\mathcal{D},\Theta_b}(\boldsymbol{x}), \widehat{\mu}_{\mathcal{D},\Theta_{b'}}(\boldsymbol{x})\right]}{\text{Var}_{\mathcal{D},\Theta_b}\left[\widehat{\mu}_{\mathcal{D},\Theta_b}(\boldsymbol{x})\right]}\end{aligned} \tag{4.3.8}$$

as $\widehat{\mu}_{\mathcal{D},\Theta_b}(\boldsymbol{x})$ and $\widehat{\mu}_{\mathcal{D},\Theta_{b'}}(\boldsymbol{x})$ are identically distributed. By the law of total covariance, the numerator in (4.3.8) can be rewritten as

$$\begin{aligned}\text{Cov}_{\mathcal{D},\Theta_b,\Theta_{b'}}\left[\widehat{\mu}_{\mathcal{D},\Theta_b}(\boldsymbol{x}), \widehat{\mu}_{\mathcal{D},\Theta_{b'}}(\boldsymbol{x})\right] &= \text{E}_{\mathcal{D}}\left[\text{Cov}_{\Theta_b,\Theta_{b'}}\left[\widehat{\mu}_{\mathcal{D},\Theta_b}(\boldsymbol{x}), \widehat{\mu}_{\mathcal{D},\Theta_{b'}}(\boldsymbol{x})|\mathcal{D}\right]\right] \\ &\quad + \text{Cov}_{\mathcal{D}}\left[\text{E}_{\Theta_b}\left[\widehat{\mu}_{\mathcal{D},\Theta_b}(\boldsymbol{x})|\mathcal{D}\right], \text{E}_{\Theta_{b'}}\left[\widehat{\mu}_{\mathcal{D},\Theta_{b'}}(\boldsymbol{x})|\mathcal{D}\right]\right] \\ &= \text{Var}_{\mathcal{D}}\left[\text{E}_{\Theta_b}\left[\widehat{\mu}_{\mathcal{D},\Theta_b}(\boldsymbol{x})|\mathcal{D}\right]\right]\end{aligned} \tag{4.3.9}$$

since conditionally to $\mathcal{D}$, estimates $\widehat{\mu}_{\mathcal{D},\Theta_b}(\boldsymbol{x})$ and $\widehat{\mu}_{\mathcal{D},\Theta_{b'}}(\boldsymbol{x})$ are independent and identically distributed. Hence, combining (4.3.5) and (4.3.9), the correlation coefficient in (4.3.8) becomes

$$\rho(\boldsymbol{x}) = \frac{\mathrm{Var}_{\mathcal{D}}\left[\mathrm{E}_{\Theta_b}\left[\widehat{\mu}_{\mathcal{D},\Theta_b}(\boldsymbol{x})|\mathcal{D}\right]\right]}{\mathrm{Var}_{\mathcal{D},\Theta_b}\left[\widehat{\mu}_{\mathcal{D},\Theta_b}(\boldsymbol{x})\right]} \tag{4.3.10}$$

$$= \frac{\mathrm{Var}_{\mathcal{D}}\left[\mathrm{E}_{\Theta_b}\left[\widehat{\mu}_{\mathcal{D},\Theta_b}(\boldsymbol{x})|\mathcal{D}\right]\right]}{\mathrm{Var}_{\mathcal{D}}\left[\mathrm{E}_{\Theta_b}\left[\widehat{\mu}_{\mathcal{D},\Theta_b}(\boldsymbol{x})\middle|\mathcal{D}\right]\right] + \mathrm{E}_{\mathcal{D}}\left[\mathrm{Var}_{\Theta_b}\left[\widehat{\mu}_{\mathcal{D},\Theta_b}(\boldsymbol{x})\middle|\mathcal{D}\right]\right]}. \tag{4.3.11}$$

The correlation coefficient $\rho(\boldsymbol{x})$ measures the correlation between a pair of predictions in the ensemble induced by repeatedly making training sample draws $\mathcal{D}$ from the population and then drawing a pair of bootstrap samples from $\mathcal{D}$.

When $\rho(\boldsymbol{x})$ is close to 1, the predictions are highly correlated, suggesting that the randomization due to the bootstrap sampling has no significant effect on the predictions. On the contrary, when $\rho(\boldsymbol{x})$ is close to 0, the predictions are uncorrelated, suggesting that the randomization due to the bootstrap sampling has a strong impact on the predictions.

One sees that $\rho(\boldsymbol{x})$ is the ratio between the variance due to the training set and the total variance. The total variance is the sum of the variance due to the training set and the variance due to randomization induced by the bootstrap samples. A correlation coefficient close to 1 and hence correlated predictions means that the total variance is mostly driven by the training set. On the contrary, a correlation coefficient close to 0 and hence de-correlated predictions means that the total variance is mostly due to the randomization induced by the bootstrap samples.

Alternatively, the variance of $\widehat{\mu}^{\mathrm{bag}}_{\mathcal{D},\Theta}(\boldsymbol{x})$ given in (4.3.4) can be re-expressed in terms of the correlation coefficient. Indeed, from (4.3.10) and (4.3.11), we have

$$\mathrm{Var}_{\mathcal{D}}\left[\mathrm{E}_{\Theta_b}\left[\widehat{\mu}_{\mathcal{D},\Theta_b}(\boldsymbol{x})|\mathcal{D}\right]\right] = \rho(\boldsymbol{x})\mathrm{Var}_{\mathcal{D},\Theta_b}\left[\widehat{\mu}_{\mathcal{D},\Theta_b}(\boldsymbol{x})\right] \tag{4.3.12}$$

and

$$\mathrm{E}_{\mathcal{D}}\left[\mathrm{Var}_{\Theta_b}\left[\widehat{\mu}_{\mathcal{D},\Theta_b}(\boldsymbol{x})|\mathcal{D}\right]\right] = (1-\rho(\boldsymbol{x}))\,\mathrm{Var}_{\mathcal{D},\Theta_b}\left[\widehat{\mu}_{\mathcal{D},\Theta_b}(\boldsymbol{x})\right], \tag{4.3.13}$$

such that (4.3.4) can be rewritten as

$$\begin{aligned}\mathrm{Var}_{\mathcal{D},\Theta}\left[\widehat{\mu}^{\mathrm{bag}}_{\mathcal{D},\Theta}(\boldsymbol{x})\right] &= \mathrm{Var}_{\mathcal{D}}\left[\mathrm{E}_{\Theta_b}\left[\widehat{\mu}_{\mathcal{D},\Theta_b}(\boldsymbol{x})|\mathcal{D}\right]\right] + \frac{1}{B}\mathrm{E}_{\mathcal{D}}\left[\mathrm{Var}_{\Theta_b}\left[\widehat{\mu}_{\mathcal{D},\Theta_b}(\boldsymbol{x})|\mathcal{D}\right]\right]\\ &= \rho(\boldsymbol{x})\mathrm{Var}_{\mathcal{D},\Theta_b}\left[\widehat{\mu}_{\mathcal{D},\Theta_b}(\boldsymbol{x})\right] + \frac{(1-\rho(\boldsymbol{x}))}{B}\mathrm{Var}_{\mathcal{D},\Theta_b}\left[\widehat{\mu}_{\mathcal{D},\Theta_b}(\boldsymbol{x})\right].\end{aligned} \tag{4.3.14}$$

As B increases, the second term disappears, but the first term remains. Hence, when $\rho(\boldsymbol{x}) < 1$, one sees that the variance of the ensemble is strictly smaller than the variance of an individual model. Let us mention that assuming $\rho(\boldsymbol{x}) < 1$ amounts to suppose that the randomization due to the bootstrap sampling influences the individual predictions.

Notice that the random perturbation introduced by the bootstrap sampling induces a higher variance for an individual prediction $\widehat{\mu}_{\mathcal{D},\Theta_b}(\boldsymbol{x})$ than for $\widehat{\mu}_{\mathcal{D}}(\boldsymbol{x})$, so that

$$\mathrm{Var}_{\mathcal{D},\Theta_b}\left[\widehat{\mu}_{\mathcal{D},\Theta_b}(\boldsymbol{x})\right] \geq \mathrm{Var}_{\mathcal{D}}\left[\widehat{\mu}_{\mathcal{D}}(\boldsymbol{x})\right]. \tag{4.3.15}$$

Therefore, bagging averages models with higher variances. Nevertheless, the bagging prediction $\widehat{\mu}^{\text{bag}}_{\mathcal{D},\Theta}(\boldsymbol{x})$ has generally a smaller variance than $\widehat{\mu}_{\mathcal{D}}(\boldsymbol{x})$. This comes from the fact that, typically, the correlation coefficient $\rho(\boldsymbol{x})$ in (4.3.14) compensates for the variance increase $\mathrm{Var}_{\mathcal{D},\Theta_b}\left[\widehat{\mu}_{\mathcal{D},\Theta_b}(\boldsymbol{x})\right] - \mathrm{Var}_{\mathcal{D}}\left[\widehat{\mu}_{\mathcal{D}}(\boldsymbol{x})\right]$, so that the combined effect of $\rho(\boldsymbol{x}) < 1$ and $\mathrm{Var}_{\mathcal{D},\Theta_b}\left[\widehat{\mu}_{\mathcal{D},\Theta_b}(\boldsymbol{x})\right] \geq \mathrm{Var}_{\mathcal{D}}\left[\widehat{\mu}_{\mathcal{D}}(\boldsymbol{x})\right]$ often leads to a variance reduction

$$\mathrm{Var}_{\mathcal{D}}\left[\widehat{\mu}_{\mathcal{D}}(\boldsymbol{x})\right] - \rho(\boldsymbol{x})\mathrm{Var}_{\mathcal{D},\Theta_b}\left[\widehat{\mu}_{\mathcal{D},\Theta_b}(\boldsymbol{x})\right] \tag{4.3.16}$$

that is positive. Because of their high variance, regression trees very likely benefit from the averaging procedure.

4.3.3 *Expected Generalization Error*

For some loss functions, such as the squared error and Poisson deviance losses, we can show that the expected generalization error for the bagging prediction $\widehat{\mu}^{\text{bag}}_{\mathcal{D},\Theta}(\boldsymbol{x})$ is smaller than the expected generalization error for an individual prediction $\widehat{\mu}_{\mathcal{D},\Theta_b}(\boldsymbol{x})$, that is,

$$\mathrm{E}_{\mathcal{D},\Theta}\left[Err\left(\widehat{\mu}^{\text{bag}}_{\mathcal{D},\Theta}(\boldsymbol{x})\right)\right] \leq \mathrm{E}_{\mathcal{D},\Theta_b}\left[Err\left(\widehat{\mu}_{\mathcal{D},\Theta_b}(\boldsymbol{x})\right)\right]. \tag{4.3.17}$$

However, while it is typically the case with bagging trees, we cannot highlight some situations where the estimate $\widehat{\mu}^{\text{bag}}_{\mathcal{D},\Theta}(\boldsymbol{x})$ performs always better than $\widehat{\mu}_{\mathcal{D}}(\boldsymbol{x})$ in the sense of the expected generalization error, even for the squared error and Poisson deviance losses.

4.3.3.1 Squared Error Loss

For the squared error loss, from (2.4.3), the expected generalization error for $\widehat{\mu}^{\text{bag}}_{\mathcal{D},\Theta}(\boldsymbol{x})$ is given by

$$\begin{aligned}\mathrm{E}_{\mathcal{D},\Theta}\left[Err\left(\widehat{\mu}^{\text{bag}}_{\mathcal{D},\Theta}(\boldsymbol{x})\right)\right] &= Err\left(\mu(\boldsymbol{x})\right) + \left(\mu(\boldsymbol{x}) - \mathrm{E}_{\mathcal{D},\Theta}\left[\widehat{\mu}^{\text{bag}}_{\mathcal{D},\Theta}(\boldsymbol{x})\right]\right)^2 \\ &\quad + \mathrm{Var}_{\mathcal{D},\Theta}\left[\widehat{\mu}^{\text{bag}}_{\mathcal{D},\Theta}(\boldsymbol{x})\right].\end{aligned} \tag{4.3.18}$$

From (4.3.3) and (4.3.6), one observes that the bias remains unchanged while the variance decreases compared to the individual prediction $\widehat{\mu}_{\mathcal{D},\Theta_b}(\boldsymbol{x})$, so that we get

$$
\begin{aligned}
&\mathrm{E}_{\mathcal{D},\Theta}\left[Err\left(\widehat{\mu}^{\text{bag}}_{\mathcal{D},\Theta}(\boldsymbol{x})\right)\right] \\
&= Err\left(\mu(\boldsymbol{x})\right) + \left(\mu(\boldsymbol{x}) - \mathrm{E}_{\mathcal{D},\Theta_b}\left[\widehat{\mu}_{\mathcal{D},\Theta_b}(\boldsymbol{x})\right]\right)^2 + \mathrm{Var}_{\mathcal{D},\Theta}\left[\widehat{\mu}^{\text{bag}}_{\mathcal{D},\Theta}(\boldsymbol{x})\right] \\
&\leq Err\left(\mu(\boldsymbol{x})\right) + \left(\mu(\boldsymbol{x}) - \mathrm{E}_{\mathcal{D},\Theta_b}\left[\widehat{\mu}_{\mathcal{D},\Theta_b}(\boldsymbol{x})\right]\right)^2 + \mathrm{Var}_{\mathcal{D},\Theta_b}\left[\widehat{\mu}_{\mathcal{D},\Theta_b}(\boldsymbol{x})\right] \\
&= \mathrm{E}_{\mathcal{D},\Theta_b}\left[Err\left(\widehat{\mu}_{\mathcal{D},\Theta_b}(\boldsymbol{x})\right)\right]. \qquad (4.3.19)
\end{aligned}
$$

For every value of $\boldsymbol{X}$, the expected generalization error of the ensemble is smaller than the expected generalization error of an individual model.

Taking the average of (4.3.19) over $\boldsymbol{X}$ leads to

$$
\mathrm{E}_{\mathcal{D},\Theta}\left[Err\left(\widehat{\mu}^{\text{bag}}_{\mathcal{D},\Theta}\right)\right] \leq \mathrm{E}_{\mathcal{D},\Theta_b}\left[Err\left(\widehat{\mu}_{\mathcal{D},\Theta_b}\right)\right]. \qquad (4.3.20)
$$

4.3.3.2 Poisson Deviance Loss

For the Poisson deviance loss, from (2.4.4) and (2.4.5), the expected generalization error for $\widehat{\mu}^{\text{bag}}_{\mathcal{D},\Theta}(\boldsymbol{x})$ is given by

$$
\mathrm{E}_{\mathcal{D},\Theta}\left[Err\left(\widehat{\mu}^{\text{bag}}_{\mathcal{D},\Theta}(\boldsymbol{x})\right)\right] = Err\left(\mu(\boldsymbol{x})\right) + \mathrm{E}_{\mathcal{D},\Theta}\left[\mathcal{E}^{\mathcal{P}}\left(\widehat{\mu}^{\text{bag}}_{\mathcal{D},\Theta}(\boldsymbol{x})\right)\right] \qquad (4.3.21)
$$

with

$$
\mathrm{E}_{\mathcal{D},\Theta}\left[\mathcal{E}^{\mathcal{P}}\left(\widehat{\mu}^{\text{bag}}_{\mathcal{D},\Theta}(\boldsymbol{x})\right)\right] = 2\mu(\boldsymbol{x})\left(\mathrm{E}_{\mathcal{D},\Theta}\left[\frac{\widehat{\mu}^{\text{bag}}_{\mathcal{D},\Theta}(\boldsymbol{x})}{\mu(\boldsymbol{x})}\right] - 1 - \mathrm{E}_{\mathcal{D},\Theta}\left[\ln\left(\frac{\widehat{\mu}^{\text{bag}}_{\mathcal{D},\Theta}(\boldsymbol{x})}{\mu(\boldsymbol{x})}\right)\right]\right). \qquad (4.3.22)
$$

We have

$$
\mathrm{E}_{\mathcal{D},\Theta}\left[\frac{\widehat{\mu}^{\text{bag}}_{\mathcal{D},\Theta}(\boldsymbol{x})}{\mu(\boldsymbol{x})}\right] = \mathrm{E}_{\mathcal{D},\Theta_b}\left[\frac{\widehat{\mu}_{\mathcal{D},\Theta_b}(\boldsymbol{x})}{\mu(\boldsymbol{x})}\right], \qquad (4.3.23)
$$

so that (4.3.22) can be expressed as

$$
\begin{aligned}
&\mathrm{E}_{\mathcal{D},\Theta}\left[\mathcal{E}^{\mathcal{P}}\left(\widehat{\mu}^{\text{bag}}_{\mathcal{D},\Theta}(\boldsymbol{x})\right)\right] \\
&= 2\mu(\boldsymbol{x})\left(\mathrm{E}_{\mathcal{D},\Theta_b}\left[\frac{\widehat{\mu}_{\mathcal{D},\Theta_b}(\boldsymbol{x})}{\mu(\boldsymbol{x})}\right] - 1 - \mathrm{E}_{\mathcal{D},\Theta_b}\left[\ln\left(\frac{\widehat{\mu}_{\mathcal{D},\Theta_b}(\boldsymbol{x})}{\mu(\boldsymbol{x})}\right)\right]\right) \\
&\quad -2\mu(\boldsymbol{x})\left(\mathrm{E}_{\mathcal{D},\Theta}\left[\ln\left(\frac{\widehat{\mu}^{\text{bag}}_{\mathcal{D},\Theta}(\boldsymbol{x})}{\mu(\boldsymbol{x})}\right)\right] - \mathrm{E}_{\mathcal{D},\Theta_b}\left[\ln\left(\frac{\widehat{\mu}_{\mathcal{D},\Theta_b}(\boldsymbol{x})}{\mu(\boldsymbol{x})}\right)\right]\right) \\
&= \mathrm{E}_{\mathcal{D},\Theta_b}\left[\mathcal{E}^{\mathcal{P}}\left(\widehat{\mu}_{\mathcal{D},\Theta_b}(\boldsymbol{x})\right)\right] \\
&\quad -2\mu(\boldsymbol{x})\left(\mathrm{E}_{\mathcal{D},\Theta}\left[\ln\left(\widehat{\mu}^{\text{bag}}_{\mathcal{D},\Theta}(\boldsymbol{x})\right)\right] - \mathrm{E}_{\mathcal{D},\Theta_b}\left[\ln\left(\widehat{\mu}_{\mathcal{D},\Theta_b}(\boldsymbol{x})\right)\right]\right). \qquad (4.3.24)
\end{aligned}
$$

Jensen's inequality implies

$$\begin{aligned}
&\mathrm{E}_{\mathcal{D},\Theta}\left[\ln \widehat{\mu}^{\mathrm{bag}}_{\mathcal{D},\Theta}(\boldsymbol{x})\right] - \mathrm{E}_{\mathcal{D},\Theta_b}\left[\ln \widehat{\mu}_{\mathcal{D},\Theta_b}(\boldsymbol{x})\right] \\
&= \mathrm{E}_{\mathcal{D},\Theta_1,\ldots,\Theta_B}\left[\ln\left(\frac{1}{B}\sum_{b=1}^{B}\widehat{\mu}_{\mathcal{D},\Theta_b}(\boldsymbol{x})\right)\right] - \mathrm{E}_{\mathcal{D},\Theta_b}\left[\ln \widehat{\mu}_{\mathcal{D},\Theta_b}(\boldsymbol{x})\right] \\
&\geq \mathrm{E}_{\mathcal{D},\Theta_1,\ldots,\Theta_B}\left[\frac{1}{B}\sum_{b=1}^{B}\ln \widehat{\mu}_{\mathcal{D},\Theta_b}(\boldsymbol{x})\right] - \mathrm{E}_{\mathcal{D},\Theta_b}\left[\ln \widehat{\mu}_{\mathcal{D},\Theta_b}(\boldsymbol{x})\right] \\
&= 0,
\end{aligned} \tag{4.3.25}$$

so that combining (4.3.24) and (4.3.25) leads to

$$\mathrm{E}_{\mathcal{D},\Theta}\left[\mathcal{E}^{\mathcal{P}}\left(\widehat{\mu}^{\mathrm{bag}}_{\mathcal{D},\Theta}(\boldsymbol{x})\right)\right] \leq \mathrm{E}_{\mathcal{D},\Theta_b}\left[\mathcal{E}^{\mathcal{P}}\left(\widehat{\mu}_{\mathcal{D},\Theta_b}(\boldsymbol{x})\right)\right] \tag{4.3.26}$$

and hence

$$\mathrm{E}_{\mathcal{D},\Theta}\left[Err\left(\widehat{\mu}^{\mathrm{bag}}_{\mathcal{D},\Theta}(\boldsymbol{x})\right)\right] \leq \mathrm{E}_{\mathcal{D},\Theta_b}\left[Err\left(\widehat{\mu}_{\mathcal{D},\Theta_b}(\boldsymbol{x})\right)\right]. \tag{4.3.27}$$

For every value of $\boldsymbol{X}$, the expected generalization error of the ensemble is smaller than the expected generalization error of an individual model.

Taking the average of (4.3.27) over $\boldsymbol{X}$ leads to

$$\mathrm{E}_{\mathcal{D},\Theta}\left[Err\left(\widehat{\mu}^{\mathrm{bag}}_{\mathcal{D},\Theta}\right)\right] \leq \mathrm{E}_{\mathcal{D},\Theta_b}\left[Err\left(\widehat{\mu}_{\mathcal{D},\Theta_b}\right)\right]. \tag{4.3.28}$$

Example Consider the example of Sect. 3.7.2. We simulate training sets $\mathcal{D}$ made of 100 000 observations and validation sets $\overline{\mathcal{D}}$ of the same size. For each simulated training set $\mathcal{D}$, we build the corresponding tree $\widehat{\mu}_{\mathcal{D}}$ with `maxdepth` $=$ `D` $= 5$, which corresponds to a reasonable size in this context (see Fig. 3.31), and we estimate its generalization error on a validation set $\overline{\mathcal{D}}$. Also, we generate bootstrap samples $\mathcal{D}^{*1}, \mathcal{D}^{*2}, \ldots$ of $\mathcal{D}$ and we produce the corresponding trees $\widehat{\mu}_{\mathcal{D}^{*1}}, \widehat{\mu}_{\mathcal{D}^{*2}}, \ldots$ with `D` $= 5$. We estimate their generalization errors on a validation set $\overline{\mathcal{D}}$, together with the generalization errors of the corresponding bagging models. Note that in this example, we use the `R` package `rpart` to build the different trees described above.

Figure 4.1 displays estimates of the expected generalization errors for $\widehat{\mu}_{\mathcal{D}}, \widehat{\mu}_{\mathcal{D}^{*b}} = \widehat{\mu}_{\mathcal{D},\Theta_b}$ and $\widehat{\mu}^{\mathrm{bag}}_{\mathcal{D},\Theta}$ for $B = 1, 2, \ldots, 10$ obtained by Monte-Carlo simulations. As expected, we notice that

$$\widehat{\mathrm{E}}_{\mathcal{D},\Theta}\left[Err\left(\widehat{\mu}^{\mathrm{bag}}_{\mathcal{D},\Theta}\right)\right] \leq \widehat{\mathrm{E}}_{\mathcal{D},\Theta_b}\left[Err\left(\widehat{\mu}_{\mathcal{D},\Theta_b}\right)\right].$$

For $B \geq 2$, bagging trees outperforms individual sample trees. Also, we note that

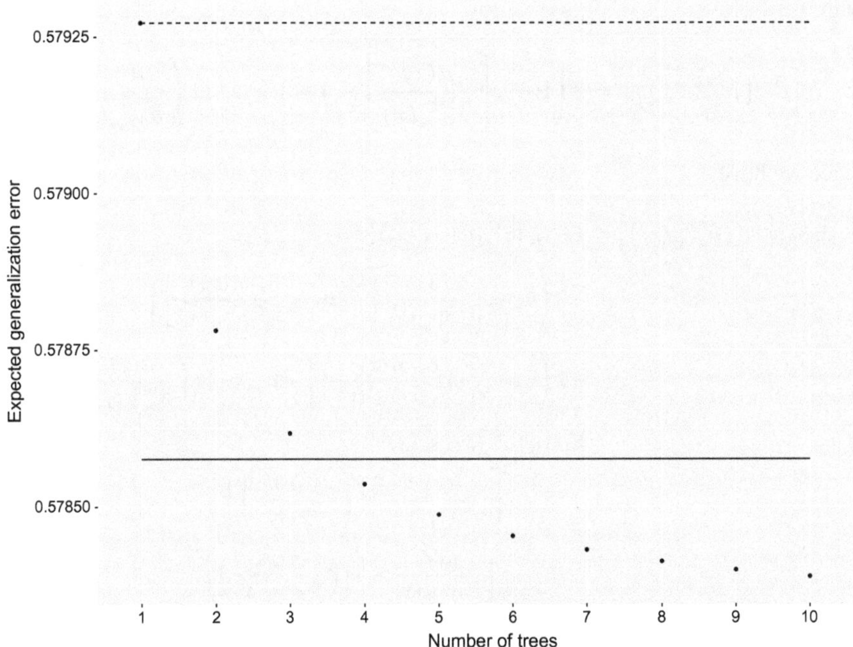

Fig. 4.1 $\widehat{\mathrm{E}}_{\mathcal{D},\boldsymbol{\Theta}}\left[Err\left(\widehat{\mu}^{\text{bag}}_{\mathcal{D},\boldsymbol{\Theta}}\right)\right]$ with respect to the number of trees B, together with $\widehat{\mathrm{E}}_{\mathcal{D},\boldsymbol{\Theta}_b}\left[Err\left(\widehat{\mu}_{\mathcal{D},\boldsymbol{\Theta}_b}\right)\right]$ (dotted line) and $\widehat{\mathrm{E}}_{\mathcal{D}}\left[Err\left(\widehat{\mu}_{\mathcal{D}}\right)\right]$ (solid line)

$$\widehat{\mathrm{E}}_{\mathcal{D}}\left[Err\left(\widehat{\mu}_{\mathcal{D}}\right)\right] \leq \widehat{\mathrm{E}}_{\mathcal{D},\Theta_b}\left[Err\left(\widehat{\mu}_{\mathcal{D},\Theta_b}\right)\right],$$

showing that the restriction imposed by the reduced sample $\mathcal{D}^{*b}$ does not allow to build trees as predictive as trees built on the entire training set $\mathcal{D}$. Finally, from $B = 4$, we note that

$$\widehat{\mathrm{E}}_{\mathcal{D},\Theta}\left[Err\left(\widehat{\mu}^{\text{bag}}_{\mathcal{D},\Theta}\right)\right] \leq \widehat{\mathrm{E}}_{\mathcal{D}}\left[Err\left(\widehat{\mu}_{\mathcal{D}}\right)\right],$$

meaning that for $B \geq 4$, bagging trees also outperforms single trees built on the entire training set.

4.3.3.3 Gamma Deviance Loss

Consider the Gamma deviance loss. From (2.4.6) and (2.4.7), the expected generalization error for $\widehat{\mu}^{\text{bag}}_{\mathcal{D},\Theta}(\boldsymbol{x})$ is given by

$$\mathrm{E}_{\mathcal{D},\Theta}\left[Err\left(\widehat{\mu}^{\text{bag}}_{\mathcal{D},\Theta}(\boldsymbol{x})\right)\right] = Err\left(\mu(\boldsymbol{x})\right) + \mathrm{E}_{\mathcal{D},\Theta}\left[\mathcal{E}^{\mathcal{G}}\left(\widehat{\mu}^{\text{bag}}_{\mathcal{D},\Theta}(\boldsymbol{x})\right)\right] \qquad (4.3.29)$$

with

$$\mathrm{E}_{\mathcal{D},\Theta}\left[\mathcal{E}^{\mathcal{G}}\left(\widehat{\mu}^{\text{bag}}_{\mathcal{D},\Theta}(\boldsymbol{x})\right)\right] = 2\left(\mathrm{E}_{\mathcal{D},\Theta}\left[\frac{\mu(\boldsymbol{x})}{\widehat{\mu}^{\text{bag}}_{\mathcal{D},\Theta}(\boldsymbol{x})}\right] - 1 - \mathrm{E}_{\mathcal{D},\Theta}\left[\ln\left(\frac{\mu(\boldsymbol{x})}{\widehat{\mu}^{\text{bag}}_{\mathcal{D},\Theta}(\boldsymbol{x})}\right)\right]\right). \tag{4.3.30}$$

Since we have

$$\begin{aligned}
\mathrm{E}_{\mathcal{D},\Theta}\left[\mathcal{E}^{\mathcal{G}}\left(\widehat{\mu}^{\text{bag}}_{\mathcal{D},\Theta}(\boldsymbol{x})\right)\right] &= \mathrm{E}_{\mathcal{D},\Theta_b}\left[\mathcal{E}^{\mathcal{G}}\left(\widehat{\mu}_{\mathcal{D},\Theta_b}(\boldsymbol{x})\right)\right] \\
&\quad +2\left(\mathrm{E}_{\mathcal{D},\Theta}\left[\frac{\mu(\boldsymbol{x})}{\widehat{\mu}^{\text{bag}}_{\mathcal{D},\Theta}(\boldsymbol{x})}\right] - \mathrm{E}_{\mathcal{D},\Theta_b}\left[\frac{\mu(\boldsymbol{x})}{\widehat{\mu}_{\mathcal{D},\Theta_b}(\boldsymbol{x})}\right]\right) \\
&\quad +2\left(\mathrm{E}_{\mathcal{D},\Theta_b}\left[\ln\left(\frac{\mu(\boldsymbol{x})}{\widehat{\mu}_{\mathcal{D},\Theta_b}(\boldsymbol{x})}\right)\right] - \mathrm{E}_{\mathcal{D},\Theta}\left[\ln\left(\frac{\mu(\boldsymbol{x})}{\widehat{\mu}^{\text{bag}}_{\mathcal{D},\Theta}(\boldsymbol{x})}\right)\right]\right) \\
&= \mathrm{E}_{\mathcal{D},\Theta_b}\left[\mathcal{E}^{\mathcal{G}}\left(\widehat{\mu}_{\mathcal{D},\Theta_b}(\boldsymbol{x})\right)\right] \\
&\quad +2\mathrm{E}_{\mathcal{D},\Theta}\left[\frac{\mu(\boldsymbol{x})}{\widehat{\mu}^{\text{bag}}_{\mathcal{D},\Theta}(\boldsymbol{x})} - \ln\left(\frac{\mu(\boldsymbol{x})}{\widehat{\mu}^{\text{bag}}_{\mathcal{D},\Theta}(\boldsymbol{x})}\right)\right] \\
&\quad -2\mathrm{E}_{\mathcal{D},\Theta_b}\left[\frac{\mu(\boldsymbol{x})}{\widehat{\mu}_{\mathcal{D},\Theta_b}(\boldsymbol{x})} - \ln\left(\frac{\mu(\boldsymbol{x})}{\widehat{\mu}_{\mathcal{D},\Theta_b}(\boldsymbol{x})}\right)\right],
\end{aligned} \tag{4.3.31}$$

we see that

$$\mathrm{E}_{\mathcal{D},\Theta}\left[\mathcal{E}^{\mathcal{G}}\left(\widehat{\mu}^{\text{bag}}_{\mathcal{D},\Theta}(\boldsymbol{x})\right)\right] \leq \mathrm{E}_{\mathcal{D},\Theta_b}\left[\mathcal{E}^{\mathcal{G}}\left(\widehat{\mu}_{\mathcal{D},\Theta_b}(\boldsymbol{x})\right)\right]$$

if and only if

$$\mathrm{E}_{\mathcal{D},\Theta}\left[\frac{\mu(\boldsymbol{x})}{\widehat{\mu}^{\text{bag}}_{\mathcal{D},\Theta}(\boldsymbol{x})} + \ln\left(\frac{\widehat{\mu}^{\text{bag}}_{\mathcal{D},\Theta}(\boldsymbol{x})}{\mu(\boldsymbol{x})}\right)\right] \leq \mathrm{E}_{\mathcal{D},\Theta_b}\left[\frac{\mu(\boldsymbol{x})}{\widehat{\mu}_{\mathcal{D},\Theta_b}(\boldsymbol{x})} + \ln\left(\frac{\widehat{\mu}_{\mathcal{D},\Theta_b}(\boldsymbol{x})}{\mu(\boldsymbol{x})}\right)\right]. \tag{4.3.32}$$

The latter inequality is fulfilled when the individual sample trees satisfy

$$\frac{\widehat{\mu}_{\mathcal{D},\Theta_b}(\boldsymbol{x})}{\mu(\boldsymbol{x})} \leq 2, \tag{4.3.33}$$

which, in turn, guarantees that $\frac{\widehat{\mu}^{\text{bag}}_{\mathcal{D},\Theta}(\boldsymbol{x})}{\mu(\boldsymbol{x})} \leq 2$. Indeed, the function $\phi : x > 0 \to \frac{1}{x} + \ln x$ is convex for $x \leq 2$, so that Jensen's inequality implies

$$
\begin{aligned}
\mathrm{E}_{\mathcal{D},\Theta}\left[\frac{\mu(\boldsymbol{x})}{\widehat{\mu}^{\text{bag}}_{\mathcal{D},\Theta}(\boldsymbol{x})}+\ln\left(\frac{\widehat{\mu}^{\text{bag}}_{\mathcal{D},\Theta}(\boldsymbol{x})}{\mu(\boldsymbol{x})}\right)\right] &= \mathrm{E}_{\mathcal{D},\Theta}\left[\phi\left(\frac{\widehat{\mu}^{\text{bag}}_{\mathcal{D},\Theta}(\boldsymbol{x})}{\mu(\boldsymbol{x})}\right)\right] \\
&= \mathrm{E}_{\mathcal{D},\Theta_1,\ldots,\Theta_B}\left[\phi\left(\frac{1}{B}\sum_{b=1}^{B}\frac{\widehat{\mu}_{\mathcal{D},\Theta_b}(\boldsymbol{x})}{\mu(\boldsymbol{x})}\right)\right] \\
&\leq \mathrm{E}_{\mathcal{D},\Theta_1,\ldots,\Theta_B}\left[\frac{1}{B}\sum_{b=1}^{B}\phi\left(\frac{\widehat{\mu}_{\mathcal{D},\Theta_b}(\boldsymbol{x})}{\mu(\boldsymbol{x})}\right)\right] \\
&= \mathrm{E}_{\mathcal{D},\Theta_b}\left[\phi\left(\frac{\widehat{\mu}_{\mathcal{D},\Theta_b}(\boldsymbol{x})}{\mu(\boldsymbol{x})}\right)\right] \\
&= \mathrm{E}_{\mathcal{D},\Theta_b}\left[\frac{\mu(\boldsymbol{x})}{\widehat{\mu}_{\mathcal{D},\Theta_b}(\boldsymbol{x})}+\ln\left(\frac{\widehat{\mu}_{\mathcal{D},\Theta_b}(\boldsymbol{x})}{\mu(\boldsymbol{x})}\right)\right]
\end{aligned}
$$

provided that condition (4.3.33) holds.

If inequality (4.3.33) is satisfied, we thus have

$$
\mathrm{E}_{\mathcal{D},\Theta}\left[Err\left(\widehat{\mu}^{\text{bag}}_{\mathcal{D},\Theta}(\boldsymbol{x})\right)\right] \leq \mathrm{E}_{\mathcal{D},\Theta_b}\left[Err\left(\widehat{\mu}_{\mathcal{D},\Theta_b}(\boldsymbol{x})\right)\right],
$$

and so

$$
\mathrm{E}_{\mathcal{D},\Theta}\left[Err\left(\widehat{\mu}^{\text{bag}}_{\mathcal{D},\Theta}\right)\right] \leq \mathrm{E}_{\mathcal{D},\Theta_b}\left[Err\left(\widehat{\mu}_{\mathcal{D},\Theta_b}\right)\right]. \tag{4.3.34}
$$

Note that condition (4.3.33) means that the individual sample prediction $\widehat{\mu}_{\mathcal{D},\Theta_b}(\boldsymbol{x})$ should be not too far from the true prediction $\mu(\boldsymbol{x})$.

4.4 Random Forests

Bagging is a technique used for reducing the variance of a prediction. Typically, it works well for high variance and low-bias procedures, such as regression trees. The procedure called random forests is a modification of bagging trees. It produces a collection of trees that are more de-correlated than in the bagging procedure, and averages them.

Recall from (4.3.14) that the variance of the bagging prediction can be expressed as

$$
\mathrm{Var}_{\mathcal{D},\Theta}\left[\widehat{\mu}^{\text{bag}}_{\mathcal{D},\Theta}(\boldsymbol{x})\right] = \rho(\boldsymbol{x})\mathrm{Var}_{\mathcal{D},\Theta_b}\left[\widehat{\mu}_{\mathcal{D},\Theta_b}(\boldsymbol{x})\right] + \frac{(1-\rho(\boldsymbol{x}))}{B}\mathrm{Var}_{\mathcal{D},\Theta_b}\left[\widehat{\mu}_{\mathcal{D},\Theta_b}(\boldsymbol{x})\right]. \tag{4.4.1}
$$

As B increases, the second term disappears while the first one remains. The correlation coefficient $\rho(\boldsymbol{x})$ in the first term limits the effect of averaging. The idea of random forests is to improve the bagging procedure by reducing the correlation coefficient $\rho(\boldsymbol{x})$ without increasing the variance $\mathrm{Var}_{\mathcal{D},\Theta_b}\left[\widehat{\mu}_{\mathcal{D},\Theta_b}(\boldsymbol{x})\right]$ too much.

Reducing correlation among trees can be achieved by adding randomness to the training procedure. The difficulty relies in modifying the training procedure without affecting too much both the bias and the variance of the individual trees.

Random forests is a combination of bagging with random feature selection at each node. Specifically, when growing a tree on a bootstrap sample, $m(\leq p)$ features are selected at random before each split and used as candidates for splitting. The random forest prediction writes

$$\widehat{\mu}^{\mathrm{rf}}_{\mathcal{D},\Theta}(\boldsymbol{x}) = \frac{1}{B}\sum_{b=1}^{B}\widehat{\mu}_{\mathcal{D},\Theta_b}(\boldsymbol{x}),$$

where $\widehat{\mu}_{\mathcal{D},\Theta_b}(\boldsymbol{x})$ denotes the prediction at point $\boldsymbol{x}$ for the bth random forest tree. Random vectors $\Theta_1, \ldots, \Theta_B$ capture not only the randomness of the bootstrap sampling, as for bagging, but also the additional randomness of the training procedure due to the random selection of m features before each split. This provides the following algorithm:

Algorithm 4.2: Random forests

For $b = 1$ to B **do**

1. Generate a bootstrap sample $\mathcal{D}^{*b}$ of $\mathcal{D}$.
2. Fit a tree on $\mathcal{D}^{*b}$.

 For each node t **do**

 (2.1) Select m ($\leq p$) features at random from the p original features.

 (2.2) Pick the best feature among the m.

 (2.3) Split the node into two daughter nodes.

 End for

 This gives prediction $\widehat{\mu}_{\mathcal{D},\Theta_b}(\boldsymbol{x})$ (use typical tree stopping criteria (but do not prune)).

End for

Output: $\widehat{\mu}^{\mathrm{rf}}_{\mathcal{D},\Theta}(\boldsymbol{x}) = \frac{1}{B}\sum_{b=1}^{B}\widehat{\mu}_{\mathcal{D},\Theta_b}(\boldsymbol{x})$.

As soon as $m < p$, random forests differs from bagging trees since the optimal split can be missed if it is not among the m features selected. Typical value of m is $\lfloor p/3 \rfloor$. However, the best value for m depends on the problem under consideration and is treated as a tuning parameter. Decreasing m reduces the correlation between any pair of trees while it increases the variance of the individual trees.

Notice that random forests is more computationally efficient on a tree-by-tree basis than bagging since the training procedure only needs to assess a part of the original features at each split. However, compared to bagging, random forests usually require more trees.

Remark 4.4.1 Obviously, results obtained in Sects. 4.3.1, 4.3.2 and 4.3.3 also hold for random forests. The only difference relies in the meaning of random vectors $\Theta_1, \ldots, \Theta_B$. For bagging, those vectors express the randomization due to the bootstrap sampling, while for random forests, they also account for the randomness due to the feature selection at each node.

4.5 Out-of-Bag Estimate

Bagging trees and random forests aggregate trees built on bootstrap samples $\mathcal{D}^{*1}, \ldots, \mathcal{D}^{*B}$. For each observation $(y_i, \boldsymbol{x}_i)$ of the training set $\mathcal{D}$, an out-of-bag prediction can be constructed by averaging only trees corresponding to bootstrap samples $\mathcal{D}^{*b}$ in which $(y_i, \boldsymbol{x}_i)$ does not appear. The out-of-bag prediction for observation $(y_i, \boldsymbol{x}_i)$ is thus given by

$$\widehat{\mu}^{\text{oob}}_{\mathcal{D},\Theta}(\boldsymbol{x}_i) = \frac{1}{\sum_{b=1}^{B} \mathrm{I}\left[(y_i, \boldsymbol{x}_i) \notin \mathcal{D}^{*b}\right]} \sum_{b=1}^{B} \widehat{\mu}_{\mathcal{D},\Theta_b}(\boldsymbol{x}_i) \mathrm{I}\left[(y_i, \boldsymbol{x}_i) \notin \mathcal{D}^{*b}\right]. \quad (4.5.1)$$

The generalization error of $\widehat{\mu}_{\mathcal{D},\Theta}$ can be estimated by

$$\widehat{Err}^{\text{oob}}(\widehat{\mu}_{\mathcal{D},\Theta}) = \frac{1}{|\mathcal{I}|} \sum_{i \in \mathcal{I}} L(y_i, \widehat{\mu}^{\text{oob}}_{\mathcal{D},\Theta}(\boldsymbol{x}_i)), \quad (4.5.2)$$

which is called the out-of-bag estimate of the generalization error.

The out-of bag estimate of the generalization error is almost identical to the $|\mathcal{I}|$-fold cross-validation estimate. However, the out-of-bag estimate does not require to fit new trees, so that bagging trees and random forests can be fit in one sequence. We stop adding new trees when the out-of-bag estimate of the generalization error stabilizes.

4.6 Interpretability

A bagged model is less interpretable than a model that is not bagged. Bagging trees and random forests are no longer a tree. However, there exist tools that enable to better understand model outcomes.

4.6.1 Relative Importances

As for a single regression tree, the relative importance of a feature can be computed for an ensemble by combining relative importances from the bootstrap trees. For the bth tree in the ensemble, denoted T_b, the relative importance of feature x_j is the sum of the deviance reductions ΔD_{χ_t} over the non-terminal nodes $t \in \widetilde{\mathcal{T}}_{(T_b)}(x_j)$ (i.e. the non-terminal nodes t of T_b for which x_j was selected as the splitting feature), that is,

$$\mathcal{I}_b(x_j) = \sum_{t \in \widetilde{\mathcal{T}}_{(T_b)}(x_j)} \Delta D_{\chi_t}. \tag{4.6.1}$$

For the ensemble, the relative importance of feature x_j is obtained by averaging the relative importances of x_j over the collection of trees, namely

$$\mathcal{I}(x_j) = \frac{1}{B} \sum_{b=1}^{B} \mathcal{I}_b(x_j). \tag{4.6.2}$$

For convenience, the relative importances are often normalized so that their sum equals to 100. Any individual number can then be interpreted as the percentage contribution to the overall model. Sometimes, the relative importances are expressed as a percent of the maximum relative importance.

An alternative to compute variable importances for bagging trees and random forests is based on out-of-bag observations. Some observations $(y_i, \boldsymbol{x}_i)$ of the training set $\mathcal{D}$ do not appear in bootstrap sample $\mathcal{D}^{*b}$. They are called the out-of-bag observations for the bth tree. Because they were not used to fit that specific tree, these observations enable to assess the predictive accuracy of $\widehat{\mu}_{\mathcal{D},\Theta_b}$, that is,

$$\widehat{Err}(\widehat{\mu}_{\mathcal{D},\Theta_b}) = \frac{1}{|\mathcal{I} \backslash \mathcal{I}^{*b}|} \sum_{i \in \mathcal{I} \backslash \mathcal{I}^{*b}} L(y_i, \widehat{\mu}_{\mathcal{D},\Theta_b}(\boldsymbol{x}_i)), \tag{4.6.3}$$

where $\mathcal{I}^{*b}$ labels the observations in $\mathcal{D}^{*b}$. The categories of feature x_j are then randomly permuted in the out-of-bag observations, so that we get perturbed observations $(y_i, \boldsymbol{x}_i^{\text{perm(j)}})$, $i \in \mathcal{I} \backslash \mathcal{I}^{*b}$, and the predictive accuracy of $\widehat{\mu}_{\mathcal{D},\Theta_b}$ is again computed as

$$\widehat{Err}^{\text{perm(j)}}(\widehat{\mu}_{\mathcal{D},\Theta_b}) = \frac{1}{|\mathcal{I} \backslash \mathcal{I}^{*b}|} \sum_{i \in \mathcal{I} \backslash \mathcal{I}^{*b}} L(y_i, \widehat{\mu}_{\mathcal{D},\Theta_b}(\boldsymbol{x}_i^{\text{perm(j)}})). \tag{4.6.4}$$

The decrease in predictive accuracy due to this permuting is averaged over all trees and is used as a measure of importance for feature x_j in the ensemble, that is

$$\mathcal{I}(x_j) = \frac{1}{B} \sum_{b=1}^{B} \left(\widehat{Err}^{\text{perm(j)}}(\widehat{\mu}_{\mathcal{D},\Theta_b}) - \widehat{Err}(\widehat{\mu}_{\mathcal{D},\Theta_b}) \right). \tag{4.6.5}$$

These importances can be normalized to improve their readability. A feature will be important if randomly permuting its values decreases the model accuracy. In such a case, it means that the model relies on the feature for the prediction.

Notice that the relative importance $\mathcal{I}(x_j)$ obtained by permutation in (4.6.5) does not measure the effect on estimate of the absence of x_j. Rather, it measures the effect of neutralizing x_j, much like setting a coefficient to zero in GLMs for instance.

4.6.2 Partial Dependence Plots

Visualizing the value of $\widehat{\mu}(\boldsymbol{x})$ as a function of $\boldsymbol{x}$ enables to understand its dependence on the joint values of the features. Such visualization is however limited to small values of p. For $p = 1$, $\widehat{\mu}(x)$ can be easily plotted, as a graph of the values of $\widehat{\mu}(x)$ against each possible value of x for single real-valued variable x, and as a bar-plot for categorical variable x, each bar corresponding to one value of the categorical variable. For $p = 2$, graphical renderings of $\widehat{\mu}(\boldsymbol{x})$ is still possible. Functions of two real-valued variables can be represented by means of contour or perspective mesh plots. Functions of a categorical variable and another variable, categorical or real, can be pictured by a sequence of plots, each one depicting the dependence of $\widehat{\mu}(\boldsymbol{x})$ on the second variable, conditioned on the values of the first variable.

For $p > 2$, representing $\widehat{\mu}(\boldsymbol{x})$ as a function of its arguments is more difficult. An alternative consists in visualizing a collection of plots, each one showing the partial dependence of $\widehat{\mu}(\boldsymbol{x})$ on selected small subsets of the features.

Consider the subvector $\boldsymbol{x}_{\mathcal{S}}$ of $l < p$ of the features $\boldsymbol{x} = (x_1, x_2, \ldots, x_p)$, indexed by $\mathcal{S} \subset \{1, 2, \ldots, p\}$. Let $\boldsymbol{x}_{\bar{\mathcal{S}}}$ be the complement subvector such that

$$\boldsymbol{x}_{\mathcal{S}} \cup \boldsymbol{x}_{\bar{\mathcal{S}}} = \boldsymbol{x}.$$

In principle, $\widehat{\mu}(\boldsymbol{x})$ depends on features in $\boldsymbol{x}_{\mathcal{S}}$ and $\boldsymbol{x}_{\bar{\mathcal{S}}}$, so that (by rearranging the order of the features if needed) we can write

$$\widehat{\mu}(\boldsymbol{x}) = \widehat{\mu}(\boldsymbol{x}_{\mathcal{S}}, \boldsymbol{x}_{\bar{\mathcal{S}}}).$$

Hence, if one conditions on specific values for the features in $\boldsymbol{x}_{\bar{\mathcal{S}}}$, then $\widehat{\mu}(\boldsymbol{x})$ can be seen as a function of the features in $\boldsymbol{x}_{\mathcal{S}}$. One way to define the partial dependence of $\widehat{\mu}(\boldsymbol{x})$ on $\boldsymbol{x}_{\mathcal{S}}$ is given by

$$\widehat{\mu}_{\mathcal{S}}(\boldsymbol{x}_{\mathcal{S}}) = \mathrm{E}_{\boldsymbol{X}_{\bar{\mathcal{S}}}}\left[\widehat{\mu}(\boldsymbol{x}_{\mathcal{S}}, \boldsymbol{X}_{\bar{\mathcal{S}}})\right]. \tag{4.6.6}$$

This average function can be used as a description of the effect of the selected subset $\boldsymbol{x}_{\mathcal{S}}$ on $\widehat{\mu}(\boldsymbol{x})$ when the features in $\boldsymbol{x}_{\mathcal{S}}$ do not have strong interactions with those in $\boldsymbol{x}_{\bar{\mathcal{S}}}$.

For instance, in the particular case where the dependence of $\widehat{\mu}(\boldsymbol{x})$ on $\boldsymbol{x}_{\mathcal{S}}$ is additive

$$\widehat{\mu}(\boldsymbol{x}) = f_{\mathcal{S}}(\boldsymbol{x}_{\mathcal{S}}) + f_{\bar{\mathcal{S}}}(\boldsymbol{x}_{\bar{\mathcal{S}}}), \tag{4.6.7}$$

so that there is no interactions between features in $\boldsymbol{x}_{\mathcal{S}}$ and $\boldsymbol{x}_{\bar{\mathcal{S}}}$, the partial dependence function (4.6.6) becomes

$$\widehat{\mu}_{\mathcal{S}}(\boldsymbol{x}_{\mathcal{S}}) = f_{\mathcal{S}}(\boldsymbol{x}_{\mathcal{S}}) + \mathrm{E}_{\boldsymbol{X}_{\bar{\mathcal{S}}}}\left[f_{\bar{\mathcal{S}}}(\boldsymbol{X}_{\bar{\mathcal{S}}})\right]. \tag{4.6.8}$$

Hence, (4.6.6) produces $f_{\mathcal{S}}(\boldsymbol{x}_{\mathcal{S}})$ up to an additive constant. In this case, one sees that (4.6.6) provides a complete description of the way $\widehat{\mu}(\boldsymbol{x})$ varies on the subset $\boldsymbol{x}_{\mathcal{S}}$.

The partial dependence function $\widehat{\mu}_{\mathcal{S}}(\boldsymbol{x}_{\mathcal{S}})$ can be estimated from the training set by

$$\frac{1}{|\mathcal{I}|}\sum_{i\in\mathcal{I}}\widehat{\mu}(\boldsymbol{x}_{\mathcal{S}}, \boldsymbol{x}_{i\bar{\mathcal{S}}}), \tag{4.6.9}$$

where $\{\boldsymbol{x}_{i\bar{\mathcal{S}}}, i \in \mathcal{I}\}$ are the values of $\boldsymbol{X}_{\bar{\mathcal{S}}}$ in the training set.

Notice that $\widehat{\mu}_{\mathcal{S}}(\boldsymbol{x}_{\mathcal{S}})$ represents the effect of $\boldsymbol{x}_{\mathcal{S}}$ on $\widehat{\mu}(\boldsymbol{x})$ after accounting for the average effects of the other features $\boldsymbol{x}_{\bar{\mathcal{S}}}$ on $\widehat{\mu}(\boldsymbol{x})$. Hence, it is different than computing the conditional expectation

$$\widetilde{\mu}_{\mathcal{S}}(\boldsymbol{x}_{\mathcal{S}}) = \mathrm{E}_{\boldsymbol{X}}\left[\widehat{\mu}(\boldsymbol{X})|\boldsymbol{X}_{\mathcal{S}} = \boldsymbol{x}_{\mathcal{S}}\right]. \tag{4.6.10}$$

Indeed, this latter expression captures the effect of $\boldsymbol{x}_{\mathcal{S}}$ on $\widehat{\mu}(\boldsymbol{x})$ ignoring the effects of $\boldsymbol{x}_{\bar{\mathcal{S}}}$. Both expressions (4.6.6) and (4.6.10) are actually equivalent only when $\boldsymbol{X}_{\mathcal{S}}$ and $\boldsymbol{X}_{\bar{\mathcal{S}}}$ are independent.

For instance, in the specific case (4.6.7) where $\widehat{\mu}(\boldsymbol{x})$ is additive, $\widetilde{\mu}_{\mathcal{S}}(\boldsymbol{x}_{\mathcal{S}})$ can be written as

$$\widetilde{\mu}_{\mathcal{S}}(\boldsymbol{x}_{\mathcal{S}}) = f_{\mathcal{S}}(\boldsymbol{x}_{\mathcal{S}}) + \mathrm{E}_{\boldsymbol{X}_{\bar{\mathcal{S}}}}\left[f_{\bar{\mathcal{S}}}(\boldsymbol{X}_{\bar{\mathcal{S}}})|\boldsymbol{X}_{\mathcal{S}} = \boldsymbol{x}_{\mathcal{S}}\right]. \tag{4.6.11}$$

One sees that $\widetilde{\mu}_{\mathcal{S}}(\boldsymbol{x}_{\mathcal{S}})$ does not produce $f_{\mathcal{S}}(\boldsymbol{x}_{\mathcal{S}})$ up to a constant, as it was the case for $\widehat{\mu}_{\mathcal{S}}(\boldsymbol{x}_{\mathcal{S}})$. This time, the behavior of $\widetilde{\mu}_{\mathcal{S}}(\boldsymbol{x}_{\mathcal{S}})$ depends on the dependence structure between $\boldsymbol{X}_{\mathcal{S}}$ and $\boldsymbol{X}_{\bar{\mathcal{S}}}$. In the case where the dependence of $\widehat{\mu}(\boldsymbol{x})$ on $\boldsymbol{x}_{\mathcal{S}}$ is multiplicative, that is

$$\widehat{\mu}(\boldsymbol{x}) = f_{\mathcal{S}}(\boldsymbol{x}_{\mathcal{S}}) \cdot f_{\bar{\mathcal{S}}}(\boldsymbol{x}_{\bar{\mathcal{S}}}), \tag{4.6.12}$$

so that the features in $\boldsymbol{x}_{\mathcal{S}}$ interact with those in $\boldsymbol{x}_{\bar{\mathcal{S}}}$, (4.6.6) becomes

$$\widehat{\mu}_{\mathcal{S}}(\boldsymbol{x}_{\mathcal{S}}) = f_{\mathcal{S}}(\boldsymbol{x}_{\mathcal{S}}) \cdot \mathrm{E}_{\boldsymbol{X}_{\bar{\mathcal{S}}}}\left[f_{\bar{\mathcal{S}}}(\boldsymbol{X}_{\bar{\mathcal{S}}})\right] \tag{4.6.13}$$

while (4.6.10) is given by

$$\widetilde{\mu}_{\mathcal{S}}(\boldsymbol{x}_{\mathcal{S}}) = f_{\mathcal{S}}(\boldsymbol{x}_{\mathcal{S}}) \cdot \mathrm{E}_{\boldsymbol{X}_{\bar{\mathcal{S}}}}\left[f_{\bar{\mathcal{S}}}(\boldsymbol{X}_{\bar{\mathcal{S}}})|\boldsymbol{X}_{\mathcal{S}} = \boldsymbol{x}_{\mathcal{S}}\right]. \tag{4.6.14}$$

In this other example, one sees that $\widehat{\mu}_S(\boldsymbol{x}_S)$ produces $f_S(\boldsymbol{x}_S)$ up to a multiplicative constant factor, so that its form does not depend neither on the dependence structure between $\boldsymbol{X}_S$ and $\boldsymbol{X}_{\bar{S}}$, while $\widetilde{\mu}_S(\boldsymbol{x}_S)$ does well.

4.7 Example

Consider the real dataset described in Sect. 3.2.4.2. We use the same training set $\mathcal{D}$ and validation set $\overline{\mathcal{D}}$ than in the example of Sect. 3.3.2.3, so that the estimates for the generalization error will be comparable.

We fit random forests with $B = 2000$ trees on $\mathcal{D}$ by means of the `R` package `rfCountData`. More precisely, we use the `R` command `rfPoisson()`, which stands for `r`andom `f`orest `Poisson`, producing random forests with Poisson deviance as loss function. The number of trees $B = 2000$ is set arbitrarily. However, we will see that $B = 2000$ is large enough (i.e. adding trees will not improve the predictive accuracy of the random forests under investigation).

The other parameters we need to fine-tune are

- the number m of features tried at each split (`mtry`);
- the size of the trees, controlled here by the minimum number of observations (`nodesize`) required in terminal nodes.

To this end, we try different values for `mtry` and `nodesize` and we split the training set $\mathcal{D}$ into five disjoint and stratified subsets $\mathcal{D}_1, \mathcal{D}_2, \ldots, \mathcal{D}_5$ of equal size. Specifically, we consider all possible values for `mtry` (from 1 to 8) together with four values for `nodesize`: 500, 1000, 5000 and 10 000. Note that the training set $\mathcal{D}$ contains 128 755 observations, so that values of 500 or 1000 for `nodesize` amounts to require at least 0.39% and 0.78% of the observations in the final nodes of the individual trees, respectively, which allows for rather large trees. Then, for each value of (`mtry`, `nodesize`), we compute the 5-fold cross-validation estimate of the generalization error from subsets $\mathcal{D}_1, \mathcal{D}_2, \ldots, \mathcal{D}_5$. The results are depicted in Fig. 4.2. We can see that the minimum 5-fold cross-validation estimate corresponds to `mtry` $= 3$ and `nodesize` $= 1000$. Notice that for any value of `nodesize`, it is never optimal to use all the features at each split (i.e. `mtry` $= 8$). Introducing a random feature selection at each node therefore improves the predictive accuracy of the ensemble. Moreover, as expected, limiting too much the size of the trees (here with `nodesize` $= 5000, 10\,000$) turns out to be counterproductive. The predictive performances for trees with `nodesize` $= 1000$ are already satisfying and comparable to the ones obtained with even smaller trees.

In Fig. 4.3, we show the out-of-bag estimate of the generalization error for random forests with `mtry` $= 3$ and `nodesize` $= 1000$ with respect to the number of trees B. We observe that $B = 2000$ is more than enough. The out-of-bag estimate is already stabilized from $B = 500$. Notice that adding more trees beyond $B = 500$ does not decrease the predictive accuracy of the random forest.

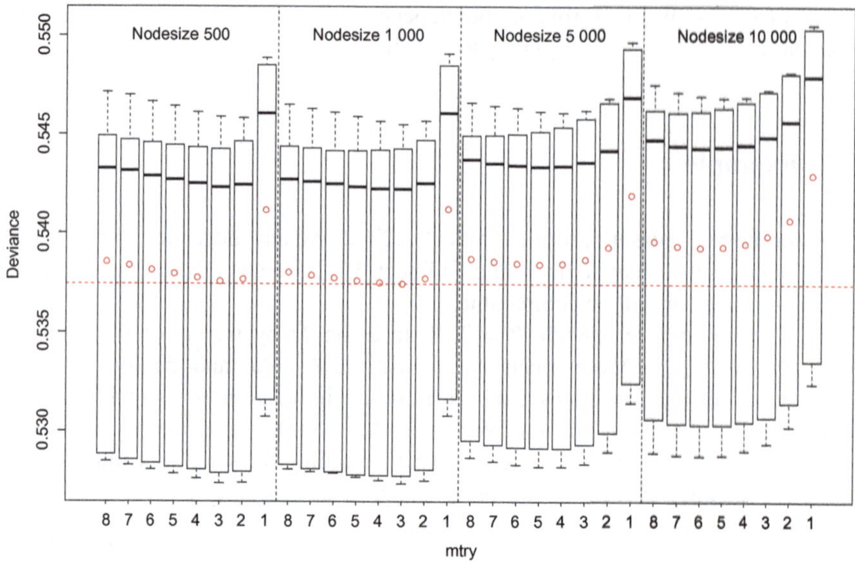

Fig. 4.2 5-fold cross-validation estimates of the generalization error for `mtry` $= 1, 2, \ldots, 8$ and `nodesize` $= 500, 1000, 5000, 10\,000$

We denote by $\widehat{\mu}^{\text{rf}^*}_{\mathcal{D},\Theta}$ the random forest fitted on the entire training set $\mathcal{D}$ with $B = 500$, `mtry` $= 3$ and `nodesize` $= 1000$. The relative importances of the features for $\widehat{\mu}^{\text{rf}^*}_{\mathcal{D},\Theta}$ are depicted in Fig. 4.4. The most important feature is AgePh followed by, in descending order, Split, Fuel, AgeCar, Cover, Gender, PowerCat and Use. Notice that this ranking of the features in terms of importance is almost identical to the one shown in Fig. 3.23 and obtained from the tree depicted in Fig. 3.17, only the order of features AgeCar and Cover is reversed (their importances are very similar here).

Figure 4.5 represents the partial dependence plots of the features for $\widehat{\mu}^{\text{rf}^*}_{\mathcal{D},\Theta}$. Specifically, one sees that the partial dependence plot for policyholder's age is relatively smooth. This is more realistic than the impact of policyholder's age deduced from regression tree $\widehat{\mu}_{T_{\alpha_{k^*}}}$ represented in Fig. 3.20, tree that is reproduced in Fig. 4.6 where we have circled the nodes using AgePh for splitting. Indeed, with only six circled nodes, $\widehat{\mu}_{T_{\alpha_{k^*}}}$ cannot reflect a smooth behavior of the expected claim frequency with respect to AgePh. While random forests enable to capture nuances of the response, regression trees suffer from a lack of smoothness.

Finally, the validation sample estimate of the generalization error of $\widehat{\mu}^{\text{rf}^*}_{\mathcal{D},\Theta}$ is given by

$$\widehat{Err}^{\text{val}}\left(\widehat{\mu}^{\text{rf}^*}_{\mathcal{D},\Theta}\right) = 0.5440970.$$

Compared to the validation sample estimate of the generalization error of $\widehat{\mu}_{T_{\alpha_{k^*}}}$, that is

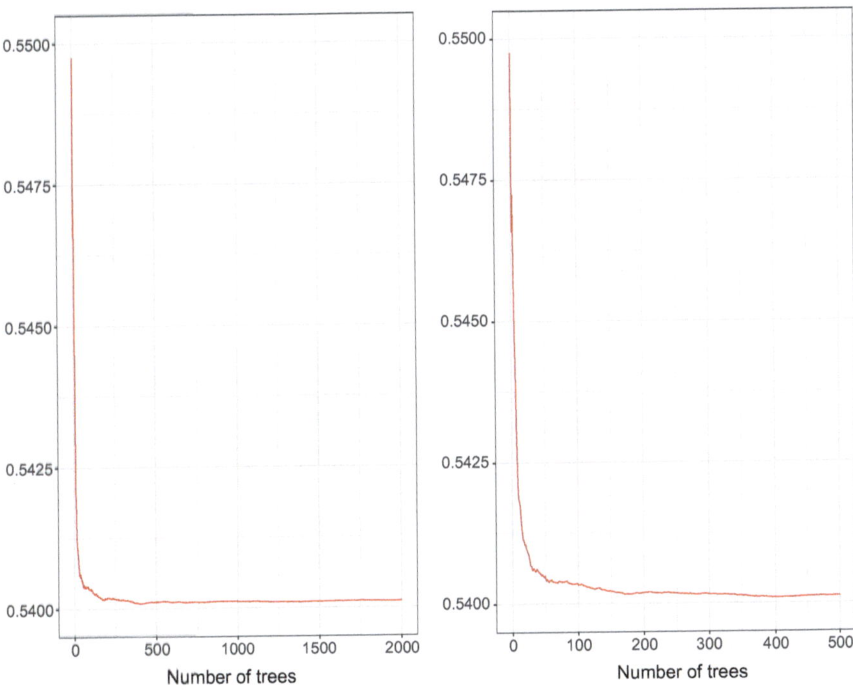

Fig. 4.3 Out-of-bag estimate of the generalization error for random forests with `mtry` = 3 and `nodesize` = 1000 with respect to the number of trees B

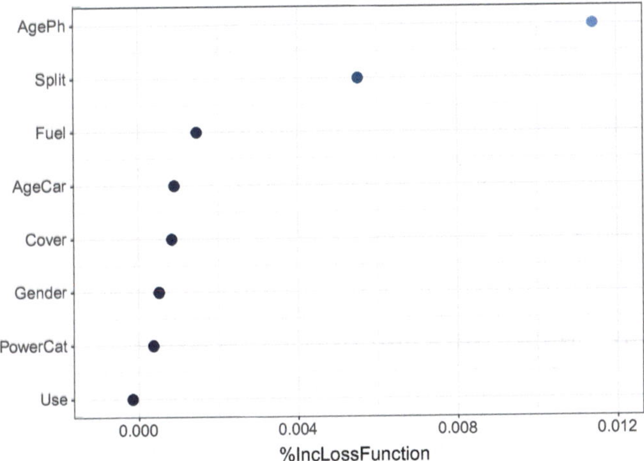

Fig. 4.4 Relative importances of the features for $\widehat{\mu}^{\text{rf}*}_{\mathcal{D},\Theta}$ obtained by permutation

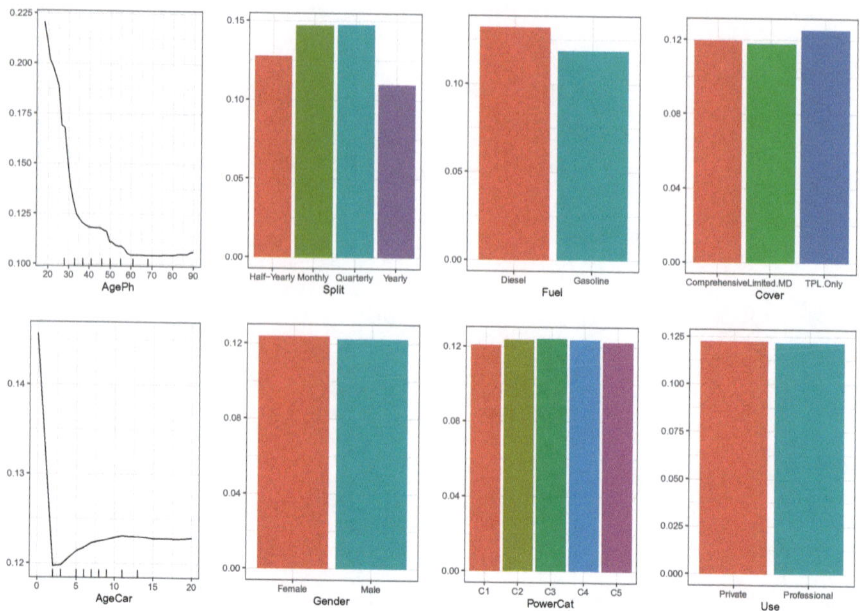

Fig. 4.5 Partial dependence plots for $\widehat{\mu}^{\text{rf}^*}_{\mathcal{D},\boldsymbol{\Theta}}$

$$\widehat{Err}^{\text{val}}\left(\widehat{\mu}_{T_{\alpha_{k^*}}}\right) = 0.5452772$$

one sees that $\widehat{\mu}^{\text{rf}^*}_{\mathcal{D},\Theta}$ improves by $1.1802\,10^{-3}$ the predictive accuracy of the single tree $\widehat{\mu}_{T_{\alpha_{k^*}}}$.

Remark 4.7.1 It is worth noticing that the selection procedure of the optimal tuning parameters may depend on the initial choice of the training set $\mathcal{D}$ and on the folds used to compute cross-validation estimates of the generalization error. This latter point can be mitigated by increasing the number of folds.

4.8 Bibliographic Notes and Further Reading

Bagging is one of the first ensemble methods proposed by Breiman (1996), who showed that aggregating multiple versions of an estimator into an ensemble improves the model accuracy. Several authors added randomness into the training procedure. Dietterich and Kong (1995) introduced the idea of random split selection. They proposed to select at each node the best split among the twenty best ones. Amit et al. (1997) rather proposed to choose the best split over a random subset of the features, and Amit and Geman (1997) also defined a large number of geometric features. Ho (1998) investigated the idea of building a decision forest whose trees are produced

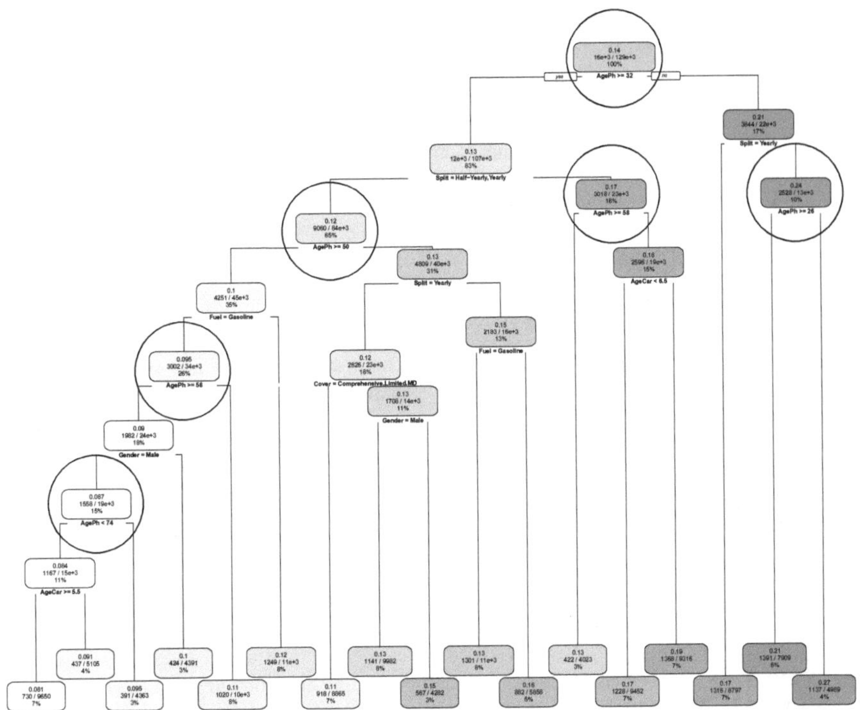

Fig. 4.6 Optimal tree $\widehat{\mu}_{T_{\alpha_{k^*}}}$ built in Chap. 3 on $\mathcal{D}$. The circled nodes use AgePh for splitting

on random subsets of the features, each tree being constructed on a random subset of the features drawn once (prior the construction of the tree). Breiman (2000) studied the addition of noise to the response in order to perturb tree structure. From these works emerged the random forests algorithm discussed in Breiman (2001).

Several authors applied random forests to insurance pricing, such as Wüthrich and Buser (2019) who adapted tree-based methods to model claim frequencies or Henckaert et al. (2020) who worked with random forests and boosted trees to develop full tariff plans built from both the frequency and severity of claims.

Note that the bias-variance decomposition of the generalization error discussed in this chapter is due to Geman et al. (1992). Also, Sects. 4.3.3.3 and 4.6.2 are largely inspired by Denuit and Trufin (2020) and by Sect. 8.2 of Friedman (2001), respectively.

Finally, the presentation is mainly inspired by Breiman (2001), Hastie et al. (2009), Wüthrich and Buser (2019) and Louppe (2014).

References

Amit Y, Geman D (1997) Shape quantization and recognition with randomized trees. Neural Comput 9(7):1545–1588

Amit Y, Geman D, Wilder K (1997) Joint induction of shape features and tree classifiers. IEEE Trans Pattern Anal Mach Intell 19(11):1300–1305

Breiman L (1996) Bagging predictors. Mach Learn 24:123–140

Breiman L (2000) Randomizing outputs to increase prediction accuracy. Mach Learn 40:229–242. ISSN 0885-6125

Breiman L (2001) Random forests. Mach Learn 45:5–32

Denuit M, Trufin J (2020) Generalization error for Tweedie models: decomposition and bagging models. Working paper

Dietterich TG, Kong EB (1995) Machine learning bias, statistical bias, and statistical variance of decision tree algorithms. Technical report, Department of Computer Science, Oregon State University

Friedman J (2001) Greedy function approximation: a gradient boosting machine. Ann Stat 29(5):1189–1232

Geman S, Bienenstock E, Doursat R (1992) Neural networks and the bias/variance dilemma. Neural comput 4(1):1–58

Hastie T, Tibshirani R, Friedman J (2009) The Elements of Statistical Learning. Data Mining, Inference, and Prediction, 2nd edn. Springer Series in Statistics

Henckaerts R, Côté M-P, Antonio K, Verbelen R (2020) Boosting insights in insurance tariff plans with tree-based machine learning methods. North Am Actuar J. https://doi.org/10.1080/10920277.2020.1745656

Ho TK (1998) The random subspace method for constructing decision forests. IEEE Trans Pattern Anal Mach Intell 13:340–354

Louppe G (2014) Understanding random forests: from theory to practice. arXiv:14077502

Wüthrich MV, Buser C (2019) Data analytics for non-life insurance pricing. Lecture notes

Chapter 5
Boosting Trees

5.1 Introduction

Bagging trees and random forests base their predictions on an ensemble of trees. In this chapter, we consider another training procedure based on an ensemble of trees, called boosting trees. However, the way the trees are produced and combined differ between random forests (and so bagging trees) and boosting trees. In random forests, the trees are created independently of each other and contribute equally to the ensemble. Moreover, the constituent trees can be quite large, even fully grown. In boosting, however, the trees are typically small, dependent on previous trees and contribute unequally to the ensemble. Both training procedures are thus different, but they produce competitive predictive performance. Note that the trees in random forests can be created simultaneously since they are independent of each other, so that computational time for random forests is in general smaller than for boosting.

5.2 Forward Stagewise Additive Modeling

Ensemble techniques assume structural models of the form

$$g(\mu(\boldsymbol{x})) = \text{score}(\boldsymbol{x}) = \sum_{m=1}^{M} \beta_m T(\boldsymbol{x}; \mathbf{a}_m), \tag{5.2.1}$$

where β_m, $m = 1, 2, \ldots, M$, are the expansion coefficients, and $T(\boldsymbol{x}; \mathbf{a}_m)$, $m = 1, 2, \ldots, M$, are usually simple functions of the features $\boldsymbol{x}$, characterized by parameters $\mathbf{a}_m$. Estimating a score of the form (5.2.1) by minimizing the corresponding training sample estimate of the generalized error

M. Denuit et al., *Effective Statistical Learning Methods for Actuaries II*,
Springer Actuarial, https://doi.org/10.1007/978-3-030-57556-4_5

$$\min_{\{\beta_m, \mathbf{a}_m\}_1^M} \sum_{i \in \mathcal{I}} L\left(y_i, g^{-1}\left(\sum_{m=1}^{M} \beta_m T(\boldsymbol{x}_i; \mathbf{a}_m)\right)\right) \tag{5.2.2}$$

is in general infeasible. It requires computationally intensive numerical optimization techniques.

One way to overcome this problem is to approximate the solution to (5.2.2) by using a greedy forward stagewise approach. Such an approach consists in sequentially fitting a single function and adding it to the expansion of prior fitted terms. Each fitted term is not readjusted as new terms are added into the expansion, contrarily to a stepwise approach where previous terms are each time readjusted when a new one is added. Specifically, we start by computing

$$\left(\widehat{\beta}_1, \widehat{\mathbf{a}}_1\right) = \underset{\{\beta_1, \mathbf{a}_1\}}{\operatorname{argmin}} \sum_{i \in \mathcal{I}} L\left(y_i, g^{-1}\left(\widehat{\text{score}}_0(\boldsymbol{x}_i) + \beta_1 T(\boldsymbol{x}_i; \mathbf{a}_1)\right)\right) \tag{5.2.3}$$

where $\widehat{\text{score}}_0(\boldsymbol{x})$ is an initial guess. Then, at each iteration $m \geq 2$, we solve the subproblem

$$\left(\widehat{\beta}_m, \widehat{\mathbf{a}}_m\right) = \underset{\{\beta_m, \mathbf{a}_m\}}{\operatorname{argmin}} \sum_{i \in \mathcal{I}} L\left(y_i, g^{-1}\left(\widehat{\text{score}}_{m-1}(\boldsymbol{x}_i) + \beta_m T(\boldsymbol{x}_i; \mathbf{a}_m)\right)\right) \tag{5.2.4}$$

with

$$\widehat{\text{score}}_{m-1}(\boldsymbol{x}) = \widehat{\text{score}}_{m-2}(\boldsymbol{x}) + \widehat{\beta}_{m-1} T(\boldsymbol{x}_i; \widehat{\mathbf{a}}_{m-1}).$$

This leads to the following algorithm.

Algorithm 5.1: Forward Stagewise Additive Modeling.

---

1. Initialize $\widehat{\text{score}}_0(\boldsymbol{x})$ to be a constant. For instance:

$$\widehat{\text{score}}_0(\boldsymbol{x}) = \underset{\beta}{\operatorname{argmin}} \sum_{i \in \mathcal{I}} L(y_i, g^{-1}(\beta)).$$

2. **For** $m = 1$ to M **do**

   2.1 Compute

$$\left(\widehat{\beta}_m, \widehat{\mathbf{a}}_m\right) = \underset{\{\beta_m, \mathbf{a}_m\}}{\operatorname{argmin}} \sum_{i \in \mathcal{I}} L\left(y_i, g^{-1}\left(\widehat{\text{score}}_{m-1}(\boldsymbol{x}_i) + \beta_m T(\boldsymbol{x}_i; \mathbf{a}_m)\right)\right). \tag{5.2.5}$$

   2.2 Update $\widehat{\text{score}}_m(\boldsymbol{x}) = \widehat{\text{score}}_{m-1}(\boldsymbol{x}) + \widehat{\beta}_m T(\boldsymbol{x}_i; \widehat{\mathbf{a}}_m)$.

   End for

3. Output: $\widehat{\mu}_{\mathcal{D}}(\boldsymbol{x}) = g^{-1}\left(\widehat{\text{score}}_M(\boldsymbol{x})\right)$.

Considering the squared-error loss together with the identity link function, step 2.1 in Algorithm 5.1 simplifies to

$$\begin{aligned}\left(\widehat{\beta}_m, \widehat{\mathbf{a}}_m\right) &= \underset{\{\beta_m, \mathbf{a}_m\}}{\operatorname{argmin}} \sum_{i \in \mathcal{I}} \left(y_i - \widehat{\text{score}}_{m-1}(\boldsymbol{x}_i) + \beta_m T(\boldsymbol{x}_i; \mathbf{a}_m)\right)^2 \\ &= \underset{\{\beta_m, \mathbf{a}_m\}}{\operatorname{argmin}} \sum_{i \in \mathcal{I}} (r_{im} + \beta_m T(\boldsymbol{x}_i; \mathbf{a}_m))^2,\end{aligned}$$

where r_{im} is the residual of the model after $m-1$ iterations on the ith observation. One sees that the term $\beta_m T(\boldsymbol{x}; \mathbf{a}_m)$ actually fits the residuals obtained after $m-1$ iterations.

The forward stagewise additive modeling described in Algorithm 5.1 is also called boosting. Boosting is thus an iterative method based on the idea that combining many simple functions should result in a powerful one. In a boosting context, the simple functions $T(\boldsymbol{x}; \mathbf{a}_m)$ are called weak learners or base learners.

There is a large variety of weak learners available for boosting models. For instance, commonly used weak learners are wavelets, multivariate adaptive regression splines, smoothing splines, classification trees and regression trees or neural networks.

Although each weak learner has advantages and disadvantages, trees are the most commonly accepted weak learners in ensemble techniques such as boosting. The nature of trees corresponds well with the concept of weak learner. At each iteration, adding a small tree will slightly improve the current predictive accuracy of the ensemble.

In this second volume, we use trees as weak learners. Boosting using trees as weak learners is then called boosting trees. As already noticed, the procedure underlying boosting trees is completely different from bagging trees and random forests.

5.3 Boosting Trees

Henceforth, we use regression trees as weak learners. That is, we consider weak learners $T(\boldsymbol{x}; \mathbf{a}_m)$ of the form

$$T(\boldsymbol{x}; \mathbf{a}_m) = \sum_{t \in \mathcal{T}_m} c_{tm} \mathrm{I}\left[\boldsymbol{x} \in \chi_t^{(m)}\right], \tag{5.3.1}$$

where $\left\{\chi_t^{(m)}\right\}_{t \in \mathcal{T}_m}$ is the partition of the feature space χ induced by the regression tree $T(\boldsymbol{x}; \mathbf{a}_m)$ and $\{c_{tm}\}_{t \in \mathcal{T}_m}$ the corresponding predictions for the score. For regression trees, the "parameters" $\mathbf{a}_m$ represent the splitting variables and their split values as well as the corresponding predictions in the terminal nodes, that is,

$$\mathbf{a}_m = \left\{c_{tm}, \chi_t^{(m)}\right\}_{t \in \mathcal{T}_m}.$$

5.3.1 Algorithm

Step 2.2 in Algorithm 5.1 becomes

$$\begin{aligned}\widehat{\text{score}}_m(\boldsymbol{x}) &= \widehat{\text{score}}_{m-1}(\boldsymbol{x}) + \widehat{\beta}_m T(\boldsymbol{x}_i; \widehat{\mathbf{a}}_m) \\ &= \widehat{\text{score}}_{m-1}(\boldsymbol{x}) + \widehat{\beta}_m \sum_{t \in \mathcal{T}_m} \widehat{c}_{tm} \mathrm{I}\left[\boldsymbol{x} \in \chi_t^{(m)}\right],\end{aligned}$$

so that it can be alternatively expressed as

$$\widehat{\text{score}}_m(\boldsymbol{x}) = \widehat{\text{score}}_{m-1}(\boldsymbol{x}) + \sum_{t \in \mathcal{T}_m} \widehat{\gamma}_{tm} \mathrm{I}\left[\boldsymbol{x} \in \chi_t^{(m)}\right] \tag{5.3.2}$$

with $\widehat{\gamma}_{tm} = \widehat{\beta}_m \widehat{c}_{tm}$. Hence, one sees that if $\left(\widehat{\beta}_m, \widehat{\mathbf{a}}_m\right)$ is solution to (5.2.5) with

$$\widehat{\mathbf{a}}_m = \left\{\widehat{c}_{tm}, \chi_t^{(m)}\right\}_{t \in \mathcal{T}_m},$$

then $\left(1, \widehat{\mathbf{b}}_m\right)$ with

$$\widehat{\mathbf{b}}_m = \left\{\widehat{\gamma}_{tm}, \chi_t^{(m)}\right\}_{t \in \mathcal{T}_m}$$

is also solution to (5.2.5). In the following, without loss of generality, we consider solutions to (5.2.5) of the form $(1, \widehat{\mathbf{a}}_m)$, which yields the following algorithm.

Algorithm 5.2: Boosting Trees.

---

1. Initialize $\widehat{\text{score}}_0(\boldsymbol{x})$ to be a constant. For instance:

$$\widehat{\text{score}}_0(\boldsymbol{x}) = \underset{\beta}{\text{argmin}} \sum_{i \in \mathcal{I}} L(y_i, g^{-1}(\beta)).$$

2. **For** $m = 1$ to M **do**

   2.1 Fit a regression tree $T(\boldsymbol{x}; \widehat{\mathbf{a}}_m)$ with

$$\widehat{\mathbf{a}}_m = \underset{\mathbf{a}_m}{\text{argmin}} \sum_{i \in \mathcal{I}} L\left(y_i, g^{-1}\left(\widehat{\text{score}}_{m-1}(\boldsymbol{x}_i) + T(\boldsymbol{x}_i; \mathbf{a}_m)\right)\right). \tag{5.3.3}$$

   2.2 Update $\widehat{\text{score}}_m(\boldsymbol{x}) = \widehat{\text{score}}_{m-1}(\boldsymbol{x}) + T(\boldsymbol{x}; \widehat{\mathbf{a}}_m)$.

   End for

3. Output: $\widehat{\mu}_{\mathcal{D}}^{\text{boost}}(\boldsymbol{x}) = g^{-1}\left(\widehat{\text{score}}_M(\boldsymbol{x})\right)$.

Notice that since

$$T(\boldsymbol{x}; \widehat{\mathbf{a}}_m) = \sum_{t \in \mathcal{T}_m} \widehat{c}_{tm} \mathrm{I}\left[\boldsymbol{x} \in \chi_t^{(m)}\right], \tag{5.3.4}$$

step 2.2 in Algorithm 5.2 can be viewed as adding $|\mathcal{T}_m|$ separate base learners $\widehat{c}_{tm}\mathrm{I}\left[\boldsymbol{x} \in \chi_t^{(m)}\right]$, $t \in \mathcal{T}_m$.

Remark 5.3.1 In Algorithm 5.2, we initialize $\widehat{\text{score}_0}(\boldsymbol{x})$ as a constant. However, if there is an existing model for the score, obtained by any training procedure, then it can also be used as a starting score $\widehat{\text{score}_0}(\boldsymbol{x})$. For instance, in actuarial pricing, practitioners traditionally use GLMs, so that they can rely on the score fitted with a GLM procedure as a starting score in Algorithm 5.2. Notice that boosting can serve as an efficient tool for model back-testing, the boosting steps being the corrections needing to be done according to the boosting procedure for the initial model.

Because the structure of regression trees is particularly powerful to account for interactions between features, a combination of a GLM score for $\widehat{\text{score}_0}(\boldsymbol{x})$ accounting for only the main effects of the features with subsequent boosting trees can be of particular interest. Indeed, the boosting trees will bring corrections to the main effects already modeled by the GLM score, if needed, but will also model the interactions between features.

5.3.2 *Particular Cases*

5.3.2.1 Squared Error Loss

For the squared-error loss with the identity link function (which is the canonical link function for the Normal distribution), we have seen that (5.3.3) simplifies to

$$\begin{aligned}
\widehat{\mathbf{a}}_m &= \underset{\mathbf{a}_m}{\operatorname{argmin}} \sum_{i \in \mathcal{I}} \left(y_i - \widehat{\text{score}}_{m-1}(\boldsymbol{x}_i) + T(\boldsymbol{x}_i; \mathbf{a}_m)\right)^2 \\
&= \underset{\mathbf{a}_m}{\operatorname{argmin}} \sum_{i \in \mathcal{I}} \left(\tilde{r}_{mi} + T(\boldsymbol{x}_i; \mathbf{a}_m)\right)^2 \\
&= \underset{\mathbf{a}_m}{\operatorname{argmin}} \sum_{i \in \mathcal{I}} L\left(\tilde{r}_{mi}, T(\boldsymbol{x}_i; \mathbf{a}_m)\right).
\end{aligned}$$

Hence, at iteration m, $T(\boldsymbol{x}_i; \mathbf{a}_m)$ is simply the best regression tree fitting the current residuals $\tilde{r}_{mi} = y_i - \widehat{\text{score}}_{m-1}(\boldsymbol{x}_i)$. Finding the solution to (5.2.5) is thus no harder than for a single tree. It amounts to fit a regression tree on the working training set

$$\mathcal{D}^{(m)} = \{(\tilde{r}_{mi}, \boldsymbol{x}_i), i \in \mathcal{I}\}.$$

5.3.2.2 Poisson Deviance Loss

Consider (5.3.3) with the Poisson deviance loss and the log-link function (which is the canonical link function for the Poisson distribution). In actuarial studies, this choice is often made to model the number of claims, so that we also account for the observation period e referred to as the exposure-to-risk. In such a case, one observation of the training set can be described by the claims count y_i, the features $\boldsymbol{x}_i$ and the exposure-to-risk e_i, so that we have

$$\mathcal{D} = \{(y_i, \boldsymbol{x}_i, e_i), i \in \mathcal{I}\}.$$

In this context, (5.3.3) becomes

$$\begin{aligned}\widehat{\mathbf{a}}_m &= \underset{\mathbf{a}_m}{\operatorname{argmin}} \sum_{i \in \mathcal{I}} L\left(y_i, e_i \exp\left(\widehat{\text{score}}_{m-1}(\boldsymbol{x}_i) + T(\boldsymbol{x}_i; \mathbf{a}_m)\right)\right) \\ &= \underset{\mathbf{a}_m}{\operatorname{argmin}} \sum_{i \in \mathcal{I}} L\left(y_i, e_i \exp\left(\widehat{\text{score}}_{m-1}(\boldsymbol{x}_i)\right) \exp\left(T(\boldsymbol{x}_i; \mathbf{a}_m)\right)\right),\end{aligned}$$

which can be expressed as

$$\widehat{\mathbf{a}}_m = \underset{\mathbf{a}_m}{\operatorname{argmin}} \sum_{i \in \mathcal{I}} L\left(y_i, e_{mi} \exp\left(T(\boldsymbol{x}_i; \mathbf{a}_m)\right)\right)$$

with

$$e_{mi} = e_i \exp\left(\widehat{\text{score}}_{m-1}(\boldsymbol{x}_i)\right)$$

for $i \in \mathcal{I}$. At iteration m, e_{mi} is a constant and can be regarded as a working exposure-to-risk applied to observation i.

Again, solving (5.3.3) is thus no harder than for a single tree. This is equivalent to building a single tree with the Poisson deviance loss and the log-link function on the working training set

$$\mathcal{D}^{(m)} = \{(y_i, \boldsymbol{x}_i, e_{mi}), i \in \mathcal{I}\}.$$

Example

Consider the example of Sect. 3.2.4.1. The optimal tree is shown in Fig. 3.8, denoted here $\widehat{\mu}_{\mathcal{D}}^{\text{tree}}$. Next to the training set $\mathcal{D}$ made of 500 000 observations, we create a validation set $\overline{\mathcal{D}}$ containing a sufficiently large number of observations to get stable results for validation sample estimates, here 1000 000 observations. The validation sample estimate of the generalization error of $\widehat{\mu}_{\mathcal{D}}^{\text{tree}}$ is then given by

$$\widehat{Err}^{\text{val}}\left(\widehat{\mu}_{\mathcal{D}}^{\text{tree}}\right) = 0.5525081. \quad (5.3.5)$$

As we have seen, the optimal tree $\widehat{\mu}_{\mathcal{D}}^{\text{tree}}$ produces the desired partition of the feature space. The difference with the true model μ comes from the predictions in the terminal nodes. We get

$$\widehat{Err}^{\text{val}}(\mu) = 0.5524827. \tag{5.3.6}$$

Let us now build Poisson boosting trees $\widehat{\mu}_{\mathcal{D}}^{\text{boost}}$ with the log-link function on $\mathcal{D}$, so that we have $\widehat{\mu}_{\mathcal{D}}^{\text{boost}}(\boldsymbol{x}) = \exp\left(\widehat{\text{score}}_M(\boldsymbol{x})\right)$. First, we initialize the algorithm with the optimal constant score

$$\widehat{\text{score}_0}(\boldsymbol{x}) = \ln\left(\frac{\sum_{i\in\mathcal{I}} y_i}{\sum_{i\in\mathcal{I}} e_i}\right) = -2.030834.$$

Note that in this simulated example, we have $e_i = 1$ for all $i \in \mathcal{I}$.

Then, we use the R command `rpart` to produce each of the M constituent trees. We control the size of the trees with the variable `maxdepth=D`. Note that specifying the number of terminal nodes J of a tree is not possible with `rpart`.

To start, we consider trees with D $= 1$, that is, we consider trees with only $J = 2$ terminal nodes. For the first iteration, the working training set is

$$\mathcal{D}^{(1)} = \{(y_i, \boldsymbol{x}_i, e_{1i}), i \in \mathcal{I}\},$$

where

$$\begin{aligned} e_{1i} &= e_i \exp(\widehat{\text{score}_0}(\boldsymbol{x}_i)) \\ &= \exp(-2.030834). \end{aligned}$$

The first tree fitted on $\mathcal{D}^{(1)}$ is

$$T(\boldsymbol{x}; \widehat{\mathbf{a}}_1) = -0.1439303\,\mathrm{I}[x_2 \geq 45] + 0.0993644\,\mathrm{I}[x_2 < 45].$$

The single split is $x_2 \geq 45$ and the predictions in the terminal nodes are -0.1439303 for $x_2 \geq 45$ and 0.0993644 for $x_2 < 45$. We then get

$$\begin{aligned} \widehat{\text{score}_1}(\boldsymbol{x}) &= \widehat{\text{score}_0}(\boldsymbol{x}) + T(\boldsymbol{x}; \widehat{\mathbf{a}}_1) \\ &= -2.030834 - 0.1439303\,\mathrm{I}[x_2 \geq 45] + 0.0993644\,\mathrm{I}[x_2 < 45]. \end{aligned}$$

The working training set at iteration $m = 2$ is

$$\mathcal{D}^{(2)} = \{(y_i, \boldsymbol{x}_i, e_{2i}), i \in \mathcal{I}\},$$

with

$$\begin{aligned} e_{2i} &= e_i \exp\left(\widehat{\text{score}_1}(\boldsymbol{x}_i)\right) \\ &= \exp\left(-2.030834 - 0.1439303\,\mathrm{I}[x_{i2} \geq 45] + 0.0993644\,\mathrm{I}[x_{i2} < 45]\right). \end{aligned}$$

The second tree fitted on $\mathcal{D}^{(2)}$ is

$$T(\boldsymbol{x}; \widehat{\mathbf{a}}_2) = -0.06514170\,\mathrm{I}[x_4 = \text{no}] + 0.06126193\,\mathrm{I}\left[x_4 = \text{yes}\right].$$

The single split is made with x_4 and the predictions in the terminal nodes are -0.06514170 for $x_4 = \text{no}$ and 0.06126193 for $x_4 = \text{yes}$. We get

$$\begin{aligned}\widehat{\text{score}_2}(\boldsymbol{x}) &= \widehat{\text{score}_1}(\boldsymbol{x}) + T(\boldsymbol{x}; \widehat{\mathbf{a}}_2) \\ &= -2.030834 \\ &\quad -0.1439303\,\mathrm{I}[x_2 \geq 45] + 0.0993644\,\mathrm{I}[x_2 < 45] \\ &\quad -0.06514170\,\mathrm{I}[x_4 = \text{no}] + 0.06126193\,\mathrm{I}\left[x_4 = \text{yes}\right].\end{aligned}$$

The working training set at iteration $m = 3$ is

$$\mathcal{D}^{(3)} = \{(y_i, \boldsymbol{x}_i, e_{3i}), i \in \mathcal{I}\},$$

with

$$\begin{aligned}e_{3i} &= e_i \exp\left(\widehat{\text{score}_2}(\boldsymbol{x}_i)\right) \\ &= \exp\left(-2.030834 - 0.1439303\,\mathrm{I}[x_{i2} \geq 45] + 0.0993644\,\mathrm{I}[x_{i2} < 45]\right) \\ &\quad \exp\left(-0.06514170\,\mathrm{I}[x_{i4} = \text{no}] + 0.06126193\,\mathrm{I}\left[x_{i4} = \text{yes}\right]\right).\end{aligned}$$

The third tree fitted on $\mathcal{D}^{(3)}$ is

$$T(\boldsymbol{x}; \widehat{\mathbf{a}}_3) = -0.03341700\,\mathrm{I}[x_2 \geq 30] + 0.08237676\,\mathrm{I}[x_2 < 30].$$

The single split is $x_2 \geq 30$ and the predictions in the terminal nodes are -0.03341700 for $x_2 \geq 30$ and 0.08237676 for $x_2 < 30$. We get

$$\begin{aligned}\widehat{\text{score}_3}(\boldsymbol{x}) &= \widehat{\text{score}_2}(\boldsymbol{x}) + T(\boldsymbol{x}; \widehat{\mathbf{a}}_3) \\ &= -2.030834 \\ &\quad -0.1439303\,\mathrm{I}[x_2 \geq 45] + 0.0993644\,\mathrm{I}[x_2 < 45] \\ &\quad -0.06514170\,\mathrm{I}[x_4 = \text{no}] + 0.06126193\,\mathrm{I}\left[x_4 = \text{yes}\right] \\ &\quad -0.03341700\,\mathrm{I}[x_2 \geq 30] + 0.08237676\,\mathrm{I}[x_2 < 30] \\ &= -2.030834 \\ &\quad +(0.0993644 + 0.08237676)\,\mathrm{I}[x_2 < 30] \\ &\quad +(0.0993644 - 0.03341700)\,\mathrm{I}[30 \leq x_2 < 45] \\ &\quad -(0.1439303 + 0.03341700)\,\mathrm{I}[x_2 \geq 45] \\ &\quad -0.06514170\,\mathrm{I}[x_4 = \text{no}] + 0.06126193\,\mathrm{I}\left[x_4 = \text{yes}\right] \\ &= -2.030834 \\ &\quad +0.1817412\,\mathrm{I}[x_2 < 30] + 0.0659474\,\mathrm{I}[30 \leq x_2 < 45] - 0.1773473\,\mathrm{I}[x_2 \geq 45] \\ &\quad -0.06514170\,\mathrm{I}[x_4 = \text{no}] + 0.06126193\,\mathrm{I}\left[x_4 = \text{yes}\right].\end{aligned}$$

The working training set at iteration $m = 4$ is

$$\mathcal{D}^{(4)} = \{(y_i, \boldsymbol{x}_i, e_{4i}), i \in \mathcal{I}\},$$

with

$$\begin{aligned}
e_{4i} &= e_i \exp\left(\widehat{\text{score}_3}(\boldsymbol{x}_i)\right) \\
&= \exp(-2.030834) \\
&\quad \exp(0.1817412\,\mathrm{I}[x_{i2} < 30] + 0.0659474\,\mathrm{I}[30 \le x_{i2} < 45] - 0.1773473\,\mathrm{I}[x_{i2} \ge 45]) \\
&\quad \exp\left(-0.06514170\,\mathrm{I}[x_{i4} = \text{no}] + 0.06126193\,\mathrm{I}\left[x_{i4} = \text{yes}\right]\right).
\end{aligned}$$

The fourth tree fitted on $\mathcal{D}^{(4)}$ is

$$T(\boldsymbol{x}; \widehat{\mathbf{a}}_4) = -0.05309399\,\mathrm{I}[x_1 = female] + 0.05029175\,\mathrm{I}[x_1 = male].$$

The single split is made with x_1 and the predictions in the terminal nodes are -0.05309399 for $x_1 = female$ and 0.05029175 for $x_1 = male$. We get

$$\begin{aligned}
\widehat{\text{score}_4}(\boldsymbol{x}) &= \widehat{\text{score}_3}(\boldsymbol{x}) + T(\boldsymbol{x}; \widehat{\mathbf{a}}_4) \\
&= -2.030834 \\
&\quad +0.1817412\,\mathrm{I}[x_2 < 30] + 0.0659474\,\mathrm{I}[30 \le x_2 < 45] - 0.1773473\,\mathrm{I}[x_2 \ge 45] \\
&\quad -0.06514170\,\mathrm{I}[x_4 = \text{no}] + 0.06126193\,\mathrm{I}\left[x_4 = \text{yes}\right] \\
&\quad -0.05309399\,\mathrm{I}[x_1 = female] + 0.05029175\,\mathrm{I}[x_1 = male].
\end{aligned}$$

The working training set at iteration $m = 5$ is

$$\mathcal{D}^{(5)} = \{(y_i, \boldsymbol{x}_i, e_{5i}), i \in \mathcal{I}\},$$

with

$$\begin{aligned}
e_{5i} &= e_i \exp\left(\widehat{\text{score}_4}(\boldsymbol{x}_i)\right) \\
&= \exp(-2.030834) \\
&\quad \exp(0.1817412\,\mathrm{I}[x_{i2} < 30] + 0.0659474\,\mathrm{I}[30 \le x_{i2} < 45] - 0.1773473\,\mathrm{I}[x_{i2} \ge 45]) \\
&\quad \exp\left(-0.06514170\,\mathrm{I}[x_{i4} = \text{no}] + 0.06126193\,\mathrm{I}\left[x_{i4} = \text{yes}\right]\right) \\
&\quad \exp(-0.05309399\,\mathrm{I}[x_{i1} = female] + 0.05029175\,\mathrm{I}[x_{i1} = male]).
\end{aligned}$$

The fifth tree fitted on $\mathcal{D}^{(5)}$ is

$$T(\boldsymbol{x}; \widehat{\mathbf{a}}_5) = -0.01979230\,\mathrm{I}[x_2 < 45] + 0.03326232\,\mathrm{I}[x_2 \ge 45].$$

The single split is $x_2 \ge 45$ and the predictions in the terminal nodes are -0.01979230 for $x_2 < 45$ and 0.03326232 for $x_2 \ge 45$. We get

$$
\begin{aligned}
\widehat{\text{score}_5}(\boldsymbol{x}) &= \widehat{\text{score}_4}(\boldsymbol{x}) + T(\boldsymbol{x}; \widehat{\mathbf{a}}_5) \\
&= -2.030834 \\
&\quad -0.05309399\,\mathrm{I}[x_1 = female] + 0.05029175\,\mathrm{I}[x_1 = male] \\
&\quad +0.1817412\,\mathrm{I}[x_2 < 30] + 0.0659474\,\mathrm{I}[30 \le x_2 < 45] - 0.1773473\,\mathrm{I}[x_2 \ge 45] \\
&\quad -0.06514170\,\mathrm{I}[x_4 = \text{no}] + 0.06126193\,\mathrm{I}\left[x_4 = \text{yes}\right] \\
&\quad -0.01979230\,\mathrm{I}[x_2 < 45] + 0.03326232\,\mathrm{I}[x_2 \ge 45] \\
&= -2.030834 \\
&\quad -0.05309399\,\mathrm{I}[x_1 = female] + 0.05029175\,\mathrm{I}[x_1 = male] \\
&\quad +(0.1817412 - 0.01979230)\,\mathrm{I}[x_2 < 30] + (0.0659474 - 0.01979230)\,\mathrm{I}[30 \le x_2 < 45] \\
&\quad +(0.03326232 - 0.1773473)\,\mathrm{I}[x_2 \ge 45] \\
&\quad -0.06514170\,\mathrm{I}[x_4 = \text{no}] + 0.06126193\,\mathrm{I}\left[x_4 = \text{yes}\right] \\
&= -2.030834 \\
&\quad -0.05309399\,\mathrm{I}[x_1 = female] + 0.05029175\,\mathrm{I}[x_1 = male] \\
&\quad +0.1619489\,\mathrm{I}[x_2 < 30] + 0.0461551\,\mathrm{I}[30 \le x_2 < 45] - 0.144085\,\mathrm{I}[x_2 \ge 45] \\
&\quad -0.06514170\,\mathrm{I}[x_4 = \text{no}] + 0.06126193\,\mathrm{I}\left[x_4 = \text{yes}\right]
\end{aligned}
$$

The working training set at iteration $m = 6$ is

$$
\mathcal{D}^{(6)} = \{(y_i, \boldsymbol{x}_i, e_{6i}), i \in \mathcal{I}\},
$$

with

$$
\begin{aligned}
e_{6i} &= e_i \exp\left(\widehat{\text{score}_5}(\boldsymbol{x}_i)\right) \\
&= \exp(-2.030834) \\
&\quad \exp(-0.05309399\,\mathrm{I}[x_{i1} = female] + 0.05029175\,\mathrm{I}[x_{i1} = male]) \\
&\quad \exp(0.1619489\,\mathrm{I}[x_{i2} < 30] + 0.0461551\,\mathrm{I}[30 \le x_{i2} < 45] - 0.144085\,\mathrm{I}[x_{i2} \ge 45]) \\
&\quad \exp\left(-0.06514170\,\mathrm{I}[x_{i4} = \text{no}] + 0.06126193\,\mathrm{I}\left[x_{i4} = \text{yes}\right]\right).
\end{aligned}
$$

The sixth tree fitted on $\mathcal{D}^{(6)}$ is

$$
T(\boldsymbol{x}; \widehat{\mathbf{a}}_6) = -0.009135574\,\mathrm{I}[x_2 \ge 30] + 0.019476688\,\mathrm{I}[x_2 < 30].
$$

The single split is $x_2 \ge 30$ and the predictions in the terminal nodes are -0.00913557 for $x_2 \ge 30$ and 0.019476688 for $x_2 < 30$. We get

$$
\begin{aligned}
\widehat{\text{score}_6}(\boldsymbol{x}) &= \widehat{\text{score}_5}(\boldsymbol{x}) + T(\boldsymbol{x}; \widehat{\mathbf{a}}_6) \\
&= -2.030834 \\
&\quad -0.05309399\,\mathrm{I}[x_1 = female] + 0.05029175\,\mathrm{I}[x_1 = male] \\
&\quad +0.1619489\,\mathrm{I}[x_2 < 30] + 0.0461551\,\mathrm{I}[30 \le x_2 < 45] - 0.144085\,\mathrm{I}[x_2 \ge 45] \\
&\quad -0.06514170\,\mathrm{I}[x_4 = \text{no}] + 0.06126193\,\mathrm{I}\left[x_4 = \text{yes}\right] \\
&\quad -0.009135574\,\mathrm{I}[x_2 \ge 30] + 0.019476688\,\mathrm{I}[x_2 < 30] \\
&= -2.030834
\end{aligned}
$$

$$-0.05309399\,\mathrm{I}\,[x_1 = female] + 0.05029175\,\mathrm{I}\,[x_1 = male]$$
$$+0.1814256\,\mathrm{I}\,[x_2 < 30] + 0.03701953\,\mathrm{I}\,[30 \leq x_2 < 45] - 0.1532206\,\mathrm{I}\,[x_2 \geq 45]$$
$$-0.06514170\,\mathrm{I}\,[x_4 = \text{no}] + 0.06126193\,\mathrm{I}\left[x_4 = \text{yes}\right].$$

Hence, for $M = 6$, i.e. after 6 iterations, we have

$$\begin{aligned}\widehat{\mu}_{\mathcal{D}}^{\text{boost}}(\boldsymbol{x}) = {} & \exp\left(-2.030834\right) \\ & \exp\left(-0.05309399\,\mathrm{I}\,[x_1 = female] + 0.05029175\,\mathrm{I}\,[x_1 = male]\right) \\ & \exp\left(0.1814256\,\mathrm{I}\,[x_2 < 30] + 0.03701953\,\mathrm{I}\,[30 \leq x_2 < 45] - 0.1532206\,\mathrm{I}\,[x_2 \geq 45]\right) \\ & \exp\left(-0.06514170\,\mathrm{I}\,[x_4 = \text{no}] + 0.06126193\,\mathrm{I}\left[x_4 = \text{yes}\right]\right).\end{aligned}$$

We see that $\widehat{\mu}_{\mathcal{D}}^{\text{boost}}$ with $M = 6$ partitions the feature space into 12 parts, as desired. In Table 5.1, we show the 12 risk classes with their corresponding expected claim frequencies $\mu(\boldsymbol{x})$ and estimated expected claim frequencies $\widehat{\mu}_{\mathcal{D}}^{\text{tree}}(\boldsymbol{x})$ and $\widehat{\mu}_{\mathcal{D}}^{\text{boost}}(\boldsymbol{x})$.

If we continue to increase the number of iterations, we overfit the training set. In Fig. 5.1, we provide the validation sample estimate $\widehat{Err}^{\text{val}}\left(\widehat{\mu}_{\mathcal{D}}^{\text{boost}}\right)$ with respect to the number of trees M, together with $\widehat{Err}^{\text{val}}(\mu)$ and $\widehat{Err}^{\text{val}}\left(\widehat{\mu}_{\mathcal{D}}^{\text{tree}}\right)$. We see that the boosting model minimizing $\widehat{Err}^{\text{val}}\left(\widehat{\mu}_{\mathcal{D}}^{\text{boost}}\right)$ contains $M = 6$ trees, and its predictive accuracy is similar to $\widehat{\mu}_{\mathcal{D}}^{\text{tree}}$. Note that when $M \leq 5$, the optimal tree $\widehat{\mu}_{\mathcal{D}}^{\text{tree}}$ performs better than $\widehat{\mu}_{\mathcal{D}}^{\text{boost}}$ while from $M = 7$, we start to overfit the training set. For instance, the seventh tree is

$$T(\boldsymbol{x}; \widehat{\mathbf{a}}_7) = -0.01147564\,\mathrm{I}\,[x_2 < 39] + 0.01109997\,\mathrm{I}\,[x_2 \geq 39].$$

Table 5.1 Risk classes with their corresponding expected claim frequencies $\mu(\boldsymbol{x})$ and estimated expected claim frequencies $\widehat{\mu}_{\mathcal{D}}^{\text{tree}}(\boldsymbol{x})$ and $\widehat{\mu}_{\mathcal{D}}^{\text{boost}}(\boldsymbol{x})$ for $M = 6$

$\boldsymbol{x}$					
x_1 (Gender)	x_2 (Age)	x_4 (Sport)	$\mu(\boldsymbol{x})$	$\widehat{\mu}_{\mathcal{D}}^{\text{tree}}(\boldsymbol{x})$	$\widehat{\mu}_{\mathcal{D}}^{\text{boost}}(\boldsymbol{x})$ (for M=6)
Female	$x_2 \geq 45$	No	0.1000	0.1005	0.1000
Male	$x_2 \geq 45$	No	0.1100	0.1085	0.1109
Female	$x_2 \geq 45$	Yes	0.1150	0.1164	0.1135
Male	$x_2 \geq 45$	Yes	0.1265	0.1285	0.1259
Female	$30 \leq x_2 < 45$	No	0.1200	0.1206	0.1210
Male	$30 \leq x_2 < 45$	No	0.1320	0.1330	0.1342
Female	$30 \leq x_2 < 45$	Yes	0.1380	0.1365	0.1373
Male	$30 \leq x_2 < 45$	Yes	0.1518	0.1520	0.1522
Female	$x_2 < 30$	No	0.1400	0.1422	0.1399
Male	$x_2 < 30$	No	0.1540	0.1566	0.1550
Female	$x_2 < 30$	Yes	0.1610	0.1603	0.1586
Male	$x_2 < 30$	Yes	0.1771	0.1772	0.1759

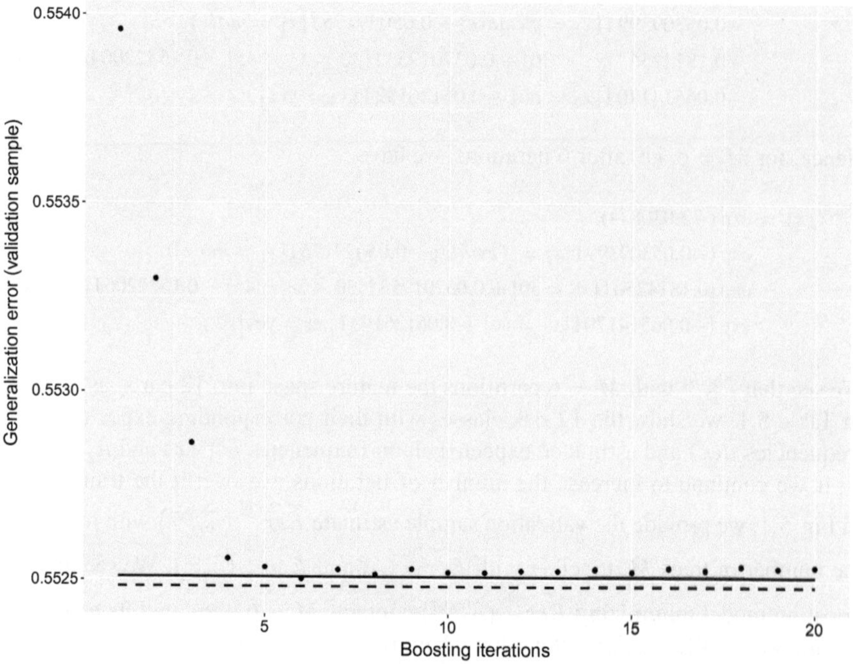

Fig. 5.1 Validation sample estimate $\widehat{Err}^{\text{val}}(\widehat{\mu}_{\mathcal{D}}^{\text{boost}})$ with respect to the number of iterations for trees with two terminal nodes, together with $\widehat{Err}^{\text{val}}(\mu)$ (dotted line) and $\widehat{Err}^{\text{val}}(\widehat{\mu}_{\mathcal{D}}^{\text{tree}})$ (solid line)

5.3.2.3 Gamma Deviance Loss

Consider the Gamma deviance loss. Using the log-link function, (5.3.3) is then given by

$$\begin{aligned}
\widehat{\mathbf{a}}_m &= \underset{\mathbf{a}_m}{\operatorname{argmin}} \sum_{i \in \mathcal{I}} L\left(y_i, \exp\left(\widehat{\text{score}}_{m-1}(\boldsymbol{x}_i) + T(\boldsymbol{x}_i; \mathbf{a}_m)\right)\right) \\
&= \underset{\mathbf{a}_m}{\operatorname{argmin}} \sum_{i \in \mathcal{I}} \Bigg\{ -2\ln\left(\frac{y_i}{\exp\left(\widehat{\text{score}}_{m-1}(\boldsymbol{x}_i) + T(\boldsymbol{x}_i; \mathbf{a}_m)\right)}\right) \\
&\qquad + 2\left(\frac{y_i}{\exp\left(\widehat{\text{score}}_{m-1}(\boldsymbol{x}_i) + T(\boldsymbol{x}_i; \mathbf{a}_m)\right)} - 1\right)\Bigg\},
\end{aligned}$$

so that we get

$$\begin{aligned}
\widehat{\mathbf{a}}_m &= \underset{\mathbf{a}_m}{\operatorname{argmin}} \sum_{i \in \mathcal{I}} \left\{ -2\ln\left(\frac{\widetilde{r}_{mi}}{\exp\left(T(\boldsymbol{x}_i; \mathbf{a}_m)\right)}\right) + 2\left(\frac{\widetilde{r}_{mi}}{\exp\left(T(\boldsymbol{x}_i; \mathbf{a}_m)\right)} - 1\right)\right\} \\
&= \underset{\mathbf{a}_m}{\operatorname{argmin}} \sum_{i \in \mathcal{I}} L\left(\widetilde{r}_{mi}, \exp\left(T(\boldsymbol{x}_i; \mathbf{a}_m)\right)\right) \qquad (5.3.7)
\end{aligned}$$

with

$$\tilde{r}_{mi} = \frac{y_i}{\exp\left(\widehat{\text{score}}_{m-1}(\boldsymbol{x}_i)\right)}$$

for $i \in \mathcal{I}$. Therefore, (5.3.3) simplifies to (5.3.7), so that finding the solution to (5.3.3) amounts to obtain the regression tree with the Gamma deviance loss and the log-link function that best predicts the working responses $\tilde{r}_{mi}$. Hence, solving (5.3.3) amounts to build the best tree on the working training set

$$\mathcal{D}^{(m)} = \{(\tilde{r}_{mi}, \boldsymbol{x}_i), i \in \mathcal{I}\},$$

using the Gamma deviance loss and the log-link function.

In practice, the choice of the log-link function is often made for the convenience of having a multiplicative model. If we rather choose the canonical link function for the Gamma distribution, that is, if we choose $g(x) = \frac{-1}{x}$ and hence $g^{-1}(x) = \frac{-1}{x}$, then (5.3.3) becomes

$$\begin{aligned}\widehat{\mathbf{a}}_m &= \underset{\mathbf{a}_m}{\text{argmin}} \sum_{i \in \mathcal{I}} L\left(y_i, \frac{-1}{\widehat{\text{score}}_{m-1}(\boldsymbol{x}_i) + T(\boldsymbol{x}_i; \mathbf{a}_m)}\right) \\ &= \underset{\mathbf{a}_m}{\text{argmin}} \sum_{i \in \mathcal{I}} \Big\{ -2\ln\left(-y_i\left(\widehat{\text{score}}_{m-1}(\boldsymbol{x}_i) + T(\boldsymbol{x}_i; \mathbf{a}_m)\right)\right) \\ &\qquad -2\left(y_i\left(\widehat{\text{score}}_{m-1}(\boldsymbol{x}_i) + T(\boldsymbol{x}_i; \mathbf{a}_m)\right) + 1\right)\Big\}. \qquad (5.3.8)\end{aligned}$$

5.3.3 *Size of the Trees*

Boosting trees have two important tuning parameters, that are the size of the trees and the number of trees M. The size of the trees can be specified in different ways, such as with the number of terminal nodes J or with the depth of the tree D.

In the boosting context, the size of the trees is controlled by the interaction depth ID. Each subsequent split can be seen as a higher-level of interaction with the previous split features. Setting ID $= 1$ produces single-split regression trees, so that no interactions are allowed. Only the main effects of the features can be captured by the score. With ID $= 2$, two-way interactions are also permitted, and for ID $= 3$, three-way interactions are also allowed, and so on. Thus, the value of ID reflects the level of interactions permitted in the score. Note that ID corresponds to the number of splits in the trees. Obviously, we have ID $= J - 1$.

In practice, the level of interactions required is often unknown, so that ID is a tuning parameter that is set by considering different values and selecting the one that minimizes the generalization error estimated on a validation set or by cross-validation. In practice, ID $= 1$ will be often insufficient, while ID > 10 is very unlikely.

Note that in the simulated example discussed in Sect. 5.3.2.2, trees with $\texttt{ID} = 1$ are big enough to get satisfying results since the true score does not contain interaction effects.

5.3.3.1 Example 1

Consider the simulated example of Sect. 3.6. In this example, using the log-link function, the true score is given by

$$\begin{aligned}\text{score}(\boldsymbol{x}) &= \ln(0.1) + \ln(1.1)\,\mathrm{I}[x_1 = male] \\ &\quad + \ln(1.4)\mathrm{I}[18 \leq x_2 < 30] + \ln(1.2)\,\mathrm{I}[30 \leq x_2 < 45] \\ &\quad + \ln(1.15)\,\mathrm{I}[x_4 = yes] + \ln(1.6/1.4)\,\mathrm{I}[x_1 = male, 18 \leq x_2 < 30].\end{aligned}$$

The first terms are functions of only one single feature while the last term is a two-variable functions, producing a second-order interaction between X_1 and X_2.

We generate 1000 000 additional observations to produce a validation set. The validation sample estimate of the generalization error of $\widehat{\mu}_{\mathcal{D}}^{\text{tree}}$, which is the optimal tree displayed in Fig. 3.25, is then given by

$$\widehat{Err}^{\text{val}}\left(\widehat{\mu}_{\mathcal{D}}^{\text{tree}}\right) = 0.5595743. \tag{5.3.9}$$

The optimal tree $\widehat{\mu}_{\mathcal{D}}^{\text{tree}}$ produces the correct partition of the feature space and differs from the true model μ by the predictions in the terminal nodes. We have

$$\widehat{Err}^{\text{val}}(\mu) = 0.5595497. \tag{5.3.10}$$

Let us fit boosting trees using the Poisson deviance loss and the log-link function. We follow the procedure described in the example of Sect. 5.3.2.2. First, we consider trees with only one split ($\texttt{ID} = 1$). Fig. 5.2 provides the validation sample estimate $\widehat{Err}^{\text{val}}\left(\widehat{\mu}_{\mathcal{D}}^{\text{boost}}\right)$ with respect to the number of trees M, together with $\widehat{Err}^{\text{val}}(\mu)$ and $\widehat{Err}^{\text{val}}\left(\widehat{\mu}_{\mathcal{D}}^{\text{tree}}\right)$. We observe that the boosting models $\widehat{\mu}_{\mathcal{D}}^{\text{boost}}$ have higher validation sample estimates of the generalization error than the optimal tree $\widehat{\mu}_{\mathcal{D}}^{\text{tree}}$, whatever the number of iterations M. Contrarily to the optimal tree, the boosting models with $\texttt{ID} = 1$ cannot account for the second-order interaction. We should consider trees with $\texttt{ID} = 2$. However, as already mentioned, specifying the number of terminal nodes J of a tree, or equivalently the interaction depth `ID`, is not possible with `rpart`.

To overcome this issue, we could rely on the `R` `gbm` package for instance, that enables to specify the interaction depth. This package implements the gradient boosting approach as described in Sect. 5.4, which is an approximation of boosting for any differentiable loss function. In Sect. 5.4.4.2, we illustrate the use of the `gbm` package on this particular example with $\texttt{ID} = 1, 2, 3, 4$.

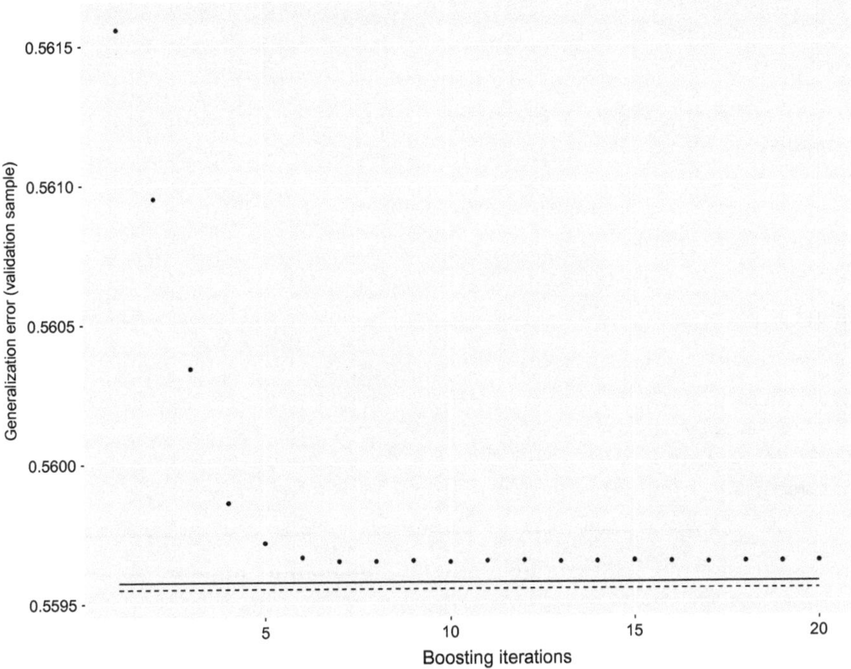

Fig. 5.2 Validation sample estimate $\widehat{Err}^{\text{val}}\left(\widehat{\mu}_{\mathcal{D}}^{\text{boost}}\right)$ with respect to the number of iterations for trees with two terminal nodes ($J = 2$), together with $\widehat{Err}^{\text{val}}(\mu)$ (dotted line) and $\widehat{Err}^{\text{val}}\left(\widehat{\mu}_{\mathcal{D}}^{\text{tree}}\right)$ (solid line)

5.3.3.2 Example 2

Consider the simulated example of Sect. 3.7.2. We generate 1000 000 additional observations to produce a validation set. The validation sample estimate of the generalization error of $\widehat{\mu}_{\mathcal{D}}^{\text{tree}}$, which is the tree minimizing the 10-fold cross validation error and shown in Fig. 3.31, is then given by

$$\widehat{Err}^{\text{val}}\left(\widehat{\mu}_{\mathcal{D}}^{\text{tree}}\right) = 0.5803903. \tag{5.3.11}$$

Contrarily to the previous example and the example in Sect. 5.3.2.2, a single tree (here $\widehat{\mu}_{\mathcal{D}}^{\text{tree}}$) cannot reproduce the desired partition of the feature space. We have seen in Sect. 3.7.2 that $\widehat{\mu}_{\mathcal{D}}^{\text{tree}}$ suffers from a lack of smoothness. The validation sample estimate of the generalization error of the true model μ is

$$\widehat{Err}^{\text{val}}(\mu) = 0.5802196, \tag{5.3.12}$$

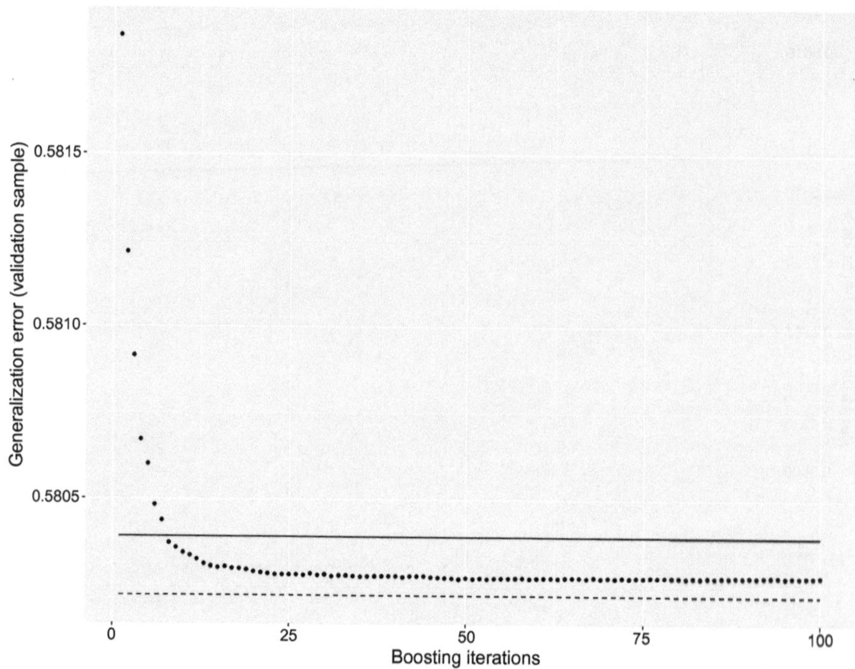

Fig. 5.3 Validation sample estimate $\widehat{Err}^{\text{val}}\left(\widehat{\mu}_{\mathcal{D}}^{\text{boost}}\right)$ with respect to the number of iterations for trees with two terminal nodes ($J = 2$), together with $\widehat{Err}^{\text{val}}(\mu)$ (dotted line) and $\widehat{Err}^{\text{val}}\left(\widehat{\mu}_{\mathcal{D}}^{\text{tree}}\right)$ (solid line)

so that the room for improvement for the validation sample estimate of the generalization error is

$$\widehat{Err}^{\text{val}}\left(\widehat{\mu}_{\mathcal{D}}^{\text{tree}}\right) - \widehat{Err}^{\text{val}}(\mu) = 0.0001706516.$$

We follow the same procedure than in example of Sect. 5.3.2.2, namely we use the Poisson deviance loss with the log-link function and we consider trees with only one split ($\mathtt{ID} = 1$) since there is no interaction effects in the true model. In Fig. 5.3, we provide the validation sample estimate $\widehat{Err}^{\text{val}}\left(\widehat{\mu}_{\mathcal{D}}^{\text{boost}}\right)$ with respect to the number of trees M, together with $\widehat{Err}^{\text{val}}(\mu)$ and $\widehat{Err}^{\text{val}}\left(\widehat{\mu}_{\mathcal{D}}^{\text{tree}}\right)$. We see that the error estimate $\widehat{Err}^{\text{val}}\left(\widehat{\mu}_{\mathcal{D}}^{\text{boost}}\right)$ becomes smaller than $\widehat{Err}^{\text{val}}\left(\widehat{\mu}_{\mathcal{D}}^{\text{tree}}\right)$ from $M = 8$ and stabilizes around $M = 25$. The smallest error $\widehat{Err}^{\text{val}}\left(\widehat{\mu}_{\mathcal{D}}^{\text{boost}}\right)$ corresponds to $M = 49$ and is given by

$$\widehat{Err}^{\text{val}}\left(\widehat{\mu}_{\mathcal{D}}^{\text{boost}}\right) = 0.5802692.$$

In Fig. 5.4, we show the validation sample estimate $\widehat{Err}^{\text{val}}\left(\widehat{\mu}_{\mathcal{D}}^{\text{boost}}\right)$ with respect to the number of trees M for constituent trees with depth $\mathtt{D} = 2, 3$. For $\mathtt{D} = 2$, the

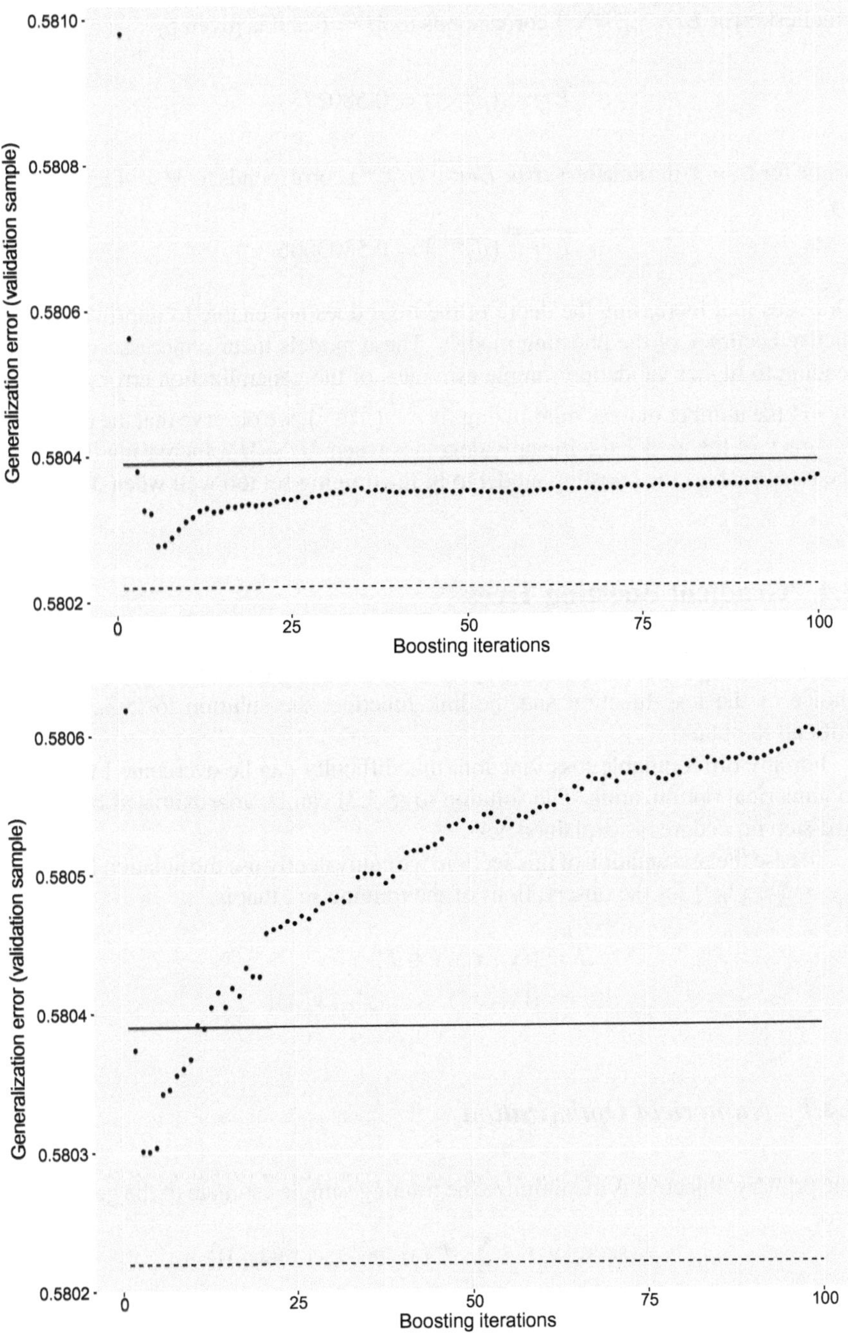

Fig. 5.4 Validation sample estimate $\widehat{Err}^{\text{val}}\left(\widehat{\mu}_{\mathcal{D}}^{\text{boost}}\right)$ with respect to the number of iterations for trees with $\mathtt{D} = 2$ (top) and $\mathtt{D} = 3$ (bottom), together with $\widehat{Err}^{\text{val}}(\mu)$ (dotted line) and $\widehat{Err}^{\text{val}}\left(\widehat{\mu}_{\mathcal{D}}^{\text{tree}}\right)$ (solid line)

smallest error $\widehat{Err}^{\text{val}}\left(\widehat{\mu}_{\mathcal{D}}^{\text{boost}}\right)$ corresponds to $M = 6$ and is given by

$$\widehat{Err}^{\text{val}}\left(\widehat{\mu}_{\mathcal{D}}^{\text{boost}}\right) = 0.5802779,$$

while for $\mathtt{D} = 3$, the smallest error $\widehat{Err}^{\text{val}}\left(\widehat{\mu}_{\mathcal{D}}^{\text{boost}}\right)$ corresponds to $M = 4$ and is given by

$$\widehat{Err}^{\text{val}}\left(\widehat{\mu}_{\mathcal{D}}^{\text{boost}}\right) = 0.5803006.$$

One sees that increasing the depth of the trees does not enable to improve the predictive accuracy of the boosting models. These models incur unnecessary variance leading to higher validation sample estimates of the generalization error. Denoting by M^* the number of trees minimizing $\widehat{Err}^{\text{val}}\left(\widehat{\mu}_{\mathcal{D}}^{\text{boost}}\right)$, we observe that the predictive accuracy of the model significantly degrades when $M > M^*$. Increasing the size of the trees enables the boosting model to fit the training set too well when $M > M^*$.

5.4 Gradient Boosting Trees

Simple fast algorithms do not always exist for solving (5.3.3). Depending on the choice of the loss function and the link function, the solution to (5.3.3) can be difficult to obtain.

For any differentiable loss function, this difficulty can be overcome by analogy to numerical optimization. The solution to (5.3.3) can be approximated by using a two-step procedure, as explained next.

To ease the presentation of this section, we equivalently use the notation $\{(y_1^*, \boldsymbol{x}_1^*), \ldots, (y_{|\mathcal{I}|}^*, \boldsymbol{x}_{|\mathcal{I}|}^*)\}$ for the observations of the training set, that is,

$$\begin{aligned} \mathcal{D} &= \{(y_i, \boldsymbol{x}_i), i \in \mathcal{I}\} \\ &= \{(y_1^*, \boldsymbol{x}_1^*), \ldots, (y_{|\mathcal{I}|}^*, \boldsymbol{x}_{|\mathcal{I}|}^*)\}. \end{aligned} \tag{5.4.1}$$

5.4.1 Numerical Optimization

The primary objective is to minimize the training sample estimate of the generalized error

$$L(\text{score}(\boldsymbol{x})) = \sum_{i \in \mathcal{I}} L\left(y_i, g^{-1}\left(\text{score}(\boldsymbol{x}_i)\right)\right) \tag{5.4.2}$$

with respect to $\text{score}(\boldsymbol{x})$, where $\text{score}(\boldsymbol{x})$ is assumed to be a sum of trees. In other words, we are interested into a function score : $\succ \to \mathbb{R}$ minimizing (5.4.2) under the constraint that it is a sum of trees.

We observe that (5.4.2) only specifies this function in the values $\boldsymbol{x}_i$, $i \in \mathcal{I}$. Hence, forgetting that we work with constrained functions, we actually try to find optimal parameters

$$\begin{aligned}\boldsymbol{\eta} &= (\eta_1, \dots, \eta_{|\mathcal{I}|})' \\ &= (\text{score}(\boldsymbol{x}_1^*), \dots, \text{score}(\boldsymbol{x}_{|\mathcal{I}|}^*))' \end{aligned} \tag{5.4.3}$$

minimizing

$$L(\boldsymbol{\eta}) = \sum_{i=1}^{|\mathcal{I}|} L\left(y_i^*, g^{-1}(\eta_i)\right). \tag{5.4.4}$$

The saturated model is of course one of the solution to (5.4.4), namely $\eta_i = g(y_i^*)$ for all $i = 1, \dots, |\mathcal{I}|$, but this solution typically leads to overfitting. Note that restricting the score to be the sum of a limited number of relatively small regression trees will prevent from overfitting.

Our problem can be viewed as the numerical optimization

$$\widehat{\boldsymbol{\eta}} = \underset{\boldsymbol{\eta}}{\text{argmin}}\, L(\boldsymbol{\eta}). \tag{5.4.5}$$

Numerical optimization methods often express the solution to (5.4.5) as

$$\widehat{\boldsymbol{\eta}} = \sum_{m=0}^{M} \mathbf{b}_m, \quad \mathbf{b}_m \in \mathbb{R}^{|\mathcal{I}|}, \tag{5.4.6}$$

where $\mathbf{b}_0$ is an initial guess and $\mathbf{b}_1, \dots, \mathbf{b}_M$ are successive increments, also called steps or boosts, each based on the preceding steps. The way each step $\mathbf{b}_m$ is computed depends on the optimization method considered.

5.4.2 Steepest Descent

One of the simplest and frequently used numerical minimization methods is the steepest descent. The steepest descent defines step $\mathbf{b}_m$ as

$$\mathbf{b}_m = -\rho_m \mathbf{g}_m,$$

where ρ_m is a scalar and

$$\mathbf{g}_m = \left(\left[\frac{\partial L(y_1^*, g^{-1}(\eta_1))}{\partial \eta_1} \right]_{\eta_1 = \widehat{\eta}_{m-1,1}}, \dots, \left[\frac{\partial L(y_{|\mathcal{I}|}^*, g^{-1}(\eta_{|\mathcal{I}|}))}{\partial \eta_{|\mathcal{I}|}} \right]_{\eta_{|\mathcal{I}|} = \widehat{\eta}_{m-1,|\mathcal{I}|}} \right)' \tag{5.4.7}$$

is the gradient of $L(\boldsymbol{\eta})$ evaluated at $\widehat{\boldsymbol{\eta}}_{m-1} = (\widehat{\eta}_{m-1,1}, \ldots, \widehat{\eta}_{m-1,|\mathcal{I}|})'$ given by

$$\widehat{\boldsymbol{\eta}}_{m-1} = \mathbf{b}_0 + \mathbf{b}_1 + \ldots + \mathbf{b}_{m-1}. \tag{5.4.8}$$

The negative gradient $-\mathbf{g}_m$ gives the local direction along with $L(\boldsymbol{\eta})$ decreases the most rapidly at $\boldsymbol{\eta} = \widehat{\boldsymbol{\eta}}_{m-1}$. The step length ρ_m is then found by

$$\rho_m = \underset{\rho>0}{\operatorname{argmin}}\, L(\widehat{\boldsymbol{\eta}}_{m-1} - \rho \mathbf{g}_m), \tag{5.4.9}$$

which provides the update

$$\widehat{\boldsymbol{\eta}}_m = \widehat{\boldsymbol{\eta}}_{m-1} - \rho_m \mathbf{g}_m. \tag{5.4.10}$$

5.4.3 Algorithm

The steepest descent is the best strategy to minimize $L(\boldsymbol{\eta})$. Unfortunately, the gradient $\mathbf{g}_m$ is only defined at the feature values $\boldsymbol{x}_i$, $i \in \mathcal{I}$, so that it cannot be generalized to other feature values whereas we would like a function defined on the entire feature space χ. Furthermore, we want to prevent from overfitting.

A way to solve these issues is to constrain the step directions to be members of a class of functions. Specifically, we reinstate the temporarily forgotten constrain that is to work with regression trees. Hence, we approximate the direction $-\mathbf{g}_m$ by a regression tree $T(\boldsymbol{x}; \widehat{\mathbf{a}}_m)$ producing

$$-\widehat{\mathbf{g}}_m = \left(T(\boldsymbol{x}_1^*; \widehat{\mathbf{a}}_m), \ldots, T(\boldsymbol{x}_{|\mathcal{I}|}^*; \widehat{\mathbf{a}}_m)\right)'$$

that is as close as possible to the negative gradient, i.e. the most parallel to $-\mathbf{g}_m$.

A common choice to measure the closeness between constrained candidates $T(\boldsymbol{x}; \mathbf{a}_m)$ for the negative gradient and the unconstrained gradient $-\mathbf{g}_m = -(g_{m1}, \ldots, g_{m|\mathcal{I}|})$ is to use the squared error, so that we get

$$\widehat{\mathbf{a}}_m = \underset{\mathbf{a}_m}{\operatorname{argmin}} \sum_{i=1}^{|\mathcal{I}|} \left(-g_{mi} - T(\boldsymbol{x}_i^*; \mathbf{a}_m)\right)^2. \tag{5.4.11}$$

The tree $T(\boldsymbol{x}; \widehat{\mathbf{a}}_m)$ is thus fitted to the negative gradient $-\mathbf{g}_m$ by least squares.

The optimal step size $\widehat{\rho}_m$ is then found by

$$\widehat{\rho}_m = \underset{\rho_m>0}{\operatorname{argmin}} \sum_{i=1}^{|\mathcal{I}|} L(y_i^*, g^{-1}\left(\widehat{\text{score}}_{m-1}(\boldsymbol{x}_i^*) + \rho_m T(\boldsymbol{x}_i^*; \widehat{\mathbf{a}}_m)\right)), \tag{5.4.12}$$

where

$$\widehat{\text{score}}_{m-1}(\boldsymbol{x}) = \widehat{\text{score}}_{m-2}(\boldsymbol{x}) + \widehat{\rho}_{m-1} T(\boldsymbol{x}; \widehat{\mathbf{a}}_{m-1})$$

and $\widehat{\text{score}}_0(\boldsymbol{x})$ is the constant function

$$\widehat{\text{score}}_0(\boldsymbol{x}) = \underset{\rho_0}{\text{argmin}} \sum_{i=1}^{|\mathcal{I}|} L(y_i^*, g^{-1}(\rho_0)).$$

As already noticed, since $T(\boldsymbol{x}; \widehat{\mathbf{a}}_m)$ can be written as

$$T(\boldsymbol{x}; \widehat{\mathbf{a}}_m) = \sum_{t \in \mathcal{T}_m} \widehat{c}_{tm} \mathrm{I}\left[\boldsymbol{x} \in \chi_t^{(m)}\right],$$

the update

$$\widehat{\text{score}}_m(\boldsymbol{x}) = \widehat{\text{score}}_{m-1}(\boldsymbol{x}) + \widehat{\rho}_m T(\boldsymbol{x}; \widehat{\mathbf{a}}_m)$$

can be viewed as adding $|\mathcal{T}_m|$ separate base learners $\widehat{c}_{tm}\mathrm{I}\left[\boldsymbol{x} \in \chi_t^{(m)}\right]$, $t \in \mathcal{T}_m$, to $\widehat{\text{score}}_{m-1}(\boldsymbol{x})$. Instead of using one coefficient for $T(\boldsymbol{x}; \widehat{\mathbf{a}}_m)$, the current procedure can be improved by using the optimal coefficients for each base learner $\widehat{c}_{tm}\mathrm{I}\left[\boldsymbol{x} \in \chi_t^{(m)}\right]$, $t \in \mathcal{T}_m$. Thus, we replace the optimal coefficient $\widehat{\rho}_m$ solution to (5.4.12) by $|\mathcal{T}_m|$ coefficients $\widehat{\rho}_{tm}$, $t \in \mathcal{T}_m$, that are solution to

$$\{\widehat{\rho}_{tm}\}_{t\in\mathcal{T}_m} = \underset{\{\rho_{tm}\}_{t\in\mathcal{T}_m}}{\text{argmin}} \sum_{i=1}^{|\mathcal{I}|} L\left(y_i^*, g^{-1}\left(\widehat{\text{score}}_{m-1}(\boldsymbol{x}_i^*) + \sum_{t\in\mathcal{T}_m} \rho_{tm}\widehat{c}_{tm}\mathrm{I}\left[\boldsymbol{x}_i^* \in \chi_t^{(m)}\right]\right)\right). \tag{5.4.13}$$

Subspaces $\{\chi_t^{(m)}\}_{t\in\mathcal{T}_m}$ of the feature space χ are disjoint, so that (5.4.13) amounts to solve

$$\widehat{\rho}_{tm} = \underset{\rho_{tm}}{\text{argmin}} \sum_{i:\boldsymbol{x}_i^* \in \chi_t^{(m)}} L\left(y_i^*, g^{-1}\left(\widehat{\text{score}}_{m-1}(\boldsymbol{x}_i^*) + \rho_{tm}\widehat{c}_{tm}\right)\right) \tag{5.4.14}$$

for each $t \in \mathcal{T}_m$. Coefficients $\widehat{\rho}_{tm}$, $t \in \mathcal{T}_m$, are thus fitted separately. Here, the least-squares coefficients $\widehat{c}_{tm}$, $t \in \mathcal{T}_m$, are simply multiplicative constants. Ignoring these latter coefficients, (5.4.14) reduces to

$$\widehat{\gamma}_{tm} = \underset{\gamma_{tm}}{\text{argmin}} \sum_{i:\boldsymbol{x}_i^* \in \chi_t^{(m)}} L\left(y_i^*, g^{-1}\left(\widehat{\text{score}}_{m-1}(\boldsymbol{x}_i^*) + \gamma_{tm}\right)\right), \tag{5.4.15}$$

meaning that $\widehat{\gamma}_{tm}$ is the best update for the score in subspace $\chi_t^{(m)}$ given the current approximation $\widehat{\text{score}}_{m-1}(\boldsymbol{x})$.

This leads to the following algorithm.

Algorithm 5.3: Gradient Boosting Trees.

---

1. Initialize $\widehat{\text{score}_0}(\boldsymbol{x})$ to be a constant. For instance:

$$\widehat{\text{score}_0}(\boldsymbol{x}) = \underset{\rho_0}{\text{argmin}} \sum_{i=1}^{|\mathcal{I}|} L(y_i^*, g^{-1}(\rho_0)). \tag{5.4.16}$$

2. **For** $m = 1$ to M **do**

2.1 For $i = 1, \ldots, |\mathcal{I}|$, compute

$$r_{mi} = -\left[\frac{\partial L(y_i^*, g^{-1}(\eta))}{\partial \eta}\right]_{\eta = \widehat{\text{score}}_{m-1}(\boldsymbol{x}_i^*)}. \tag{5.4.17}$$

2.2 Fit a regression tree $T(\boldsymbol{x}; \widehat{\mathbf{a}}_m)$ to the working responses r_{mi} by least squares, i.e.

$$\widehat{\mathbf{a}}_m = \underset{\mathbf{a}_m}{\text{argmin}} \sum_{i=1}^{\mathcal{I}} \left(r_{mi} - T(\boldsymbol{x}_i^*; \mathbf{a}_m)\right)^2, \tag{5.4.18}$$

giving the partition $\{\chi_t^{(m)}, t \in \mathcal{T}_m\}$ of the feature space.

2.3 For $t \in \mathcal{T}_m$, compute

$$\widehat{\gamma}_{tm} = \underset{\gamma_{tm}}{\text{argmin}} \sum_{i:\boldsymbol{x}_i^* \in \chi_t^{(m)}} L\left(y_i^*, g^{-1}\left(\widehat{\text{score}}_{m-1}(\boldsymbol{x}_i^*) + \gamma_{tm}\right)\right). \tag{5.4.19}$$

2.4 Update $\widehat{\text{score}}_m(\boldsymbol{x}) = \widehat{\text{score}}_{m-1}(\boldsymbol{x}) + \sum_{t \in \mathcal{T}_m} \widehat{\gamma}_{tm} \mathrm{I}\left[\boldsymbol{x} \in \chi_t^{(m)}\right]$.

End for

3. Output: $\widehat{\mu}_{\mathcal{D}}^{\text{grad boost}}(\boldsymbol{x}) = g^{-1}\left(\widehat{\text{score}}_M(\boldsymbol{x})\right)$.

Step 2.2 in Algorithm 5.3 determines the partition $\{\chi_t^{(m)}, t \in \mathcal{T}_m\}$ of the feature space for the mth iteration. Then, given the partition $\{\chi_t^{(m)}, t \in \mathcal{T}_m\}$, step 2.3 estimates the predictions in the terminal nodes $t \in \mathcal{T}_m$ at the mth iteration based on the current score $\widehat{\text{score}}_{m-1}(\boldsymbol{x})$.

Instead of fitting directly one tree as in step 2.1 in Algorithm 5.2, we first fit a tree on working responses by least squares (step 2.2 in Algorithm 5.3) to get the structure of the tree and then we compute the predictions in the final nodes (step 2.3 in Algorithm 5.3).

5.4.4 Particular Cases

5.4.4.1 Squared Error Loss

For the squared-error loss with the identity link function, i.e.

$$L\left(y, g^{-1}(\text{score}(\boldsymbol{x}))\right) = (y - \text{score}(\boldsymbol{x}))^2\,,$$

we directly get

$$\widehat{\text{score}}_0(\boldsymbol{x}) = \frac{\sum_{i=1}^{|\mathcal{I}|} y_i^*}{|\mathcal{I}|}$$

for (5.4.16) and the working responses (5.4.17) become

$$\begin{aligned} r_{mi} &= -\left[\frac{\partial(y_i^* - \eta)^2}{\partial\eta}\right]_{\eta=\widehat{\text{score}}_{m-1}(\boldsymbol{x}_i^*)} \\ &= 2\left(y_i^* - \widehat{\text{score}}_{m-1}(\boldsymbol{x}_i^*)\right). \end{aligned} \tag{5.4.20}$$

Therefore, the working responses r_{mi} are just the ordinary residuals, as in Sect. 5.3.2.1, so that gradient boosting trees is here equivalent to boosting trees. The predictions (5.4.19) are given by

$$\widehat{\gamma}_{tm} = \frac{\sum_{i:\boldsymbol{x}_i^* \in \chi_t^{(m)}} \left(y_i^* - \widehat{\text{score}}_{m-1}(\boldsymbol{x}_i^*)\right)}{\text{card}\{i : \boldsymbol{x}_i^* \in \chi_t^{(m)}\}}. \tag{5.4.21}$$

5.4.4.2 Poisson Deviance Loss

Consider the Poisson deviance loss together with the log-link function, and the training set

$$\begin{aligned} \mathcal{D} &= \{(y_i, \boldsymbol{x}_i, e_i), i \in \mathcal{I}\} \\ &= \{(y_1^*, \boldsymbol{x}_1^*, e_1^*), \ldots, (y_{|\mathcal{I}|}^*, \boldsymbol{x}_{|\mathcal{I}|}^*, e_{|\mathcal{I}|}^*)\}. \end{aligned} \tag{5.4.22}$$

Since

$$L\left(y, eg^{-1}(\text{score}(\boldsymbol{x}))\right) = 2y\left[\ln\left(\frac{y}{e\exp(\text{score}(\boldsymbol{x}))}\right) - 1 + \frac{e\exp(\text{score}(\boldsymbol{x}))}{y}\right],$$

(5.4.16) and (5.4.17) become

$$\widehat{\text{score}}_0(\boldsymbol{x}) = \ln\left(\frac{\sum_{i=1}^{|\mathcal{I}|} y_i^*}{\sum_{i=1}^{|\mathcal{I}|} e_i^*}\right)$$

and

$$\begin{aligned} r_{mi} &= -2y_i^* \left[\frac{\partial\left\{\ln\left(\frac{y_i^*}{e_i^* \exp(\eta)}\right) - 1 + \frac{e_i^* \exp(\eta)}{y_i^*}\right\}}{\partial\eta}\right]_{\eta=\widehat{\text{score}}_{m-1}(\boldsymbol{x}_i^*)} \\ &= 2\left(y_i^* - e_i^* \exp(\widehat{\text{score}}_{m-1}(\boldsymbol{x}_i^*))\right), \end{aligned} \tag{5.4.23}$$

respectively. One sees that (5.4.23) can be rewritten as

$$r_{mi} = 2\left(y_i^* - e_{mi}^*\right), \tag{5.4.24}$$

with $e_{mi}^* = e_i^* \exp(\widehat{\text{score}}_{m-1}(\boldsymbol{x}_i^*))$. At iteration m, step 2.2 in Algorithm 5.3 amounts to fit a regression tree $T(\boldsymbol{x}; \widehat{\mathbf{a}}_m)$ on the residuals (5.4.24). The predictions (5.4.19) are given by

$$\widehat{\gamma}_{tm} = \ln\left(\frac{\sum_{i:\boldsymbol{x}_i^* \in \mathcal{X}_t^{(m)}} y_i^*}{\sum_{i:\boldsymbol{x}_i^* \in \mathcal{X}_t^{(m)}} e_{mi}^*}\right). \tag{5.4.25}$$

As we have seen, boosting trees is fully manageable in this case since fitting the mth tree in the sequence is no harder than fitting a single regression tree. Thus, gradient boosting trees is a bit artificial here, as noted in Wüthrich and Buser (2019). At each iteration, the structure of the new tree is obtained by least squares while the Poisson deviance loss can be used without any problem. Hence, gradient boosting trees introduces an extra step which is artificial, leading to an unnecessary approximation.

Example

Consider the example in Sect. 5.3.3.1. We use the `R` package `gbm` to build the gradient boosting models $\widehat{\mu}_{\mathcal{D}}^{\text{grad boost}}$ with the Poisson deviance loss and the log-link function. Note that the command `gbm` enables to specify the interaction depth of the constituent trees.

Figure 5.5 displays the validation sample estimate $\widehat{Err}^{\text{val}}\left(\widehat{\mu}_{\mathcal{D}}^{\text{grad boost}}\right)$ with respect to the number of trees M for $\texttt{ID} = 1, 2, 3, 4$, together with $\widehat{Err}^{\text{val}}(\mu)$ and $\widehat{Err}^{\text{val}}\left(\widehat{\mu}_{\mathcal{D}}^{\text{tree}}\right)$.

The gradient boosting models with $\texttt{ID} = 1$ produce results similar to those depicted in Fig. 5.2. As expected, because of the second-order interaction between X_1 and X_2, $\texttt{ID} = 2$ is the optimal choice in this example. The gradient boosting model $\widehat{\mu}_{\mathcal{D}}^{\text{grad boost}}$ with $\texttt{ID} = 2$ and $M = 6$ produces the lowest validation sample estimate of the generalization error that is similar to the single optimal tree $\widehat{\mu}_{\mathcal{D}}^{\text{tree}}$. Note that gradient boosting models with $\texttt{ID} = 3, 4$ have lower validation sample estimates of the generalization error as long as $M \leq 5$. This is due to the fact that models with $\texttt{ID} = 3, 4$ learn faster than those with $\texttt{ID} = 2$. From $M = 6$, gradient boosting mod-

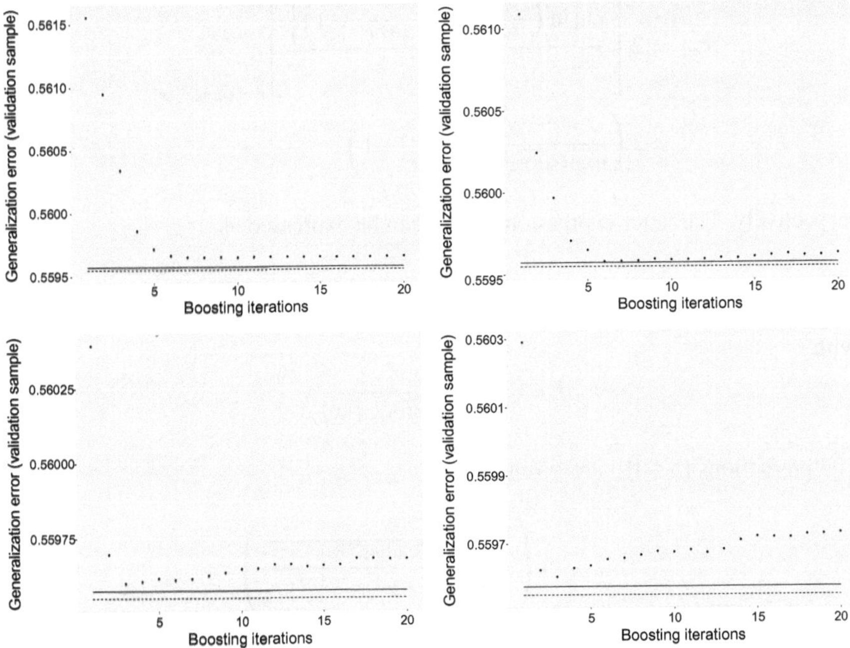

Fig. 5.5 Validation sample estimate $\widehat{Err}^{\text{val}}\left(\widehat{\mu}_{\mathcal{D}}^{\text{grad boost}}\right)$ with respect to the number of iterations for trees with ID $=1$ (top left), ID $=2$ (top right), ID $=3$ (bottom left), ID $=4$ (bottom right), together with $\widehat{Err}^{\text{val}}(\mu)$ (dotted line) and $\widehat{Err}^{\text{val}}\left(\widehat{\mu}_{\mathcal{D}}^{\text{tree}}\right)$ (solid line)

els with ID $=2$ outperform models with ID $=3,4$. The learning capacity of trees with ID $=3,4$ is too strong, so that they reduce the training sample estimate of the generalization error too quickly.

5.4.4.3 Gamma Deviance Loss

For the Gamma deviance loss with the log-link function, i.e.

$$L\left(y, g^{-1}\left(\operatorname{score}(\boldsymbol{x})\right)\right)=-2 \ln \left(\frac{y}{\exp \left(\operatorname{score}(\boldsymbol{x})\right)}\right)+2\left(\frac{y}{\exp \left(\operatorname{score}(\boldsymbol{x})\right)}-1\right), \tag{5.4.26}$$

(5.4.16) and (5.4.17) become

$$\widehat{\operatorname{score}}_0(\boldsymbol{x})=\ln \left(\frac{\sum_{i=1}^{|\mathcal{I}|} y_i^*}{|\mathcal{I}|}\right)$$

and

$$r_{mi} = 2 \left[\frac{\partial \left\{ \ln \left(\frac{y_i^*}{\exp(\eta)} \right) - \left(\frac{y_i^*}{\exp(\eta)} - 1 \right) \right\}}{\partial \eta} \right]_{\eta = \widehat{\text{score}}_{m-1}(\boldsymbol{x}_i^*)}$$
$$= 2 \left(\frac{y_i^*}{\exp(\widehat{\text{score}}_{m-1}(\boldsymbol{x}_i^*))} - 1 \right), \tag{5.4.27}$$

respectively. The latter expression for r_{mi} can be expressed as

$$r_{mi} = 2 \left(\tilde{r}_{mi} - 1 \right) \tag{5.4.28}$$

with

$$\tilde{r}_{mi} = \frac{y_i^*}{\exp(\widehat{\text{score}}_{m-1}(\boldsymbol{x}_i^*))}. \tag{5.4.29}$$

The predictions (5.4.19) are given by

$$\widehat{\gamma}_{tm} = \ln \left(\frac{\sum_{i:\boldsymbol{x}_i^* \in \chi_t^{(m)}} \frac{y_i^*}{\exp(\widehat{\text{score}}_{m-1}(\boldsymbol{x}_i^*))}}{\text{card}\{i : \boldsymbol{x}_i^* \in \chi_t^{(m)}\}} \right). \tag{5.4.30}$$

Again, as for the Poisson deviance loss with the log-link function, we have seen that boosting trees for the Gamma deviance loss is easy to implement with the log-link function, so that gradient boosting trees is also a bit artificial here.

If we consider the canonical link function $g(x) = \frac{-1}{x}$, i.e.

$$L\left(y, g^{-1}\left(\text{score}(\boldsymbol{x})\right)\right) = -2 \ln \left(-y \text{score}(\boldsymbol{x})\right) - 2 \left(y \text{score}(\boldsymbol{x}) + 1\right), \tag{5.4.31}$$

we get for (5.4.16) and (5.4.17)

$$\widehat{\text{score}}_0(\boldsymbol{x}) = \frac{-|\mathcal{I}|}{\sum_{i=1}^{|\mathcal{I}|} y_i^*}$$

and

$$r_{mi} = 2 \left[\frac{\partial \left\{ \ln \left(-y_i^* \eta\right) + y_i^* \eta + 1 \right\}}{\partial \eta} \right]_{\eta = \widehat{\text{score}}_{m-1}(\boldsymbol{x}_i^*)}$$
$$= 2 \left(\frac{1}{\widehat{\text{score}}_{m-1}(\boldsymbol{x}_i^*)} + y_i^* \right), \tag{5.4.32}$$

respectively. The predictions (5.4.19) satisfy

$$\sum_{i:\boldsymbol{x}_i^* \in \chi_t^{(m)}} \frac{-1}{\widehat{\text{score}}_{m-1}(\boldsymbol{x}_i^*) + \widehat{\gamma}_{tm}} = \sum_{i:\boldsymbol{x}_i^* \in \chi_t^{(m)}} y_i^*. \tag{5.4.33}$$

5.5 Boosting Versus Gradient Boosting

In practice, actuaries often use distributions that belong to the Tweedie class together with the log-link function to approximate $\mu(\boldsymbol{x})$. The log-link function is chosen mainly because of the multiplicative structure it produces for the resulting model. The Tweedie class regroups the members of the ED family having power variance functions $V(\mu) = \mu^{\xi}$ for some ξ.

Specifically, the Tweedie class includes continuous distributions such as the Normal, Gamma and Inverse Gaussian distributions. It also includes the Poisson and compound Poisson-Gamma distributions. Compound Poisson-Gamma distributions can be used for modeling annual claim amounts, having positive probability at zero and a continuous distribution on the positive real numbers. In practice, annual claim amounts are often decomposed into claim numbers and claim severities and separate analyses of these quantities are conducted. Typically, the Poisson distribution is used for modeling claim counts and the Gamma or Inverse Gaussian distributions for claim severities.

The following table gives a list of all Tweedie distributions.

	Type	Name
$\xi < 0$	Continuous	-
$\xi = 0$	Continuous	Normal
$0 < \xi < 1$	Non existing	-
$\xi = 1$	Discrete	Poisson
$1 < \xi < 2$	Mixed, non-negative	Compound Poisson-Gamma
$\xi = 2$	Continuous, positive	Gamma
$2 < \xi < 3$	Continuous, positive	-
$\xi = 3$	Continuous, positive	Inverse Gaussian
$\xi > 3$	Continuous, positive	-

Negative values of ξ gives continuous distributions on the whole real axis. For $0 < \xi < 1$, no ED member exists. Only the cases $\xi \geq 1$ are thus interesting for application in insurance. The corresponding deviance loss function is

$$L(y, \widehat{\mu}) = \begin{cases} (y - \widehat{\mu})^2 \text{ for } \xi = 0 \\ 2\left(y \ln \frac{y}{\widehat{\mu}} - (y - \widehat{\mu})\right) \text{ for } \xi = 1 \\ 2\left(-\ln \frac{y}{\widehat{\mu}} + \frac{y}{\widehat{\mu}} - 1\right) \text{ for } \xi = 2 \\ 2\left(\frac{\max(y,0)^{2-\xi}}{(1-\xi)(2-\xi)} - \frac{y\widehat{\mu}^{1-\xi}}{1-\xi} + \frac{\widehat{\mu}^{2-\xi}}{2-\xi}\right) \text{ else.} \end{cases} \tag{5.5.1}$$

Notice that we consider non-negative responses in our applications, so that we could write $\max(y, 0) = y$. However, for the sake of completeness, we let $\max(y, 0)$.

The Poisson and Gamma distributions with the log-link function give rise to simple boosting algorithms, as discussed previously. In the Gamma case, we have seen that

$$L\left(y_i, \exp\left(\widehat{\text{score}}_{m-1}(\boldsymbol{x}_i) + T(\boldsymbol{x}_i; \mathbf{a}_m)\right)\right) = L\left(\tilde{r}_{mi}, \exp\left(T(\boldsymbol{x}_i; \mathbf{a}_m)\right)\right) \tag{5.5.2}$$

with

$$\tilde{r}_{mi} = \frac{y_i}{\exp\left(\widehat{\text{score}}_{m-1}(\boldsymbol{x}_i)\right)},$$

while in the Poisson case, we have noticed that

$$L\left(y_i, e_i \exp\left(\widehat{\text{score}}_{m-1}(\boldsymbol{x}_i) + T(\boldsymbol{x}_i; \mathbf{a}_m)\right)\right) = L\left(y_i, e_{mi} \exp\left(T(\boldsymbol{x}_i; \mathbf{a}_m)\right)\right) \quad (5.5.3)$$

with

$$e_{mi} = e_i \exp\left(\widehat{\text{score}}_{m-1}(\boldsymbol{x}_i)\right).$$

For a Poisson response Y, we know that the actuary is allowed to work either with the observed claim count Y or with the observed claim rate $\widetilde{Y} = \frac{Y}{e}$ provided the weight $\nu = e$ enters the analysis. The distribution of the claim rate $\widetilde{Y}$ still belongs to the Tweedie class and is called the Poisson rate distribution. See Property 2.5.1 in Denuit, Hainaut and Trufin (2019) for more details. This is reflected by the fact that the Poisson deviance loss satisfies

$$L\left(y_i, e_i \exp\left(\widehat{\text{score}}_{m-1}(\boldsymbol{x}_i) + T(\boldsymbol{x}_i; \mathbf{a}_m)\right)\right) = \nu_i L\left(\tilde{y}_i, \exp\left(\widehat{\text{score}}_{m-1}(\boldsymbol{x}_i) + T(\boldsymbol{x}_i; \mathbf{a}_m)\right)\right) \quad (5.5.4)$$

with $\nu_i = e_i$ and $\tilde{y}_i = \frac{y_i}{e_i}$. Working with the claim rates $\tilde{y}_i$ then yields

$$\begin{aligned}
&\nu_i L(\tilde{y}_i, \exp(\widehat{\text{score}}_{m-1}(\boldsymbol{x}_i) + T(\boldsymbol{x}_i; \mathbf{a}_m))) \\
&= \nu_i 2\left[\tilde{y}_i \ln\left(\frac{\tilde{y}_i}{\exp(\widehat{\text{score}}_{m-1}(\boldsymbol{x}_i) + T(\boldsymbol{x}_i; \mathbf{a}_m))}\right) - (\tilde{y}_i - \exp(\widehat{\text{score}}_{m-1}(\boldsymbol{x}_i) + T(\boldsymbol{x}_i; \mathbf{a}_m)))\right] \\
&= \nu_i \exp(\widehat{\text{score}}_{m-1}(\boldsymbol{x}_i)) \\
&\quad 2\left[\frac{\tilde{y}_i}{\exp(\widehat{\text{score}}_{m-1}(\boldsymbol{x}_i))} \ln\left(\frac{\tilde{y}_i}{\exp(\widehat{\text{score}}_{m-1}(\boldsymbol{x}_i)) \exp(T(\boldsymbol{x}_i; \mathbf{a}_m))}\right)\right. \\
&\quad \left. - \left(\frac{\tilde{y}_i}{\exp(\widehat{\text{score}}_{m-1}(\boldsymbol{x}_i))} - \exp(T(\boldsymbol{x}_i; \mathbf{a}_m))\right)\right] \\
&= \nu_i \exp(\widehat{\text{score}}_{m-1}(\boldsymbol{x}_i)) L\left(\frac{\tilde{y}_i}{\exp(\widehat{\text{score}}_{m-1}(\boldsymbol{x}_i))}, \exp(T(\boldsymbol{x}_i; \mathbf{a}_m))\right) \\
&= \nu_{mi} L\left(\tilde{r}_{mi}, \exp(T(\boldsymbol{x}_i; \mathbf{a}_m))\right)
\end{aligned} \quad (5.5.5)$$

with

$$\nu_{mi} = \nu_i \exp(\widehat{\text{score}}_{m-1}(\boldsymbol{x}_i))$$

and

$$\tilde{r}_{mi} = \frac{\tilde{y}_i}{\exp(\widehat{\text{score}}_{m-1}(\boldsymbol{x}_i))}.$$

The mth iteration of the boosting procedure reduces to build a single tree on the working training set

$$\mathcal{D}^{(m)} = \{(\nu_{mi}, \tilde{r}_{mi}, \boldsymbol{x}_i), i \in \mathcal{I}\},$$

using the Poisson deviance loss and the log-link function. The weights are each time updated together with the responses that are assumed to follow Poisson rate distributions.

While the weights are updated differently in the Poisson (rate) and Gamma cases (they remain constant through the boosting procedure in the Gamma case), it is interesting to notice that the working responses at the mth iteration are in both cases the original ones divided by the current predictions $\exp(\widehat{\text{score}}_{m-1}(\boldsymbol{x}_i))$.

The next result shows that the latter observation is also true for any member of the Tweedie class with the log-link function. Actually, any member of the Tweedie class with the log-link function gives rise to a simple boosting algorithm.

Proposition 5.5.1 *Consider the deviance loss function (5.5.1). Then, (5.3.3) with the log-link function, that is*

$$\widehat{\boldsymbol{a}}_m = \underset{\boldsymbol{a}_m}{\operatorname{argmin}} \sum_{i \in \mathcal{I}} \nu_i L\left(y_i, \exp\left(\widehat{score}_{m-1}(\boldsymbol{x}_i) + T(\boldsymbol{x}_i; \boldsymbol{a}_m)\right)\right)$$

can be rewritten as

$$\widehat{\boldsymbol{a}}_m = \underset{\boldsymbol{a}_m}{\operatorname{argmin}} \sum_{i \in \mathcal{I}} \nu_{mi} L\left(\tilde{r}_{mi}, \exp\left(T(\boldsymbol{x}_i; \boldsymbol{a}_m)\right)\right) \tag{5.5.6}$$

with

$$\nu_{mi} = \nu_i \exp(\widehat{score}_{m-1}(\boldsymbol{x}_i))^{2-\xi}$$

and

$$\tilde{r}_{mi} = \frac{y_i}{\exp(\widehat{score}_{m-1}(\boldsymbol{x}_i))}.$$

Proof The cases $\xi = 1$ and $\xi = 2$ have already been discussed. Turning to the Normal case $\xi = 0$, we have

$$\begin{aligned}
&\nu_i L(y_i, \exp(\widehat{\text{score}}_{m-1}(\boldsymbol{x}_i) + T(\boldsymbol{x}_i; \mathbf{a}_m))) \\
&= \nu_i [y_i - \exp(\widehat{\text{score}}_{m-1}(\boldsymbol{x}_i) + T(\boldsymbol{x}_i; \mathbf{a}_m))]^2 \\
&= \nu_i [y_i - \exp(\widehat{\text{score}}_{m-1}(\boldsymbol{x}_i)) \exp(T(\boldsymbol{x}_i; \mathbf{a}_m))]^2 \\
&= \nu_i \exp(2\widehat{\text{score}}_{m-1}(\boldsymbol{x}_i)) \left[\frac{y_i}{\exp(\widehat{\text{score}}_{m-1}(\boldsymbol{x}_i))} - \exp(T(\boldsymbol{x}_i; \mathbf{a}_m))\right]^2 \\
&= \nu_i \exp(2\widehat{\text{score}}_{m-1}(\boldsymbol{x}_i)) L\left(\frac{y_i}{\exp(\widehat{\text{score}}_{m-1}(\boldsymbol{x}_i))}, \exp(T(\boldsymbol{x}_i; \mathbf{a}_m))\right) \\
&= \nu_{mi} L\left(\tilde{r}_{mi}, \exp(T(\boldsymbol{x}_i; \mathbf{a}_m))\right).
\end{aligned}$$

Finally, when $\xi \notin \{0, 1, 2\}$, it comes

$$\begin{aligned}
&\nu_i L(y_i, \exp(\widehat{\text{score}}_{m-1}(\boldsymbol{x}_i) + T(\boldsymbol{x}_i; \mathbf{a}_m))) \\
&= \nu_i 2 \left(\frac{\max(y_i, 0)^{2-\xi}}{(1-\xi)(2-\xi)} - \frac{y_i \exp(\widehat{\text{score}}_{m-1}(\boldsymbol{x}_i) + T(\boldsymbol{x}_i; \mathbf{a}_m))^{1-\xi}}{1-\xi} + \frac{\exp(\widehat{\text{score}}_{m-1}(\boldsymbol{x}_i) + T(\boldsymbol{x}_i; \mathbf{a}_m))^{2-\xi}}{2-\xi} \right) \\
&= \nu_i \exp(\widehat{\text{score}}_{m-1}(\boldsymbol{x}_i))^{2-\xi} 2 \left(\frac{\max(\tilde{r}_{mi}, 0)^{2-\xi}}{(1-\xi)(2-\xi)} - \frac{\tilde{r}_{mi} \exp(T(\boldsymbol{x}_i; \mathbf{a}_m))^{1-\xi}}{1-\xi} + \frac{\exp(T(\boldsymbol{x}_i; \mathbf{a}_m))^{2-\xi}}{2-\xi} \right) \\
&= \nu_{mi} L\left(\tilde{r}_{mi}, \exp(T(\boldsymbol{x}_i; \mathbf{a}_m))\right),
\end{aligned}$$

which completes the proof. □

This result shows that when we work with the log-link function and a response that belongs to the Tweedie class (and so with a loss function of the form (5.5.1)), solving (5.3.3) amounts to build a single regression tree on the working training set

$$\mathcal{D}^{(m)} = \{(\nu_{mi}, \tilde{r}_{mi}, \boldsymbol{x}_i), i \in \mathcal{I}\}.$$

Therefore, in these cases, that are the most relevant for application in insurance, boosting trees should be preferred to gradient boosting trees since the latter procedure introduces an extra step which is unnecessary on the one hand and that leads to an approximation that can be easily avoided with boosting trees on the other hand.

5.6 Regularization and Randomness

5.6.1 Shrinkage

Boosting (and gradient boosting) models are susceptible to overfitting. They employ the greedy strategy of selecting the optimal weak learner at each step. Such a strategy produces an optimal solution at each stage of the training procedure. However, it does not find the optimal global solution and often fits the training set too closely when M is large: after a certain number of iterations, reducing the training sample estimate of the generalization error starts to increase the generalization error.

Regularization methods aim to prevent such overfitting by constraining the training procedure. Controlling the value of M is a natural regularization strategy. For a certain size of the constituent trees, that can be specified with `ID`, there is an optimal number of trees M^* minimizing the generalization error. In practice, M^* can be estimated as the value of M that minimizes the validation sample estimate (or the cross validation estimate) of the generalization error.

Another regularization strategy consists in adding only a fraction of the prediction produced by the new tree to the current one. This fraction is often referred to as the learning rate or shrinkage factor and takes its values between 0 and 1. That is, line 2.2 in Algorithm 5.2 and line 2.4 in Algorithm 5.3 are replaced by

$$\text{Update } \widehat{\text{score}}_m(\boldsymbol{x}) = \widehat{\text{score}}_{m-1}(\boldsymbol{x}) + \tau\, T(\boldsymbol{x}; \widehat{\mathbf{a}}_m) \tag{5.6.1}$$

and

$$\text{Update } \widehat{\text{score}}_m(\boldsymbol{x}) = \widehat{\text{score}}_{m-1}(\boldsymbol{x}) + \tau \sum_{t \in \mathcal{T}_m} \widehat{\gamma}_{tm} \mathrm{I}\left[\boldsymbol{x} \in \chi_t^{(m)}\right], \tag{5.6.2}$$

respectively, where $0 < \tau \leq 1$ is the shrinkage parameter. The shrinkage parameter is another parameter to fine-tune. Small values for τ work best, but result in larger computation time since more iterations are necessary. The optimal number of iterations M^* and τ are closely related: smaller values of τ lead to larger values of M^*. Empirically, small values for the shrinkage parameter $\tau (< 0.1)$ yield dramatic improvements for regression estimation. Because the trees are small trees built with no pruning, many iterations are generally computationally manageable.

5.6.2 *Randomness*

The bootstrap sampling procedure in bagging offers a reduction in variance for bagging models. Stochastic (gradient) boosting exploits the same device to improve both the computational time and the predictive accuracy of (gradient) boosting. The boosting algorithm is updated with a random sampling scheme. At each iteration, a random sample of the training set is taken without replacement before building the next tree on that subsample. The fraction of training set used at each iteration to produce the random samples is known as the bagging fraction, denoted α, and becomes another parameter to fine-tune. A typical value for α is 0.5.

The boosting and gradient boosting predictions then writes

$$\widehat{\mu}_{\mathcal{D},\Theta}^{\text{boost}}(\boldsymbol{x}) = g^{-1}\left(\widehat{\text{score}}_M(\boldsymbol{x})\right)$$

and

$$\widehat{\mu}_{\mathcal{D},\Theta}^{\text{grad boost}}(\boldsymbol{x}) = g^{-1}\left(\widehat{\text{score}}_M(\boldsymbol{x})\right)$$

respectively, where

$$\Theta = (\Theta_1, \ldots, \Theta_M),$$

Θ_m expressing the randomization due to the random sampling at iteration m.

5.7 Interpretability

As for bagging trees and random forests, boosting trees are less interpretable than a single tree. Nevertheless, we can also rely on tools such as relative importances and partial dependences to better understand model outcomes. Moreover, Friedman's H-statistics, introduced in Friedman and Popescu (2008), enable to know which features are involved in interactions with other features, the identities of the other features with which they interact, as well as the order and strength of the respective interaction effects. Note that Friedman's H-statistics can be computed for any regression model, including bagging trees and random forests.

5.7.1 Relative Importances

For constituent tree m, the relative importance of feature x_j, denoted $\mathcal{I}_m(x_j)$, is the sum of the deviance reductions over the non-terminal nodes of the mth tree for which x_j was selected as the splitting feature. The relative importance of feature x_j is then obtained by summing the relative importances of x_j over the collection of trees, that is,

$$\mathcal{I}(x_j) = \sum_{m=1}^{M} \mathcal{I}_m(x_j). \tag{5.7.1}$$

To improve their readability, the relative importances are often normalized so that their sum equals to 100. Any individual number can then be interpreted as the percentage contribution to the overall model. Sometimes, the relative importances are expressed as a percent of the maximum relative importance.

5.7.2 Partial Dependence Plots

Also, the partial dependence of $\widehat{\mu}_{\mathcal{D},\Theta}^{\text{boost}}(\boldsymbol{x})$ or $\widehat{\mu}_{\mathcal{D},\Theta}^{\text{grad boost}}(\boldsymbol{x})$ on selected small subsets of the features helps the analyst to improve its model understanding. We refer the reader to Sect. 4.6.2 for more details about partial dependence plots.

5.7.3 Friedman's H-Statistics

Tree-based models are praised for their ability to account for interaction effects between features. Knowing which features are involved in interactions with other features, the identities of the other features with which they interact, as well as the order and strength of the respective interaction effects provide useful information to the analyst.

Consider two features x_j and x_k. If these two variables do not interact, then the function $\widehat{\mu}(\boldsymbol{x})$ can be written as

$$\widehat{\mu}(\boldsymbol{x}) = f_{\backslash j}(\boldsymbol{x}_{\backslash j}) + f_{\backslash k}(\boldsymbol{x}_{\backslash k}), \tag{5.7.2}$$

where $\boldsymbol{x}_{\backslash j}$ and $\boldsymbol{x}_{\backslash k}$ represent all the features except x_j and x_k, respectively. If a given feature x_j interacts with none of the other features, then $\widehat{\mu}(\boldsymbol{x})$ can be expressed as

$$\widehat{\mu}(\boldsymbol{x}) = f_{\backslash j}(\boldsymbol{x}_{\backslash j}) + f_j(x_j). \tag{5.7.3}$$

Considering a third feature x_l, $\widehat{\mu}(\boldsymbol{x})$ can be written as

$$\widehat{\mu}(\boldsymbol{x}) = f_{\backslash j}(\boldsymbol{x}_{\backslash j}) + f_{\backslash k}(\boldsymbol{x}_{\backslash k}) + f_{\backslash l}(\boldsymbol{x}_{\backslash l}) \tag{5.7.4}$$

if there is no three-variable interaction between x_j, x_k and x_l, where $\boldsymbol{x}_{\backslash l}$ represents all the features except x_l. In the same way, similar expressions can be defined for the absence of higher order interaction effects.

Recall that the partial dependence of $\widehat{\mu}(\boldsymbol{x})$ on the subvector $\boldsymbol{x}_{\mathcal{S}}$ of $\boldsymbol{x}$ is defined by

$$\widehat{\mu}_{\mathcal{S}}(\boldsymbol{x}_{\mathcal{S}}) = \mathrm{E}_{\boldsymbol{X}_{\bar{\mathcal{S}}}}\left[\widehat{\mu}(\boldsymbol{x}_{\mathcal{S}}, \boldsymbol{X}_{\bar{\mathcal{S}}})\right] \tag{5.7.5}$$

and can be estimated from the training set by

$$\widehat{\widehat{\mu}}_{\mathcal{S}}(\boldsymbol{x}_{\mathcal{S}}) = \frac{1}{|\mathcal{I}|}\sum_{i\in\mathcal{I}} \widehat{\mu}(\boldsymbol{x}_{\mathcal{S}}, \boldsymbol{x}_{i\bar{\mathcal{S}}}), \tag{5.7.6}$$

where $\{\boldsymbol{x}_{i\bar{\mathcal{S}}}, i \in \mathcal{I}\}$ are the values of $\boldsymbol{X}_{\bar{\mathcal{S}}}$ in the training set. In this section, we consider that all partial dependence functions and the regression model $\widehat{\mu}(\boldsymbol{x})$ are centered to have a mean of zero.

If there is no interaction between x_j and x_k, then, from (5.7.2), the partial dependence of $\widehat{\mu}(\boldsymbol{x})$ on $\boldsymbol{x}_{\mathcal{S}} = (x_j, x_k)$ can be decomposed into the sum of the respective partial dependences on each feature separately, that is,

$$\begin{aligned}
\widehat{\mu}_{j,k}(x_j, x_k) &= \mathrm{E}_{\boldsymbol{X}_{\backslash j,k}}\left[f_{\backslash j}(x_k, \boldsymbol{X}_{\backslash j,k})\right] + \mathrm{E}_{\boldsymbol{X}_{\backslash j,k}}\left[f_{\backslash k}(x_j, \boldsymbol{X}_{\backslash j,k})\right] \\
&= \widehat{\mu}_k(x_k) - \mathrm{E}_{\boldsymbol{X}_{\backslash k}}\left[f_{\backslash k}(\boldsymbol{X}_{\backslash k})\right] + \widehat{\mu}_j(x_j) - \mathrm{E}_{\boldsymbol{X}_{\backslash j}}\left[f_{\backslash j}(\boldsymbol{X}_{\backslash j})\right] \\
&= \widehat{\mu}_k(x_k) + \widehat{\mu}_j(x_j) - \mathrm{E}_{\boldsymbol{X}}\left[\widehat{\mu}(\boldsymbol{X})\right] \\
&= \widehat{\mu}_k(x_k) + \widehat{\mu}_j(x_j).
\end{aligned} \tag{5.7.7}$$

Moreover, if a given feature x_j interacts with none of the other features, then, from (5.7.3) and since $\mathrm{E}_{\boldsymbol{X}}\left[\widehat{\mu}(\boldsymbol{X})\right] = \mathrm{E}_{\boldsymbol{X}_{\backslash j}}\left[f_{\backslash j}(\boldsymbol{X}_{\backslash j})\right] + \mathrm{E}_{X_j}\left[f_j(X_j)\right] = 0$, we have

$$\begin{aligned}\widehat{\mu}(\boldsymbol{x}) &= \widehat{\mu}_{\setminus j}(\boldsymbol{x}_{\setminus j}) - \mathrm{E}_{\boldsymbol{X}_{\setminus j}}\left[f_{\setminus j}(\boldsymbol{X}_{\setminus j})\right] + \widehat{\mu}_j(x_j) - \mathrm{E}_{X_j}\left[f_j(X_j)\right] \\ &= \widehat{\mu}_{\setminus j}(\boldsymbol{x}_{\setminus j}) + \widehat{\mu}_j(x_j), \end{aligned} \tag{5.7.8}$$

where $\widehat{\mu}_{\setminus j}(\boldsymbol{x}_{\setminus j})$ is the partial dependence of $\widehat{\mu}(\boldsymbol{x})$ on all features except x_j.

If there is no three-variable interaction between x_j, x_k and x_l, then, from (5.7.4), we have

$$\begin{aligned}\widehat{\mu}_{j,k,l}(x_j, x_k, x_l) &= \mathrm{E}_{\boldsymbol{X}_{\setminus j,k,l}}\left[f_{\setminus j}(x_k, x_l, \boldsymbol{X}_{\setminus j,k,l})\right] + \mathrm{E}_{\boldsymbol{X}_{\setminus j,k,l}}\left[f_{\setminus k}(x_j, x_l, \boldsymbol{X}_{\setminus j,k,l})\right] \\ &\quad + \mathrm{E}_{\boldsymbol{X}_{\setminus j,k,l}}\left[f_{\setminus l}(x_j, x_k, \boldsymbol{X}_{\setminus j,k,l})\right] \\ &= \widehat{\mu}_{k,l}(x_k, x_l) + \widehat{\mu}_{j,l}(x_j, x_l) + \widehat{\mu}_{j,k}(x_j, x_k) \\ &\quad - \widehat{\mu}_j(x_j) - \widehat{\mu}_k(x_k) - \widehat{\mu}_l(x_l) \end{aligned} \tag{5.7.9}$$

since

$$\widehat{\mu}_j(x_j) = \mathrm{E}_{\boldsymbol{X}_{\setminus j}}\left[f_{\setminus j}(\boldsymbol{X}_{\setminus j})\right] + \mathrm{E}_{\boldsymbol{X}_{\setminus j,k}}\left[f_{\setminus k}(x_j, \boldsymbol{X}_{\setminus j,k})\right] + \mathrm{E}_{\boldsymbol{X}_{\setminus j,l}}\left[f_{\setminus l}(x_j, \boldsymbol{X}_{\setminus j,l})\right]$$

and

$$\begin{aligned}\widehat{\mu}_{j,k}(x_j, x_k) &= \mathrm{E}_{\boldsymbol{X}_{\setminus j,k}}\left[f_{\setminus j}(x_k, \boldsymbol{X}_{\setminus j,k})\right] + \mathrm{E}_{\boldsymbol{X}_{\setminus j,k}}\left[f_{\setminus k}(x_j, \boldsymbol{X}_{\setminus j,k})\right] \\ &\quad + \mathrm{E}_{\boldsymbol{X}_{\setminus j,k,l}}\left[f_{\setminus l}(x_j, x_k, \boldsymbol{X}_{\setminus j,k,l})\right]\end{aligned}$$

with

$$\mathrm{E}_{\boldsymbol{X}_{\setminus j}}\left[f_{\setminus j}(\boldsymbol{X}_{\setminus j})\right] + \mathrm{E}_{\boldsymbol{X}_{\setminus k}}\left[f_{\setminus k}(\boldsymbol{X}_{\setminus k})\right] + \mathrm{E}_{\boldsymbol{X}_{\setminus l}}\left[f_{\setminus l}(\boldsymbol{X}_{\setminus l})\right] = \mathrm{E}_{\boldsymbol{X}}\left[\widehat{\mu}(\boldsymbol{X})\right] = 0.$$

Similar expressions can be obtained for the absence of higher order interactions.

Expressions (5.7.7), (5.7.8) and (5.7.9) can be used to test for the presence of interaction effects. Specifically, to test a potential interaction between two given features x_j and x_k, we can rely on the statistic

$$\hat{H}^2_{j,k} = \frac{\sum_{i\in\mathcal{I}}\left[\widehat{\widehat{\mu}}_{j,k}(x_{ji}, x_{ki}) - \widehat{\widehat{\mu}}_j(x_{ji}) - \widehat{\widehat{\mu}}_k(x_{ki})\right]^2}{\sum_{i\in\mathcal{I}}\widehat{\widehat{\mu}}^2_{j,k}(x_{ji}, x_{ki})}, \tag{5.7.10}$$

where x_{ji} indicates that the estimated partial dependence function is evaluated at the observed value of x_j for policyholder i. Considering that all partial dependence functions are centered at zero, the numerator measures the variance of the interaction and the denominator measures the total variance, so that the statistic (5.7.12) measures the interaction strength as the amount of variance explained by the interaction. A value of 0 for $H^2_{j,k}$ means that there is no interaction between x_j and x_k while a value of 1 means that the effect of x_j and x_k on the model is exclusively explained by the interaction.

In the same way, for a given feature x_j, we can use the statistic

$$H_j^2 = \frac{\sum_{i\in\mathcal{I}} \left[\widehat{\mu}(\boldsymbol{x}_i) - \widehat{\widehat{\mu}}_j(x_{ji}) - \widehat{\widehat{\mu}}_{\backslash j}(\boldsymbol{x}_{i\backslash j})\right]^2}{\sum_{i\in\mathcal{I}} \widehat{\mu}^2(\boldsymbol{x}_i)} \tag{5.7.11}$$

to test whether x_j interacts with any other variable. The statistic (5.7.11) differs from zero to the extent that x_j interacts with one or more other features.

In the case where x_j interacts with more than one other feature, say with at least x_k and x_l, it is interesting to determine whether these interactions represent separate two-way interactions between (x_j, x_k) and (x_j, x_l) only, or whether there is an additional three-way interaction between (x_j, x_k, x_l). This alternative can be tested by means of the statistic

$$H_{j,k,l}^2 = \sum_{i\in\mathcal{I}} \left[\widehat{\widehat{\mu}}_{j,k,l}(x_{ji}, x_{ki}, x_{li}) - \widehat{\widehat{\mu}}_{j,k}(x_{ji}, x_{ki}) - \widehat{\widehat{\mu}}_{j,l}(x_{ji}, x_{li}) - \widehat{\widehat{\mu}}_{k,l}(x_{ki}, x_{li}) \right.$$
$$\left. + \widehat{\widehat{\mu}}_j(x_{ji}) + \widehat{\widehat{\mu}}_k(x_{ki}) + \widehat{\widehat{\mu}}_k(x_{li})\right]^2 \Big/ \sum_{i\in\mathcal{I}} \widehat{\widehat{\mu}}^2_{j,k,l}(x_{ji}, x_{ki}, x_{li}). \tag{5.7.12}$$

which measures the fraction of variance of $\widehat{\widehat{\mu}}_{j,k,l}(x_{ji}, x_{ki}, x_{li})$ not explained by the lower order interaction effects among these features. Similarly, additional statistics for higher order interactions can be built, if needed.

5.8 Example

Consider the real dataset described in Sect. 3.2.4.2. We use the same training set $\mathcal{D}$ and validation set $\overline{\mathcal{D}}$ than in the examples of Sects. 3.3.2.3 and 4.7.

We fit gradient boosting trees on $\mathcal{D}$ with the Poisson deviance loss and the log-link function by means of the R package gbm. The parameters we need to fine-tune are

- the number of trees M;
- the size of the trees ID;
- the bagging fraction α;
- the shrinkage parameter τ.

To this end, we consider different values for the tuning parameters ID, α and τ, namely ID $= 1, 2, 3, 4, 5, 6$, $\alpha = 1, 0.75, 0.5$ and $\tau = 1, 0.1, 0.01$, and we split the training set $\mathcal{D}$ into five disjoint and stratified subsets $\mathcal{D}_1, \mathcal{D}_2, \ldots, \mathcal{D}_5$ of equal size. Then, for each value of (ID, α, τ), depicted in Table 5.2, we compute the 5-fold cross-validation estimates of the generalization error (from subsets $\mathcal{D}_1, \mathcal{D}_2, \ldots, \mathcal{D}_5$) for models including up to 4000 trees, and we select the optimal number of trees as the number of trees corresponding to the model with the smallest 5-fold cross-validation estimate.

In Fig. 5.6, we show the optimal numbers of trees obtained for the different values of (ID, α, τ) under consideration. We notice that considering 4000 trees was more than enough in this example, even for ID $= 1$ and $\tau = 0.01$. Also, one sees that we

Table 5.2 Values for (ID, α, τ)

	ID	α	τ
1–06	1,2,3,4,5,6	1.00	1.00
07–12	1,2,3,4,5,6	0.75	1.00
13–18	1,2,3,4,5,6	0.50	1.00
19–24	1,2,3,4,5,6	1.00	0.10
25–30	1,2,3,4,5,6	0.75	0.10
31–36	1,2,3,4,5,6	0.50	0.10
37–42	1,2,3,4,5,6	1.00	0.01
43–48	1,2,3,4,5,6	0.75	0.01
49–54	1,2,3,4,5,6	0.50	0.01

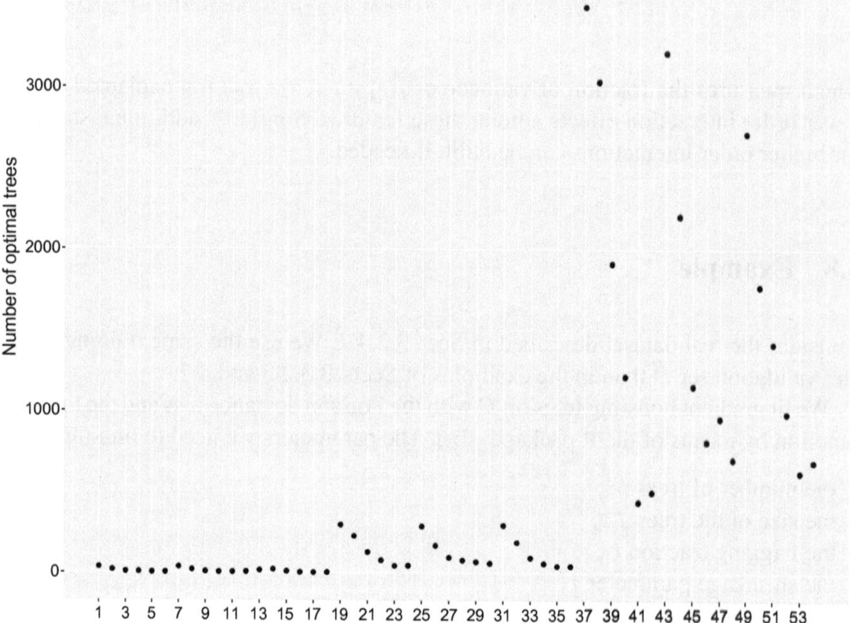

Fig. 5.6 Optimal numbers of trees for the values of (ID, α, τ) summarized in Table 5.2

need more trees when the shrinkage parameter decreases or when the size of the trees decreases, highlighting the interplay between these tuning parameters.

Figure 5.7 displays 5-fold cross-validation estimates of the generalization error for the models built with their corresponding optimal number of trees. One sees that the introduction of a shrinkage parameter enables to improve the predictive accuracy of the boosting procedure. Also, results with $\tau = 0.01$ are slightly better than the ones obtained with $\tau = 0.1$. To a lesser extent, adding randomness into the training

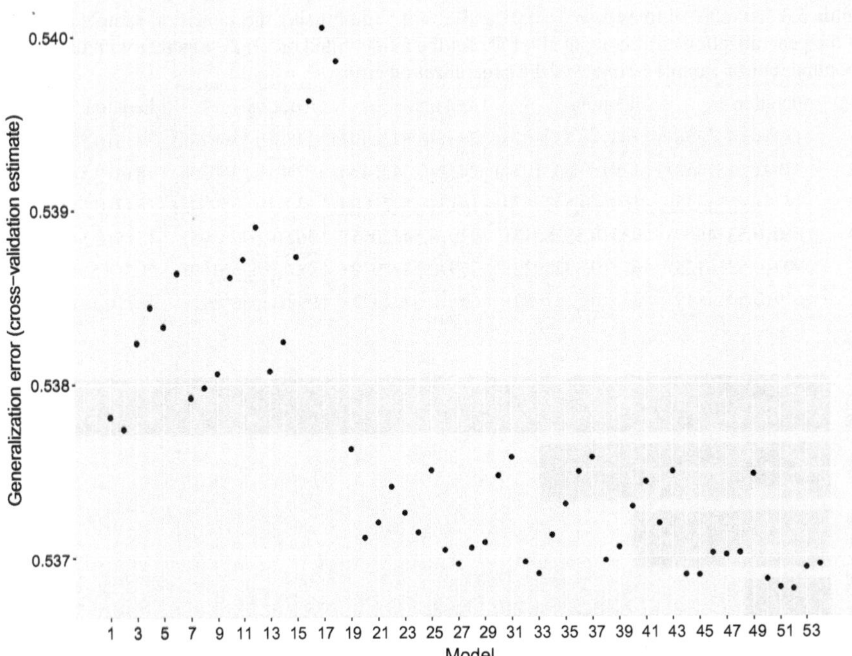

Fig. 5.7 5-fold cross-validation estimates of the generalization error for the best models corresponding to the values of $(\mathtt{ID}, \alpha, \tau)$ summarized in Table 5.2

procedure appears to be relevant in this example. For instance, one sees that for any given value of ID, the model with $(\tau = 0.01, \alpha = 0.5)$ performs slightly better than the model with $(\tau = 0.01, \alpha = 0.75)$, which, in turn, performs slightly better than the model with $(\tau = 0.01, \alpha = 1)$. Finally, if we look at models with $\tau = 0.01$, i.e. models 37 to 54, we observe that the interaction depths minimizing the 5-fold cross-validation estimate of the generalization error are $\mathtt{ID} = 2$ for $\alpha = 1$, $\mathtt{ID} = 3$ for $\alpha = 0.75$ and $\mathtt{ID} = 4$ for $\alpha = 0.5$, so that the best value for ID ranges from 2 to 4. Based on the results depicted in Fig. 5.7, we decide to select $M = 986$, $\mathtt{ID} = 4$, $\alpha = 0.5$ and $\tau = 0.01$ as the optimal tuning parameters, which correspond to the values minimizing the 5-fold cross-validation estimate (i.e. model 52). We denote by $\widehat{\mu}^{\text{grad boost}^*}_{\mathcal{D},\Theta}$ the gradient boosting model fitted on the entire training set with these optimal parameters.

Remark 5.8.1 Remark 4.7.1 about the instability related to the selection procedure of the optimal tuning parameters still holds for (gradient) boosting trees. As an illustration, Table 5.3 provides, for each iteration j of the 5 cross-validation, the number of trees for $\tau = 0.01$ and $\alpha = 0.5$ minimizing the validation-sample estimate of the generalization error computed on $\mathcal{D}_j$ for models fitted on $\mathcal{D}\backslash\mathcal{D}_j$, together with the corresponding out-of-sample estimate of the generalization error. We can see that the optimal tuning parameters for ID and M are unstable over the

Table 5.3 Number of trees for $\tau = 0.01$ and $\alpha = 0.5$ minimizing the validation-sample estimate of the generalization error computed on $\mathcal{D}_j$ for the model fitted on $\mathcal{D}\backslash\mathcal{D}_j$ together with the corresponding out-of-sample estimate of the generalization error

ID	Iteration 1	Iteration 2	Iteration 3	Iteration 4	Iteration 5
1	1841 (0.5323761)	1847 (0.5336023)	2895 (0.5424494)	2356 (0.5368788)	3475 (0.5417317)
2	3194 (0.5314581)	1291 (0.5331031)	2422 (0.5422430)	1740 (0.5360356)	1814 (<u>0.5412181</u>)
3	1422 (<u>0.5314159</u>)	1004 (0.5329624)	1413 (<u>0.5421080</u>)	1401 (0.5359842)	1642 (0.5413473)
4	860 (0.5314600)	933 (0.5328013)	919 (0.5422055)	962 (0.5358395)	1315 (0.5415447)
5	907 (0.5316125)	615 (0.5328922)	779 (0.5424979)	987 (<u>0.5356169</u>)	702 (0.5416647)
6	619 (0.5314849)	738 (<u>0.5327782</u>)	652 (0.5424877)	992 (0.5360672)	707 (0.5418545)

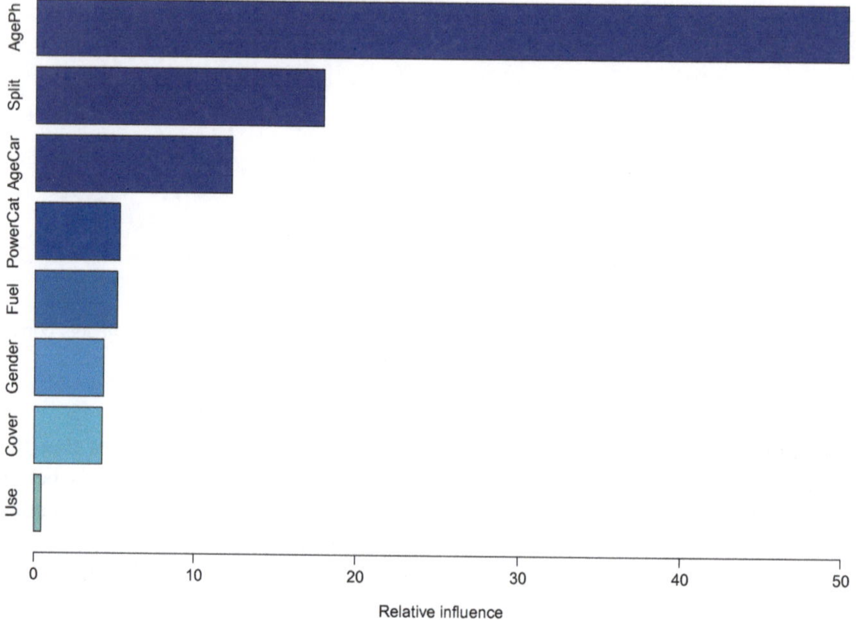

Fig. 5.8 Relative importances of the features for $\widehat{\mu}^{\text{grad boost}*}_{\mathcal{D},\Theta}$

iterations, getting ($M = 1422$, $\mathtt{ID} = 3$) for iteration 1, ($M = 738$, $\mathtt{ID} = 6$) for iteration 2, ($M = 1413$, $\mathtt{ID} = 3$) for iteration 3, ($M = 987$, $\mathtt{ID} = 5$) for iteration 4 and ($M = 1814$, $\mathtt{ID} = 2$) for iteration 5.

The relative importances of the features for $\widehat{\mu}^{\text{grad boost}*}_{\mathcal{D},\Theta}$ are depicted in Fig. 5.8. The most important feature is AgePh followed by, in descending order, Split, AgeCar, PowerCat, Fuel, Gender, Cover and Use. Notice that compared to Fig. 4.4, AgePh, Split, Gender and Use keep the same ranking in terms of importance.

Figures 5.9 and 5.10 represent the partial dependence plots of the features and the H-statistics $H^2_{j,k}$ for $\widehat{\mu}^{\text{grad boost}*}_{\mathcal{D},\Theta}$, respectively. Here we focus on two-way interactions.

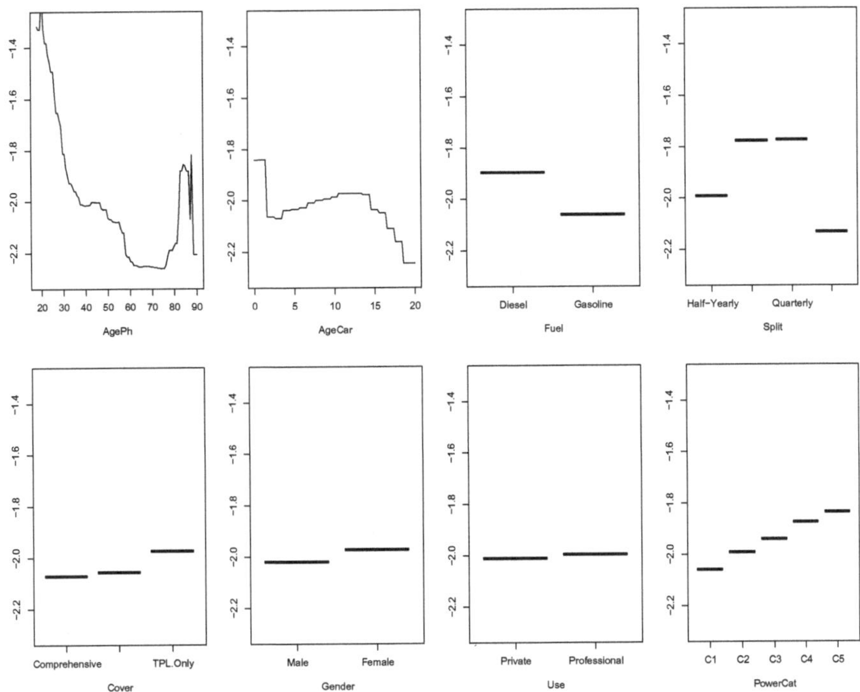

Fig. 5.9 Partial dependence plots for $\widehat{\mu}^{\text{grad boost*}}_{\mathcal{D},\Theta}$ (on the score scale)

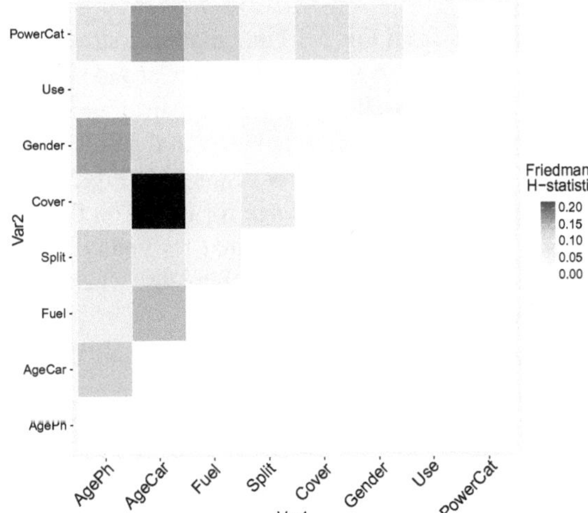

Fig. 5.10 H-statistic $H^2_{j,k}$ for $\widehat{\mu}^{\text{grad boost*}}_{\mathcal{D},\Theta}$

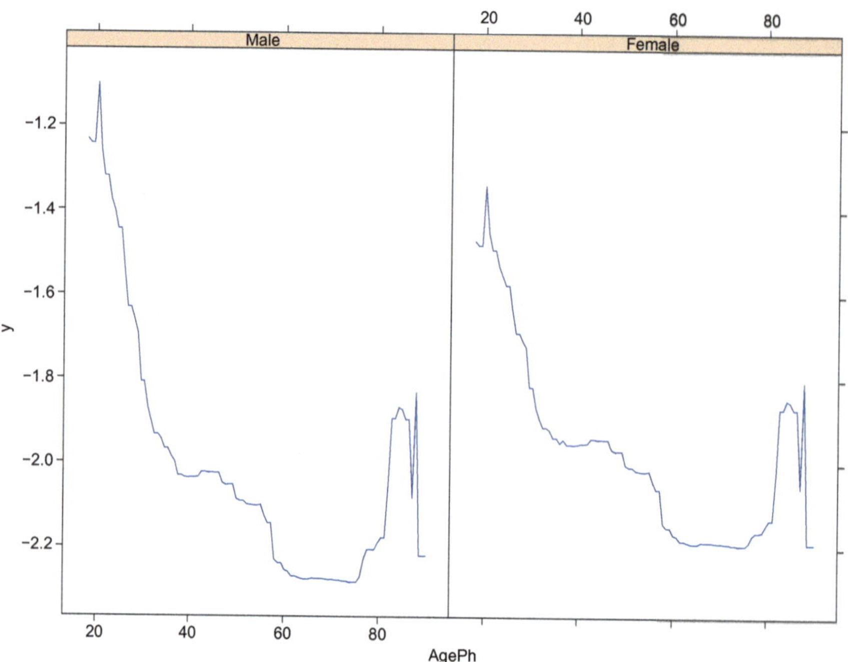

Fig. 5.11 Effect of the policyholder's age on the score for males (left-hand side) and females (right-hand side)

One observes from Fig. 5.10 that the three strongest interactions are found between Agecar and Cover, AgeCar and PowerCat and between AgePh and Gender, this latter interaction being well-know in MTPL insurance. The H-statistic $H_{j,k}^2$ informs us on the strength of the interaction between features x_j and x_k but does not give any clue on how the effect behaves. For instance, in Fig. 5.11, we show the effect (on the score scale) of the policyholder's age for males on the left-hand side and for females on le right-hand side. We observe that for young policyholders, males are on average more risky drivers compared to females, whereas at older ages female drivers are perceived as more risky than males.

Finally, the validation sample estimate of the generalization error of $\widehat{\mu}_{\mathcal{D},\Theta}^{\text{grad boost*}}$ (computed on $\overline{\mathcal{D}}$) is given by

$$\widehat{Err}^{\text{val}}\left(\widehat{\mu}_{\mathcal{D},\Theta}^{\text{grad boost*}}\right) = 0.5431231.$$

Compared to the validation sample estimates of the generalization error of $\widehat{\mu}_{T_{\alpha_{k^*}}}$ and $\widehat{\mu}_{\mathcal{D},\Theta}^{\text{rf*}}$ (also computed on $\overline{\mathcal{D}}$) fitted in Sects. 3.3.2.3 and 4.7, that are

$$\widehat{Err}^{\text{val}}\left(\widehat{\mu}_{T_{\alpha_{k^*}}}\right) = 0.5452772$$

and

$$\widehat{Err}^{\text{val}}\left(\widehat{\mu}^{\text{rf}^*}_{\mathcal{D},\Theta}\right) = 0.5440970,$$

one sees that $\widehat{\mu}^{\text{rf}^*}_{\mathcal{D},\Theta}$ improves by $2.1541\ 10^{-3}$ the predictive accuracy of the single tree $\widehat{\mu}_{T_{\alpha_{k^*}}}$ and by $0.9739\ 10^{-3}$ the predictive accuracy of the random forest $\widehat{\mu}^{\text{rf}^*}_{\mathcal{D},\Theta}$.

5.9 Bibliographic Notes and Further Reading

Boosting was originally designed for classification problems. Valiant (1984) and Kearns and Valiant (1989) introduced the concept of combining weak classifiers into a strong classifier. These works influenced Schapire, who developed the first simple boosting procedure (Schapire 1990). The performance of the simple boosting algorithm of Schapire was improved by Freund (1995). Freund and Schapire collaborated to produce the AdaBoost algorithm (Freund and Schapire 1996a, 1997). To support their algorithms, Freund and Schapire (1996a) and Schapire and Singer (1999) derived some upper bounds on the generalization error. Other theories attempting to explain boosting come from game theory (Freund and Schapire 1996b,Breiman 1998, 1998) and Vapnik-Chervonenkis theory (Schapire et al. 1998). In particular, Breiman (1998) explained the algorithm as a gradient descent approach with numerical optimization and statistical estimation. In practice, the AdaBoost algorithm was shown to be a powerful prediction tool, far beyond the expectations implied by the bounds and the theoretical developments.

Friedman et al. (2000) made the link between the AdaBoost algorithm and the statistical concepts of loss functions, additive modeling and logistic regression. They showed that boosting can be viewed as a forward stagewise additive model that minimizes exponential loss. Friedman (2001) proposed a boosting method called Gradient Boosting Machine for regression and classification problems, which combines weak learners. Bühlmann and Hothorn (2007) adopted penalty splines, linear regressors, and trees in various scenarios. Ridgeway (2007) uses only trees as the base learners. Gradient boosting machine with neural networks can be found in Denuit et al. (2019).

Friedman and Popescu (2008) presented techniques to identify the variables that are involved in interactions with other variables, the strength and degree of those interactions, as well as the identities of the other variables with which they interact. Tree-based models are known for their ability to account for interaction effects between features, as illustrated in Buchner et al. (2017) and Schiltz et al. (2018).

Several authors applied boosting and gradient boosting to insurance pricing. Guelman (2012) proposed gradient boosted trees for predicting auto insurance loss. Liu et al. (2014) treated the claim frequency prediction problem by using multi class AdaBoost trees. Wüthrich and Buser (2019) adapted tree-based methods to model claim frequencies. Yang et al. (2018) predicted insurance premiums by using a gradient boosted tree algorithm to Tweedie models. Lee and Lin (2018) introduced

Delta Boosting Machine as a new member of the boosting gamily with application to general insurance. Pesantez-Narvaez et al. (2019) employed XGBoost to predict the occurrence of claims using telematics data. Henckaert et al. (2020) worked with random forests and boosted trees to develop full tariff plans built from both the frequency and severity of claims.

We mainly based our presentation on Hastie et al. (2009), Friedman (2000) and Friedman and Popescu (2008). Sect. 5.5 is inspired by Denuit et al. (2020). Hastie et al. (2009) and Kuhn and Johnson (2013) made a good overview of the existing literature.

References

Breiman L (1998) Arcing classifiers (with discussion). Ann Stat 26(3):801–849

Breiman L (1999) Prediction games and arcing algorithms. Neural Comput 11(7):1493–1517

Buchner F, Wasem J, Schillo S (2017) Regression trees identify relevant interactions: can this improve the predictive performance of risk adjustment? Health Econ 26(1):74–85

Bühlmann P, Hothorn T (2007) Boosting algorithms: regularization, prediction and model fitting. Stat Sci 22(4):477–505

Denuit M, Hainaut D, Trufin J (2019) Effective statistical learning methods for actuaries III: neural networks and extensions. Springer actuarial lecture notes

Denuit M, Hainaut D, Trufin J (2020) Boosting versus gradient boosting for Tweedie models with log-link function. Working paper

Freund Y (1995) Boosting a weak learning algorithm by majority. Inf Comput 121(2):256–285

Freund Y, Schapire R (1996a) Experiments with a new boosting algorithm. In: Machine learning: proceedings of the thirteenth international conference, Morgan Kauffman, San Francisco, pp 148–156

Freund Y, Schapire R (1996b) Game theory, on-line prediction and boosting. In: Proceedings of the ninth annual conference on computational learning theory, Desenzano del Garda, Italy, pp 325–332

Freund Y, Schapire R (1997) A decision-theoretic generalization of on-line learning and an application to boosting. J Comput Syst Sci 55(1):119–139

Friedman J, Hastie T, Tibshirani R (2000) Additive logistic regression: a statistical view of boosting (with discussion and a rejoinder by the authors). Ann Stat 28(2):337–407

Friedman J (2001) Greedy function approximation: a gradient boosting machine. Ann Stat 29(5):1189–1232

Friedman J, Popescu B (2008) Predictive learning via rule ensembles. Ann Appl Stat 2(3):916–954

Guelman L (2012) Gradient boosting trees for auto insurance loss cost modeling and prediction. Expert Syst Appl 39(3):3659–3667

Hastie T, Tibshirani R, Friedman J (2009) The elements of statistical learning. Data mining, inference, and prediction. Springer series in statistics, 2nd edn

Henckaerts R, Côté M-P, Antonio K, Verbelen R (2020) Boosting insights in insurance tariff plans with tree-based machine learning methods. North Am Actu J. https://doi.org/10.1080/10920277.2020.1745656

Kearns M, Valiant LG (1989) Cryptographic limitations on learning Boolean formulae and finite automata. In: Proceedings of the twenty-first annual ACM symposium on theory of computing, Seattle, pp 433–444

Kuhn M, Johnson K (2013) Applied predictive modeling. Springer, New York

Lee SCK, Lin S (2018) Delta boosting machine with application to general insurance. N Am Actu J 22(3):405–425

Liu Y, Wang B, Lv S (2014) Using multi-class AdaBoost tree for prediction frequency of auto insurance. J Appl Financ Bank 4(5):45–53

Pesantez-Narvaez J, Guillen M, Alcañiz M (2019) Predicting motor insurance claims using telematics data XGBoost versus logistic regression. Risks 7(2):1–16

Ridgeway G (2007) Generalized boosted models: a guide to the GBM package. Update 1(1):

Schapire R (1990) The strength of weak learnability. Mach Learn 5:197–227

Schapire R, Freund Y, Bartlett P, Lee W (1998) Boosting the margin: a new explanation for the effectiveness of voting methods. Ann Stat 26(5):1651–1686

Schapire R, Singer Y (1999) Improved boosting algorithms using confidence-rated predictions. Mach Learn 37:297–336

Schiltz F, Masci C, Agasisti T, Horn D (2018) Using regression tree ensembles to model interaction effects: a graphical approach. Appl Econ 50(58):6341–6354

Valiant LG (1984) A theory of the learnable. Commun ACM 27(11):1134–1142

Wüthrich MV, Buser C (2019) Data analytics for non-life insurance pricing. Lecture notes

Yang Y, Qian W, Zou H (2018) Insurance premium prediction via gradient tree-boosted Tweedie compound Poisson models. J Bus Econ Stat 36(3):456–470

Chapter 6
Other Measures for Model Comparison

6.1 Introduction

Actuarial pricing models are generally calibrated so that they minimize the generalization error computed with an appropriate loss function. Model selection is based on the generalization error. Regression models are then evaluated by assessing their generalization error with the same loss function, which is done by comparing their generalization errors computed on a validation set.

Model selection and model assessment are thus based on the same objective function (the deviance in our ED family setting), which provides consistency in the approach. However, in our ED family setting, using the deviance as a tool for model assessment has some drawbacks. As we have seen throughout the different examples made in the previous chapters, the deviance only slightly reacts to a model improvement. Moreover, a decrease of a certain amount of the deviance is difficult to interpret for the analyst. There is a need for additional measures to assess pricing models, notably more economic criteria. In this chapter, we describe complementary measures frequently used by practitioners for model assessment.

Remark 6.1.1 Training a model with an objective function (say objective function 1) and assessing it with another one (say objective function 2) is not without criticism. If the ultimate goal of the analyst is to minimize the second objective function, then one could wonder why not directly training the model using this second objective function as well.

M. Denuit et al., *Effective Statistical Learning Methods for Actuaries II*, Springer Actuarial, https://doi.org/10.1007/978-3-030-57556-4_6

6.2 Measures of Association

6.2.1 Context

The merits of a regression model can be assessed using the pairs $(Y, \widehat{\mu}(\boldsymbol{X}))$. Besides using validation sample estimates for the generalization error to assess the predictive power of a model $\widehat{\mu}$, measures of association for the pairs $(Y, \widehat{\mu}(\boldsymbol{X}))$ are also frequently used by practitioners. In insurance, the most popular ones are Kendall's tau and Spearman's rho, which are based on concordance probabilities.

Kendall's tau and Spearman's rho, defined in the following, are efficient tools for measuring the strength of dependence between continuous outcomes. They can be expressed in terms of the corresponding copula only and are thus independent of the marginal distributions. When they are applied to discrete variables, they are no more distribution-free so that their ranges are restricted to sub-intervals of $[-1, 1]$. This makes their interpretation more difficult: relatively small values for Kendall's tau and Spearman's rho may in fact strongly support the fitted model if their maximal possible values are small as well. In case the response variable Y is discrete, such as the number of claims, correlation indices are thus often restricted to a sub-interval $[-1, 1]$. That is why positive values of Kendall's tau and Spearman's rho for the pairs $(Y, \widehat{\mu}(\boldsymbol{X}))$ must be compared to their highest attainable values and not 1.

Notice that, even for discrete responses, predictors $\widehat{\mu}(\boldsymbol{X})$ are generally continuous random variables. This is the case when there is at least one continuous feature comprised in the available information $\boldsymbol{X}$ (so that the score is continuous) and the function $\widehat{\mu}$ is a continuously increasing function of the score. Of course, predictors $\widehat{\mu}(\boldsymbol{X})$ can still be discrete when all the features are discrete or when they are piecewise constant predictors (such as a single tree, for instance). However, this is unlikely to be the case since actuarial pricing is nowadays based on more sophisticated models than piecewise constant predictors (trees being combined into random forests, for instance) and uses more and more features. Even though predictors $\widehat{\mu}(\boldsymbol{X})$ are generally continuous, we also consider the discrete case for $\widehat{\mu}(\boldsymbol{X})$. For ease of exposition, we mean by the random variable Z a predictor $\widehat{\mu}(\boldsymbol{X})$ and we denote by p_k the probabilities $\mathrm{P}[Y = k]$, $k \in \mathbb{N}$. When Z is discrete, we assume it is valued in $\{z_1, z_2, \ldots, z_m\}$ with $z_1 < z_2 < \ldots < z_m$, and we define

$$j_k = \min\{j \in \{1, 2, \ldots, m\} : F_Z(z_j) \geq F_Y(k)\}, \quad k \in \mathbb{N}.$$

This section aims to derive the best possible upper bounds for Kendall's tau and Spearman's rho when the response takes its values in $\mathbb{N} = \{0, 1, 2, \ldots\}$.

6.2.2 Probability of Concordance

6.2.2.1 Definition

Consider independent copies (Y_1, Z_1) and (Y_2, Z_2) of (Y, Z). Then, (Y_1, Z_1) and (Y_2, Z_2) are said to be concordant if $(Y_1 - Y_2)(Z_1 - Z_2) > 0$ holds true whereas they are said to be discordant when $(Y_1 - Y_2)(Z_1 - Z_2) < 0$.

Tied pairs (that is, pairs of observations that have equal values of Y or Z) may occur in practice. Specifically, the probability that a tie occurs is given by

$$\mathrm{P}[(Y_1 - Y_2)(Z_1 - Z_2) = 0] = \begin{cases} \mathrm{P}[Y_1 = Y_2 \text{ or } Z_1 = Z_2] \text{ if } Z \text{ is discrete} \\ \\ \mathrm{P}[Y_1 = Y_2] \text{ if } Z \text{ is continuous.} \end{cases}$$

The concordance probabilities can be expressed as follows.

Proposition 6.2.1 *If H denotes the joint distribution function of the pair (Y, Z), then*

$$\begin{aligned} P[(Y_1 - Y_2)(Z_1 - Z_2) > 0] &= 2E[H(Y, Z)] - P[Y_1 = Y_2] - P[Z_1 = Z_2] \\ &= 2\sum_{k=0}^{\infty}\sum_{l=1}^{\infty} P[Y_1 = k, Y_2 = k + l, Z_1 < Z_2]. \end{aligned}$$

Proof As Z_1 and Z_2 are independent and identically distributed, we have

$$\begin{aligned} \mathrm{P}[Z_1 \leq Z_2] &= \mathrm{P}[Z_1 < Z_2] + \mathrm{P}[Z_1 = Z_2] \\ &= \mathrm{P}[Z_1 > Z_2] + \mathrm{P}[Z_1 = Z_2] \\ &= \frac{1 - \mathrm{P}[Z_1 = Z_2]}{2} + \mathrm{P}[Z_1 = Z_2] \\ &= \frac{1 + \mathrm{P}[Z_1 = Z_2]}{2}. \end{aligned}$$

This allows us to write

$$\begin{aligned} \mathrm{P}[Y_1 \leq Y_2, Z_1 \leq Z_2] &= 1 - \mathrm{P}[Y_1 > Y_2] - \mathrm{P}[Z_1 > Z_2] + \mathrm{P}[Y_1 < Y_2, Z_1 < Z_2] \\ &\quad - \mathrm{P}[Y_1 < Y_2, Z_1 < Z_2] + \frac{\mathrm{P}[Y_1 = Y_2]}{2} + \frac{\mathrm{P}[Z_1 = Z_2]}{2}. \end{aligned}$$

The concordance probability can finally be expressed as

$$\begin{aligned} \mathrm{P}[(Y_1 - Y_2)(Z_1 - Z_2) > 0] &= 2\mathrm{P}[Y_1 < Y_2, Z_1 < Z_2] \\ &= 2\mathrm{P}[Y_1 \leq Y_2, Z_1 \leq Z_2] - \mathrm{P}[Y_1 = Y_2] - \mathrm{P}[Z_1 = Z_2] \\ &= 2\mathrm{E}[H(Y, Z)] - \mathrm{P}[Y_1 = Y_2] - \mathrm{P}[Z_1 = Z_2] \end{aligned}$$

as announced. Now, since

$$P[(Y_1 - Y_2)(Z_1 - Z_2) > 0] = 2P[Y_1 < Y_2, Z_1 < Z_2],$$

it suffices to notice that

$$P[Y_1 < Y_2, Z_1 < Z_2] = \sum_{k=0}^{\infty}\sum_{l=1}^{\infty} P[Y_1 = k, Y_2 = k + l, Z_1 < Z_2],$$

which gives the second announced equality and ends the proof. □

Proposition 6.2.1 shows that concordance probabilities get higher when we replace the joint distribution function H with H_+ whose graph lies everywhere above H, provided the marginals are kept unchanged. A natural candidate for H_+ is the Fréchet–Höffding upper bound H^u defined as

$$H^u(y, z) = \min\{F_Y(y), F_Z(z)\}.$$

Proposition 6.2.2 *We have*

$$P[(Y_1 - Y_2)(Z_1 - Z_2) > 0] \leq P[(Y_1^u - Y_2^u)(Z_1^u - Z_2^u) > 0] \qquad (6.2.1)$$

where (Y_1^u, Z_1^u) and (Y_2^u, Z_2^u) are independent copies of the random pair (Y^u, Z^u) obeying the Fréchet–Höffding upper bound H^u, i.e.

$$Z^u = F_Z^{-1}(U) \text{ and } Y^u = \sum_{k=0}^{\infty} kI[F_Y(k-1) \leq U < F_Y(k)] \qquad (6.2.2)$$

with U being uniformly distributed over the unit interval $[0, 1]$.

Proof The joint distribution function of the random pair (Y, Z) satisfies

$$H(y, z) \leq \min\{F_Y(y), F_Z(z)\} \text{ for all } y \text{ and } z.$$

This ensures that

$$E[H(Y, Z)] \leq E[\min\{F_Y(Y), F_Z(Z)\}]$$

holds true. Now, the inequality $E[g(Y, Z)] \leq E[g(Y^u, Z^u)]$ is known to be valid for every supermodular function g (see e.g. Denuit et al. 2005, Sect. 6.2.4). As every joint distribution function is supermodular, we also have

$$E[\min\{F_Y(Y), F_Z(Z)\}] \leq E[\min\{F_Y(Y^u), F_Z(Z^u)\}],$$

so that

$$E[H(Y, Z)] \leq E[\min\{F_Y(Y^u), F_Z(Z^u)\}]$$

is true. Hence, as

$$P[Y^u \leq y, Z^u \leq z] = \min\{F_Y(y), F_Z(y)\},$$

we have the announced result by Proposition 6.2.1. □

6.2.2.2 Upper Bounds

Based on Proposition 6.2.2, we can establish upper bounds for concordance probabilities.

Proposition 6.2.3 *If Z is continuous, then*

$$P[(Y_1 - Y_2)(Z_1 - Z_2) > 0] \leq 2E[F_Y(Y-)]. \tag{6.2.3}$$

Proof By (6.2.1), since $P[(Y_1^u - Y_2^u)(Z_1^u - Z_2^u) > 0] = 2P[Y_1^u < Y_2^u, Z_1^u < Z_2^u]$, it suffices to show that $P[Y_1^u < Y_2^u, Z_1^u < Z_2^u] = E[F_Y(Y-)]$ with

$$Z_i^u = F_Z^{-1}(U_i) \text{ and } Y_i^u = \sum_{k=0}^{\infty} k\mathrm{I}\left[F_Y(k-1) \leq U_i < F_Y(k)\right]$$

for $i = 1, 2$, where U_1 and U_2 are independent random variables, uniformly distributed over the unit interval $[0, 1]$. We get

$$\begin{aligned} P[Y_1^u < Y_2^u, Z_1^u < Z_2^u] &= \sum_{k=0}^{\infty} P[Y_1^u = k, Y_2^u > k, Z_1^u < Z_2^u] \\ &= \sum_{k=0}^{\infty} P[F_Y(k-1) \leq U_1 < F_Y(k), F_Y(k) \leq U_2, U_1 < U_2] \\ &= \sum_{k=0}^{\infty} P[F_Y(k-1) \leq U_1 < F_Y(k), F_Y(k) \leq U_2] \\ &= \sum_{k=0}^{\infty} \big(F_Y(k) - F_Y(k-1)\big)\bar{F}_Y(k) \\ &= \sum_{k=0}^{\infty} p_k \bar{F}_Y(k) \\ &= E[\bar{F}_Y(Y)], \end{aligned}$$

which ends the proof since $E[F_Y(Y)] + E[F_Y(Y-)] = 1$. □

Proposition 6.2.4 *If Z is discrete, then*

$$P[(Y_1 - Y_2)(Z_1 - Z_2) > 0]$$
$$\leq 2E[F_Y(Y-)] - 2\sum_{k=0}^{\infty}\left(F_Y(k) - \max\{F_Y(k-1), F_Z(z_{j_k-1})\}\right)\left(F_Z(z_{j_k}) - F_Y(k)\right).$$

Proof From (6.2.1), it amounts to show that

$$\mathrm{P}[Y_1^u < Y_2^u, Z_1^u < Z_2^u]$$
$$= \mathrm{E}[F_Y(Y-)] - \sum_{k=0}^{\infty}\left(F_Y(k) - \max\{F_Y(k-1), F_Z(z_{j_k-1})\}\right)\left(F_Z(z_{j_k}) - F_Y(k)\right)$$

with

$$Z_i^u = \sum_{j=1}^{m} z_j \mathrm{I}\left[F_Z(z_{j-1}) \leq U_i < F_Z(z_j)\right] \text{ and } Y_i^u = \sum_{k=0}^{\infty} k\mathrm{I}\left[F_Y(k-1) \leq U_i < F_Y(k)\right]$$

for $i = 1, 2$, where U_1 and U_2 are independent random variables, uniformly distributed over the unit interval $[0, 1]$ and $z_0 < z_1$ such that $F_Z(z_0) = 0$. We then have

$$\mathrm{P}[Y_1^u < Y_2^u, Z_1^u < Z_2^u] = \sum_{k=0}^{\infty} \mathrm{P}[Y_1^u = k, Y_2^u > k, Z_1^u < Z_2^u]$$
$$= \sum_{k=0}^{\infty} \mathrm{P}[Z_1^u < Z_2^u | Y_1^u = k, Y_2^u > k]\mathrm{P}[Y_1^u = k, Y_2^u > k]$$
$$= \sum_{k=0}^{\infty} \mathrm{P}[Z_1^u < Z_2^u | (U_1, U_2) \in \mathcal{A}_k] p_k \bar{F}_Y(k)$$
$$= \sum_{k=0}^{\infty} \left(1 - \mathrm{P}[Z_1^u = Z_2^u | (U_1, U_2) \in \mathcal{A}_k]\right) p_k \bar{F}_Y(k)$$

where

$$\mathcal{A}_k = [F_Y(k-1), F_Y(k)[\times[F_Y(k), 1], \quad k \in \mathbb{N}.$$

Define

$$\mathcal{B}_j = [F_Z(z_{j-1}), F_Z(z_j)[\times[F_Z(z_{j-1}), F_Z(z_j)[, \quad j \in \{1, \ldots, m\}.$$

Then,

$$
\begin{aligned}
\mathrm{P}[Z_1^u = Z_2^u | (U_1, U_2) \in \mathcal{A}_k] &= \sum_{j=1}^{m} \mathrm{P}[(U_1, U_2) \in \mathcal{B}_j | (U_1, U_2) \in \mathcal{A}_k] \\
&= \frac{1}{p_k \bar{F}_Y(k)} \sum_{j=1}^{m} \mathrm{P}[(U_1, U_2) \in \mathcal{A}_k \cap \mathcal{B}_j] \\
&= \frac{1}{p_k \bar{F}_Y(k)} \sum_{j=1}^{m} \alpha_{k,j} \beta_{k,j}
\end{aligned}
$$

with

$$
\alpha_{k,j} = \big(\min\{F_Y(k), F_Z(z_j)\} - \max\{F_Y(k-1), F_Z(z_{j-1})\}\big)_+
$$

and

$$
\beta_{k,j} = \big(F_Z(z_j) - \max\{F_Y(k), F_Z(z_{j-1})\}\big)_+ ,
$$

where, for any real number r, we let r_+ denote the positive part of r; that is, $r_+ = r$ if $r \geq 0$ and $r_+ = 0$ if $r < 0$.

For $j < j_k$, we get $\beta_{k,j} = 0$ since $F_Y(k) \geq F_Z(z_j) \geq F_Z(z_{j-1})$. Also, for $j > j_k$, we have $\alpha_{k,j} = 0$ since $F_Z(z_j) \geq F_Z(z_{j-1}) \geq F_Y(k) \geq F_Y(k-1)$. Now, in the remaining case $j = j_k$, it comes $F_Z(z_{j_k}) \geq F_Y(k) \geq F_Z(z_{j_k-1})$ and hence

$$
\alpha_{k,j_k} = F_Y(k) - \max\{F_Y(k-1), F_Z(z_{j_k-1})\} \text{ and } \beta_{k,j_k} = F_Z(z_{j_k}) - F_Y(k).
$$

Finally, we then get

$$
\begin{aligned}
&\mathrm{P}[Y_1^u < Y_2^u, Z_1^u < Z_2^u] \\
&= \mathrm{E}[\bar{F}_Y(Y)] - \sum_{k=0}^{\infty} \Big(F_Y(k) - \max\{F_Y(k-1), F_Z(z_{j_k-1})\}\Big)\big(F_Z(z_{j_k}) - F_Y(k)\big),
\end{aligned}
$$

which completes the proof. □

6.2.3 Kendall's Tau

6.2.3.1 Definition

Kendall's tau is a widely used measure of dependence between Y and $Z = \widehat{\mu}(\boldsymbol{X})$, defined as

$$
\tau[Y, Z] = \mathrm{P}[(Y_1 - Y_2)(Z_1 - Z_2) > 0] - \mathrm{P}[(Y_1 - Y_2)(Z_1 - Z_2) < 0].
$$

With continuous random variables, Kendall's tau is completely determined by the copula and unrelated to the marginal distributions. This is no more true in general for random variables that are not necessarily continuous (see for instance Nešlehová 2007).

6.2.3.2 Upper Bounds

In the following, we derive the best possible upper bounds on Kendall's tau for discrete responses Y. We start with the case of a continuous Z.

Proposition 6.2.5 *If Z is continuous, then*

$$\tau[Y, Z] \leq 2E[F_Y(Y-)]. \tag{6.2.4}$$

Proof If the random pair (Y, Z) obeys the upper Fréchet–Höffding bound, i.e. under (6.2.2), we get

$$\begin{aligned} \mathrm{P}[(Y_1 - Y_2)(Z_1 - Z_2) < 0] &= 2\mathrm{P}[Y_1 > Y_2, Z_1 < Z_2] \\ &= 2\sum_{k=0}^{\infty} \mathrm{P}[Y_1 > k, Y_2 = k, Z_1 < Z_2] \\ &= 2\sum_{k=0}^{\infty} \mathrm{P}[F_Y(k) \leq U_1, F_Y(k-1) \leq U_2 < F_Y(k), U_1 < U_2] \\ &= 0. \end{aligned}$$

The maximal value of Kendall's tau corresponds to a random pair distributed according to the Fréchet–Höffding upper bound because this distribution simultaneously maximizes the concordance probability and leads to zero discordance probability. This ends the proof. □

We now turn to the case where Z is discrete.

Proposition 6.2.6 *If Z is discrete, then*

$$\tau[Y, Z] \leq 2E[F_Y(Y-)] - 2\sum_{k=0}^{\infty} \left(F_Y(k) - \max\{F_Y(k-1), F_Z(z_{j_k-1})\}\right)\left(F_Z(z_{j_k}) - F_Y(k)\right). \tag{6.2.5}$$

Proof Similarly to the continuous case for Z, when the random pair (Y, Z) obeys the upper Fréchet–Höffding bound, we obviously have $\mathrm{P}[(Y_1 - Y_2)(Z_1 - Z_2) < 0] = 0$, so that we directly get the desired result. □

We notice that the latter upper bound (6.2.5) is smaller than the upper bound (6.2.4) obtained when Z is continuous. Let us define

$$k^* = \min\{k \in \mathbb{N} : F_Y(k) > 0\}.$$

The difference between the upper bounds (6.2.4) and (6.2.5) is then given by

$$\begin{aligned}
&2\sum_{k=0}^{\infty}\left(F_Y(k) - \max\{F_Y(k-1), F_Z(z_{j_k-1})\}\right)\left(F_Z(z_{j_k}) - F_Y(k)\right)\\
&= 2\sum_{k=k^*}^{\infty}\left(F_Y(k) - \max\{F_Y(k-1), F_Z(z_{j_k-1})\}\right)\left(F_Z(z_{j_k}) - F_Y(k)\right)\\
&= 2\left(F_Y(k^*) - F_Z(z_{j_{k^*}-1})\right)\left(F_Z(z_{j_{k^*}}) - F_Y(k^*)\right)\\
&\quad +2\sum_{k=k^*+1}^{\infty}\left(F_Y(k) - \max\{F_Y(k-1), F_Z(z_{j_k-1})\}\right)\left(F_Z(z_{j_k}) - F_Y(k)\right).
\end{aligned}$$

One sees that if Z is such that $F_Z(z_{m-1}) < F_Y(k^*)$, then this difference becomes

$$\begin{aligned}
&2\left(F_Y(k^*) - F_Z(z_{m-1})\right)\left(1 - F_Y(k^*)\right) + 2\sum_{k=k^*+1}^{\infty}\left(F_Y(k) - F_Y(k-1)\right)\left(1 - F_Y(k)\right)\\
&= -2F_Z(z_{m-1})\left(1 - F_Y(k^*)\right) + 2\sum_{k=k^*}^{\infty}\left(F_Y(k) - F_Y(k-1)\right)\left(1 - F_Y(k)\right)\\
&= -2F_Z(z_{m-1})\left(1 - F_Y(k^*)\right) + 2\mathrm{E}[F_Y(Y-)],
\end{aligned}$$

which is larger as $F_Z(z_{m-1})$ decreases. By letting $F_Z(z_{m-1}) \to 0$, the difference even tends to its maximum $2\mathrm{E}[F_Y(Y-)]$, such that the upper bound (6.2.5) tends to zero as well.

6.2.4 Spearman's Rho

6.2.4.1 Definition

Consider independent copies (Y_1, Z_1), (Y_2, Z_2) and (Y_3, Z_3) of (Y, Z). It is well known that the population version of Spearman's rho is defined in terms of the probability of concordance minus the probability of discordance for the random pairs (Y_1, Z_1) and (Y_2, Z_3), that is,

$$\rho[Y, Z] = 3\left(\mathrm{P}[(Y_1 - Y_2)(Z_1 - Z_3) > 0] - \mathrm{P}[(Y_1 - Y_2)(Z_1 - Z_3) < 0]\right).$$

Proposition 6.2.7 *$\rho[Y, Z]$ can be expressed in terms of the joint distribution H of (Y, Z) as*

$$\rho[Y, Z] = 3\left(E[H(Y^*-, Z^*)] + E[H(Y^*, Z^*-)] + E[H(Y^*-, Z^*-)] + E[H(Y^*, Z^*)]\right. \tag{6.2.6}$$

where the random variables Y^* and Z^* are independent and distributed as Y and Z, respectively.

Proof We refer the reader to Mesfioui and Tajar (2005) for a formal proof. □

With continuous random variables, Spearman's rho is entirely described by the copula and unrelated to the marginal distributions. This is no more true in general for random variables that are not necessarily continuous. (see for instance Nešlehová 2007). Tied pairs may occur. Specifically, for the random pairs (Y_1, Z_1) and (Y_2, Z_3), the probability that a tie occurs is given by

$$\mathrm{P}[(Y_1 - Y_2)(Z_1 - Z_3) = 0] = \begin{cases} \mathrm{P}[Y_1 = Y_2 \text{ or } Z_1 = Z_3] \text{ when } Z \text{ is discrete} \\ \\ \mathrm{P}[Y_1 = Y_2] \text{ when } Z \text{ is continuous.} \end{cases}$$

6.2.4.2 Upper Bounds

Upper bounds on Spearman's rho can be obtained by replacing the joint distribution function H in (6.2.6) with the Fréchet–Höffding upper bound

$$H^u(y, z) = \min\{F_Y(y), F_Z(z)\}.$$

Spearman's rho $\rho[Y, Z]$ is then bounded by

$$\begin{aligned} \rho_{\max} &= 3\left(\mathrm{E}[\min\left\{F_Y(Y^*-), F_Z(Z^*)\right\}] + \mathrm{E}[\min\left\{F_Y(Y^*), F_Z(Z^*-)\right\}]\right) \\ &\quad +3\left(\mathrm{E}[\min\left\{F_Y(Y^*-), F_Z(Z^*-)\right\}] + \mathrm{E}[\min\left\{F_Y(Y^*), F_Z(Z^*)\right\}] - 1\right). \end{aligned} \tag{6.2.7}$$

For a continuous Z, upper bound (6.2.7) can be particularized as follows.

Proposition 6.2.8 *If Z is continuous, then*

$$\rho[Y, Z] \leq 3\left(1 - E\left[F_Y^2(Y)\right] - E\left[F_Y^2(Y-)\right]\right). \tag{6.2.8}$$

Proof Since Z is continuous, $\rho_{\max}$ can be expressed as

$$\rho_{\max} = 3\left(2\mathrm{E}[\min\left\{F_Y(Y), U\right\}] + 2\mathrm{E}[\min\left\{F_Y(Y-), U\right\}] - 1\right) \tag{6.2.9}$$

where U is a random variable uniformly distributed over the unit interval $[0, 1]$ and independent of Y. We get

$$\begin{aligned}\mathrm{E}[\min\{F_Y(Y-), U\}] &= \mathrm{E}[\mathrm{E}[\min\{F_Y(Y-), U\}|Y]] \\ &= \mathrm{E}\left[\int_0^{F_Y(Y-)} u du + \int_{F_Y(Y-)}^1 F_Y(Y-) du\right] \\ &= \frac{1}{2}\mathrm{E}[2F_Y(Y-) - F_Y^2(Y-)]. \end{aligned} \tag{6.2.10}$$

Similarly, we have

$$\mathrm{E}[\min\{F_Y(Y), U\}] = \frac{1}{2}\mathrm{E}[2F_Y(Y) - F_Y^2(Y)]. \tag{6.2.11}$$

Inserting (6.2.10) and (6.2.11) in (6.2.9), we get the desired result. □

Turning to the case of a discrete Z, we define

$$\psi(z_j) = \begin{cases} \min\{k \in \mathbb{N} : F_Y(k) > F_Z(z_j)\} \text{ for } j \in \{1, 2, \ldots, m-1\} \\ +\infty \text{ for } j = m. \end{cases}$$

Using the notation $\phi(k)$ for j_k, $k \in \mathbb{N}$, we then get the following result.

Proposition 6.2.9 *If Z is discrete, then*

$$\begin{aligned}\rho[Y, Z] \le 9 - 3\left(E[F_Y(Y)(F_Z(z_{\phi(Y)}) + F_Z(z_{\phi(Y)}-))] + E[F_Y(Y-)(F_Z(z_{\phi(Y-)}) + F_Z(z_{\phi(Y-)}-))]\right) \\ -3\left(E[F_Z(Z)(F_Y(\psi(Z)) + F_Y(\psi(Z)-))] + E[F_Z(Z-)(F_Y(\psi(Z-)) + F_Y(\psi(Z-)-))]\right).\end{aligned} \tag{6.2.12}$$

Proof The best upper bound for $\rho[Y, Z]$ is given by (6.2.7), namely

$$\begin{aligned}\rho_{\max} = 3\left(\mathrm{E}[\min\{F_Y(Y^*-), F_Z(Z^*)\}] + \mathrm{E}[\min\{F_Y(Y^*), F_Z(Z^*-)\}]\right) \\ +3\left(\mathrm{E}[\min\{F_Y(Y^*-), F_Z(Z^*-)\}] + \mathrm{E}[\min\{F_Y(Y^*), F_Z(Z^*)\}] - 1\right).\end{aligned} \tag{6.2.13}$$

First, we note that

$$\begin{aligned}\min\{F_Y(Y^*), F_Z(Z^*)\} &= F_Y(Y^*)I[F_Z(Z^*) \ge F_Y(Y^*)] + F_Z(Z^*)I[F_Z(Z^*) < F_Y(Y^*)] \\ &= F_Y(Y^*)I[Z^* \ge z_{\phi(Y^*)}] + F_Z(Z^*)I[Y^* \ge \psi(Z^*)].\end{aligned}$$

Hence, we get

$$\begin{aligned}\mathrm{E}[\min\{F_Y(Y^*), F_Z(Z^*)\}] &= \mathrm{E}[\mathrm{E}[F_Y(Y^*)I[Z^* \ge z_{\phi(Y^*)}]|Y^*]] + \mathrm{E}[\mathrm{E}[F_Z(Z^*)I[Y^* \ge \psi(Z^*)]|Z^*]] \\ &= \mathrm{E}[F_Y(Y^*)(1 - F_Z(z_{\phi(Y^*)}-))] + \mathrm{E}[F_Z(Z^*)(1 - F_Y(\psi(Z^*)-))].\end{aligned} \tag{6.2.14}$$

Also, we have

$$\begin{aligned}\min\{F_Y(Y^*-), F_Z(Z^*)\} &= F_Y(Y^*-)I[F_Z(Z^*) \geq F_Y(Y^*-)] + F_Z(Z^*)I[F_Z(Z^*) < F_Y(Y^*-)] \\ &= F_Y(Y^*-)I[Z^* \geq z_{\phi(Y^*-)}] + F_Z(Z^*)I[Y^* \geq \psi(Z^*)+],\end{aligned}$$

which leads to

$$\begin{aligned}\mathrm{E}[\min\{F_Y(Y^*-), F_Z(Z^*)\}] = \mathrm{E}[F_Y(Y^*-)(1 - F_Z(z_{\phi(Y^*-)}-))] \\ +\mathrm{E}[F_Z(Z^*)(1 - F_Y(\psi(Z^*)))].\end{aligned} \quad (6.2.15)$$

Similarly, we get

$$\begin{aligned}\mathrm{E}[\min\{F_Y(Y^*), F_Z(Z^*-)\}] = \mathrm{E}[F_Y(Y^*)(1 - F_Z(z_{\phi(Y^*)}))] \\ +\mathrm{E}[F_Z(Z^*-)(1 - F_Y(\psi(Z^*-)-))]\end{aligned} \quad (6.2.16)$$

and

$$\begin{aligned}\mathrm{E}[\min\{F_Y(Y^*-), F_Z(Z^*-)\}] = \mathrm{E}[F_Y(Y^*-)(1 - F_Z(z_{\phi(Y^*-)}))] \\ +\mathrm{E}[F_Z(Z^*-)(1 - F_Y(\psi(Z^*-)))].\end{aligned} \quad (6.2.17)$$

Finally, the announced upper bound for $\rho[Y, Z]$ is obtained by inserting (6.2.14), (6.2.15), (6.2.16) and (6.2.17) in (6.2.13). □

6.2.5 *Numerical Example*

Let us illustrate the computation of the upper bounds (6.2.4), (6.2.5), (6.2.8) and (6.2.12) in a situation of practical relevance. To this end, we consider the motor third-party liability insurance portfolio introduced in Sect. 3.2.4.2. Specifically, we restrict our example to the 124 524 insurance policies of the portfolio that have been observed during the whole year. Figure 6.1 displays for each feature the number of policies by category/value, and Table 6.1 shows the observed numbers of claims.

Let $n = 124\,524$ be the number of policies considered in this example and let us denote by Y_i the number of claims of policy i ($i = 1, \ldots, n$). The probabilities p_k can be directly estimated using the empirical proportions

$$\widehat{p}_k = \frac{\sum_{i=1}^n \mathrm{I}[Y_i = k]}{n}, \quad k \in \mathbb{N}.$$

Hence, based on the observations for the number of claims summarized in Table 6.1, we get the empirical proportions depicted in Table 6.2.

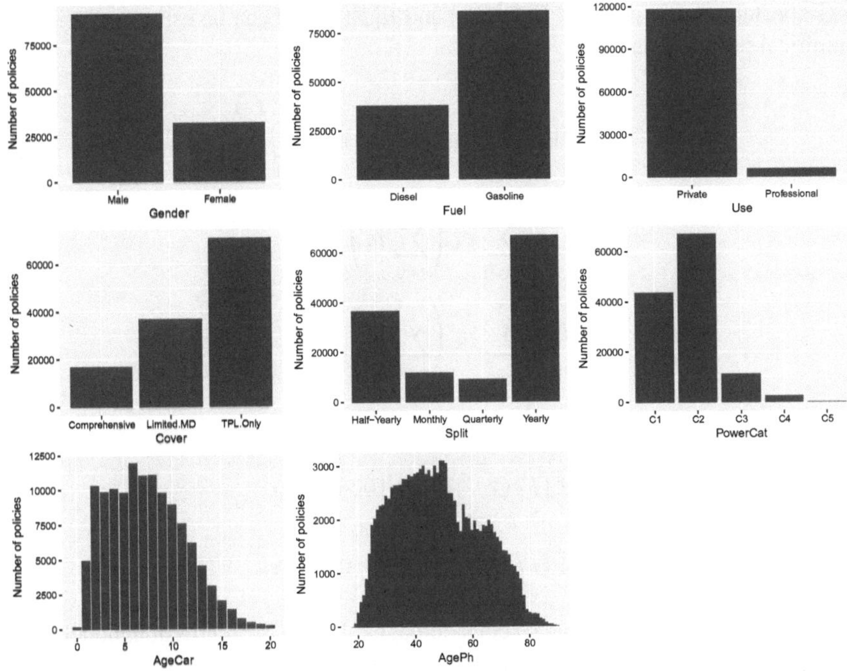

Fig. 6.1 The categories/values of the explanatory variables and their corresponding numbers of policies

Table 6.1 Descriptive statistics for the number of claims

Number of claims	Number of policies
0	110 231
1	13 024
2	1 149
3	115
4	5
≥5	0

Table 6.2 Empirical proportions $\widehat{p}_k$ displayed as percentages

k	0	1	2	3	4	≥5
$\widehat{p}_k$	88.52189	10.45903	0.92271	0.09235	0.00402	0

Expectations $\mathrm{E}[F_Y(Y-)]$, $\mathrm{E}[F_Y^2(Y)]$ and $\mathrm{E}[F_Y^2(Y-)]$ can be estimated using the empirical proportions $\widehat{p}_k$,

$$\begin{aligned}\mathrm{E}[F_Y(Y-)] &= \sum_{k=1}^{\infty} p_k F_Y(k-1) = \sum_{k=1}^{\infty}\sum_{l=0}^{k-1} p_k p_l \\ \mathrm{E}[F_Y^2(Y)] &= \sum_{k=0}^{\infty} p_k \left(\sum_{l=0}^{k} p_l\right)^2 \\ \mathrm{E}[F_Y^2(Y-)] &= \sum_{k=1}^{\infty} p_k \left(\sum_{l=0}^{k-1} p_l\right)^2 .\end{aligned}$$

With the $\widehat{p}_k$ displayed in Table 6.2, we get

$$\begin{aligned}\widehat{\mathrm{E}}[F_Y(Y-)] &= 0.1026812 \\ \widehat{\mathrm{E}}[F_Y^2(Y)] &= 0.8063110 \\ \widehat{\mathrm{E}}[F_Y^2(Y-)] &= 0.0919602.\end{aligned}$$

Of course, other estimators can be considered, exploiting the regression model structure (whereas the proposed one is purely nonparametric).

Therefore, when Z is continuous, which is the case when AgePh and AgeCar are treated as continuous variables using GAM for instance, upper bounds (6.2.4) on Kendall's tau and (6.2.8) on Spearman's rho can be estimated by

$$2\widehat{\mathrm{E}}[F_Y(Y-)] = 0.2053624$$

and

$$3\left(1 - \widehat{\mathrm{E}}\left[F_Y^2(Y)\right] - \widehat{\mathrm{E}}\left[F_Y^2(Y-)\right]\right) = 0.3051865,$$

respectively. The actuarial analyst wishing to assess the predictive performances of the regression model for claim counts using Kendall's tau or Spearman's rho must therefore compare the obtained value to the upper bound 20.5% for Kendall's tau and 30.5% for Spearman's rho, and not to 1 (which cannot be attained with the data under consideration).

In case Z is discrete, we need the distribution function of Z in order to compute upper bounds (6.2.5) and (6.2.12). Consider that the number of claims Y is predicted using a regression tree. For the illustration, we fit several regression trees, controlling the size of the trees by the minimum number of observations required in the terminal nodes. The results are shown in Tables 6.3 and 6.4. The corresponding regression trees and distribution functions of Z are displayed in Figs. 6.2, 6.3 and 6.4.

From Table 6.3, we notice that the upper bound (6.2.5) increases with m to ultimately tend to the upper bound (6.2.4) derived in the continuous case. Also, the values for Kendall's tau are rather small (around 2%), as it is generally the case with

Table 6.3 Upper bound (6.2.5), Kendall's tau and its normalized version $\left(\frac{\text{Kendall's tau}}{\text{Upper bound}}\right)$ as well as the corresponding m (i.e. number of values taken by Z) for trees whose sizes are controlled by the minimum number of observations (Min. nb of obs.) required in the terminal nodes

Min. nb of obs.	m	Kendall's tau	Upper bound	$\frac{\text{Kendall's tau}}{\text{Upper bound}}$
35 000	2	0.0167340	0.1645063	0.1017228
30 000/25 000	3	0.0227923	0.1695944	0.1343933
20 000	4	0.0257349	0.1893385	0.1359203
15 000	5	0.0281253	0.1941500	0.1448638
10 000	7	0.0287892	0.1941500	0.1482834
5000	14	0.0313762	0.2000066	0.1568756
1000	66	0.0355663	0.2050653	0.1734387

Table 6.4 Upper bound (6.2.12), Spearman's rho and its normalized version $\left(\frac{\text{Spearman's rho}}{\text{Upper bound}}\right)$ as well as the corresponding m (i.e. number of values taken by Z) for trees whose sizes are controlled by the minimum number of observations (Min. nb of obs.) required in the terminal nodes

Min. nb of obs.	m	Spearman's rho	Upper bound	$\frac{\text{Spearman's rho}}{\text{Upper bound}}$
35 000	2	0.0582118	0.2467594	0.2359052
30 000/25 000	3	0.0665856	0.2543916	0.2617447
20 000	4	0.0723529	0.2840077	0.2547567
15 000	5	0.0784423	0.2912251	0.2693530
10 000	7	0.0790575	0.2912251	0.2714653
5000	14	0.0854499	0.2992722	0.2855257
1000	66	0.0965594	0.3050679	0.3165177

claim frequency data. However, they do not compare to 1 (that cannot be attained) but with the upper bounds ranging from 16% to 20% (depending on m). This may lead the analyst to a different conclusion than the one deduced from its normalized version obtained by dividing Kendall's tau by its corresponding upper bound.

Notice that the values for Kendall's tau depicted in Table 6.3 are training sample estimates, so that they are overly optimistic and favor larger regression trees (i.e. larger m). Hence, validation sample estimates for Kendall's tau should be smaller than the already small values presented in Table 6.3, showing even more the need to compare values of Kendall's tau with the highest attainable values and not 1.

Remark 6.2.10 In Table 6.3, training sample estimates for Kendall's tau more than double from the simplest model ($m = 2$) to the most complex one ($m = 66$) while training sample estimates for its normalized version only increase by 70.5%. This is due to the fact that the values of the upper bound also increase with m, like the training sample estimates for Kendall's tau. In a way, one can say that training sample estimates for normalized Kendall's tau penalize the model's complexity.

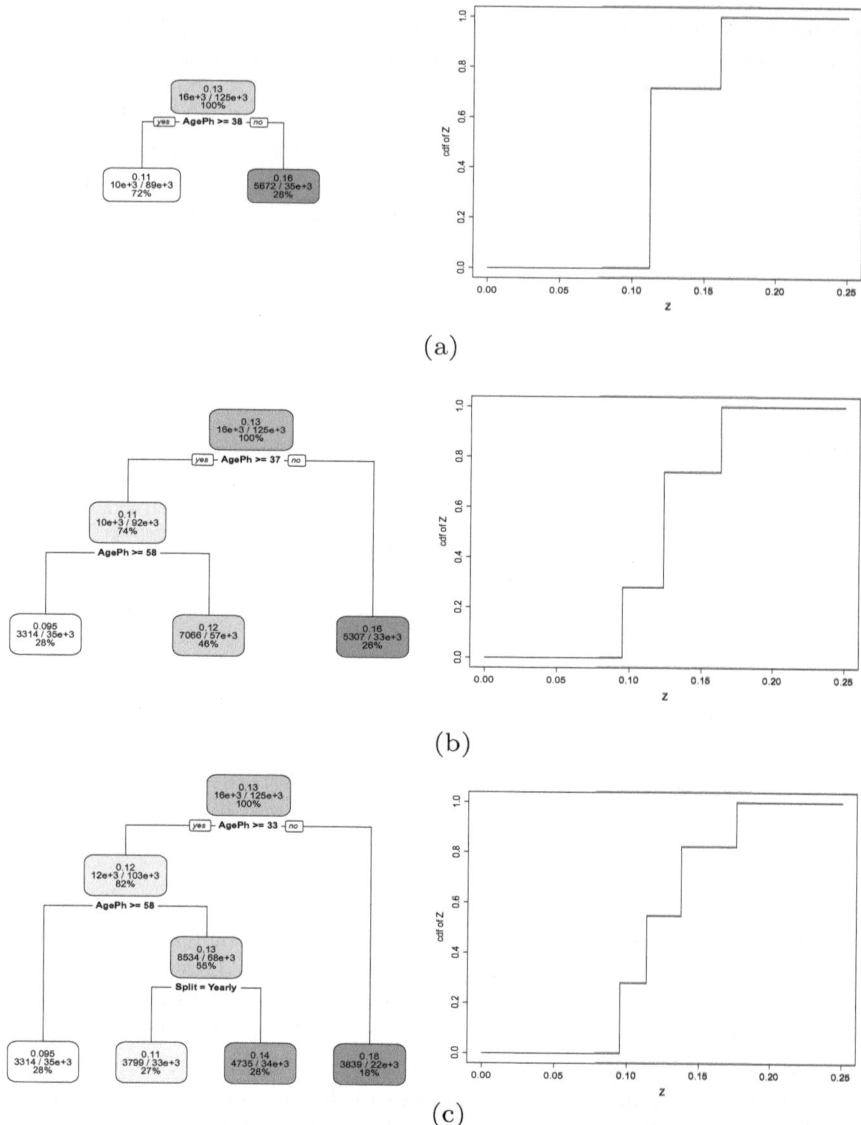

Fig. 6.2 **a** Regression tree (on the left) and distribution function of Z (on the right) when the minimum number of observations per final node is set to 35 000. **b** Regression tree (on the left) and distribution function of Z (on the right) when the minimum number of observations per final node is set to 30 000 (same results when minimum number of observations per node set to 25 000). **c** Regression tree (on the left) and distribution function of Z (on the right) when the minimum number of observations per final node is set to 20 000

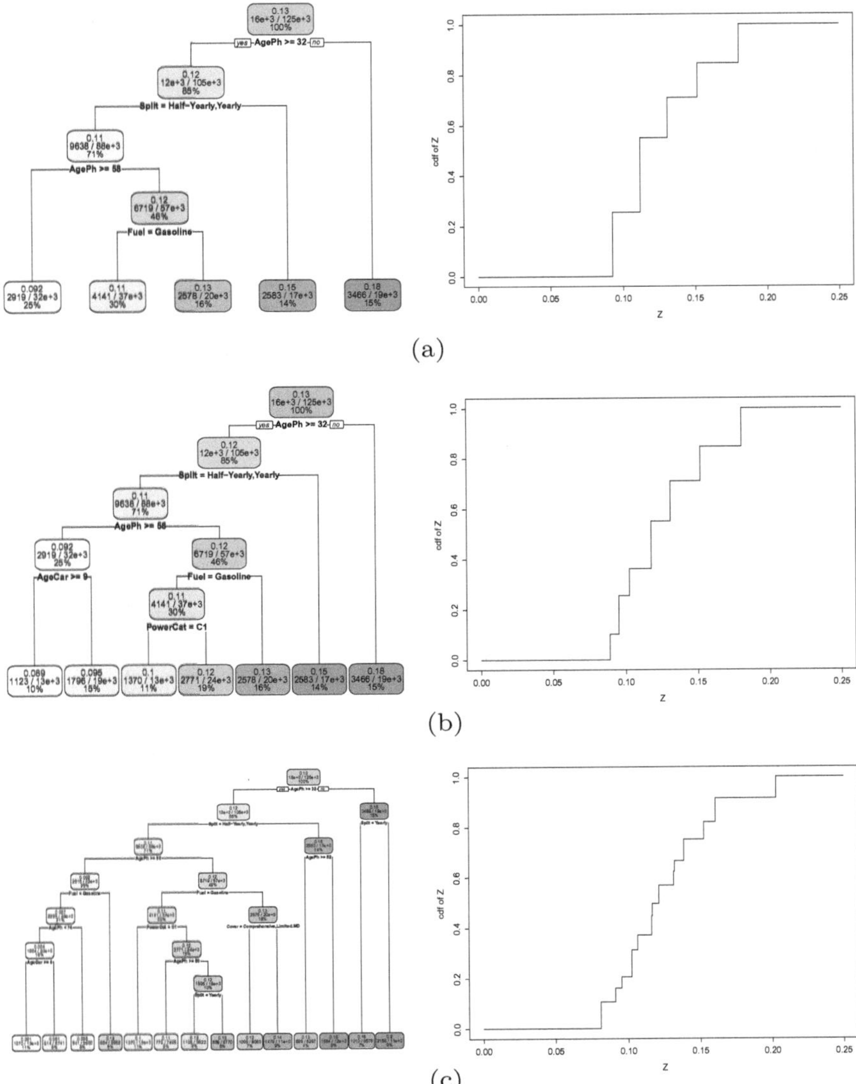

Fig. 6.3 **a** Regression tree (on the left) and distribution function of Z (on the right) when the minimum number of observations per final node is set to 15 000. **b** Regression tree (on the left) and distribution function of Z (on the right) when the minimum number of observations per final node is set to 10 000. **c** Regression tree (on the left) and distribution function of Z (on the right) when the minimum number of observations per final node is set to 5 000

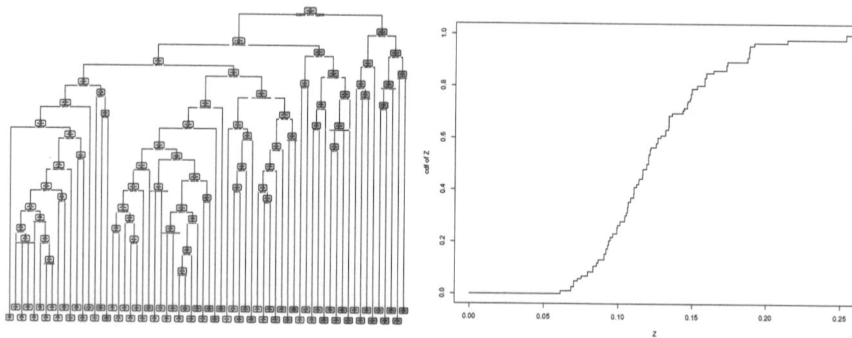

Fig. 6.4 Regression tree (on the left) and cdf of Z (on the right) when minimum number of observations = 1000

Considering the values displayed in Table 6.4, we see that the values for Spearman's rho are between 5% and 10%, which are also rather small compared to 1. Again, this may lead the analyst to a different conclusion than the one deduced from its normalized version, whose values range between 23% and 32%. As for Kendall's tau, the upper bound increases with the number of values m taken by Z to finally converge to the upper bound we get when Z is continuous. Also, the values for Spearman's rho shown in Table 6.4 are training sample estimates, so that they are overly optimistic and favor larger regression trees.

6.3 Measuring Lift

6.3.1 Motivation

We need actuarial measures of quality, especially for comparing different pricing models. In insurance applications, more economic criteria must be used to decide whether a model is worth to be implemented or not. A very accurate, and thus costly to maintain pricing model offering limited gains compared to the existing, simple one, will certainly be abandoned. The notion of lift proves to be useful in that respect.

Actuaries now resort to the concept of lift to evaluate a given model, or to compare competing models. Lift measures the model's ability to prevent adverse selection. Precisely, it quantifies the model's ability to charge each insured an actuarially fair rate, thereby minimizing the potential for loosing business attracted by competitors using finer price lists.

Measuring lift is an important component of model validation: once a predictive model has been built, it is essential to determine its performance of predicting the true premium $\mu(\boldsymbol{X})$ given the available features $\boldsymbol{X}$. Notice that the response Y itself does not play a direct role in the determination of the premium (beyond the definition of

$\mu(\boldsymbol{X})$ and the calibration of the supervised regression model delivering the prices $\widehat{\mu}$), as departures $Y - \mu(\boldsymbol{X})$ cancel out when averaged over a sufficiently large portfolio (this is the very essence of insurance). The premium $\widehat{\mu}(\boldsymbol{X})$ has to be as close as possible to the true premium $\mu(\boldsymbol{X})$. The very aim of ratemaking is not to predict the actual losses Y but to create accurate estimates of $\mu(\boldsymbol{X})$, which is unobserved. Goodness-of-lift must be measured with the validation set, not with the training set (else, a model over-fitting the training data may appear to provide a high lift).

Let $\widehat{\mu}_1$ and $\widehat{\mu}_2$ be two predictors based on the information contained in $\boldsymbol{X}$. Both are candidate premiums and attempt to predict the true premium $\mu(\boldsymbol{X})$. There are many methods to obtain such predictors, ranging from the classical GLMs to neural networks. Lift charts sort data based on the ratio $R = \widehat{\mu}_1/\widehat{\mu}_2$ to compare the two predictors. This ratio is called the relativity. The procedure is as follows. First, calculate the ratio R for each observation in the validation set and sort the data according to it. Second, bucket the data into equally populated classes. If $\widehat{\mu}_2$ is the new model, to be compared with the current one $\widehat{\mu}_1$, then the superiority of $\widehat{\mu}_2$ over $\widehat{\mu}_1$ is demonstrated by plotting loss ratios corresponding to the old model $\widehat{\mu}_1$. If the buckets with low relativities have lower loss ratios then we have lift, that is, if loss ratios exhibit an increasing trend then $\widehat{\mu}_2$ is preferred over $\widehat{\mu}_1$.

6.3.2 *Predictors Characteristics*

In actuarial pricing, the actuary aims to predict the technical premium $\mu(\boldsymbol{X})$. To this end, a predictor $\widehat{\mu}(\boldsymbol{X})$ is built from the available information $\boldsymbol{X}$. To ease the exposition, we assume that all predictors $\widehat{\mu}(\boldsymbol{X})$ under consideration, as well as the conditional expectation $\mu(\boldsymbol{X})$ are continuous random variables admitting probability density functions. This is generally the case when there is at least one continuous feature comprised in the available information $\boldsymbol{X}$ and the function $\widehat{\mu}$ is a continuously increasing function of a real score. However, this may rule out predictions based on discrete features only, as well as piecewise constant predictors, e.g., a single tree, since in those cases, $\widehat{\mu}(\boldsymbol{X})$ takes only a limited number of values. Now, as actuarial pricing is nowadays based on more sophisticated models (trees being combined into random forests, for instance), this assumption does not really restrict the generality of the approach. Notice that the response Y may be discrete (such as the number of claims, for instance) as the continuity only concerns $\mu(\boldsymbol{X})$ and $\widehat{\mu}(\boldsymbol{X})$.

The predictor is also supposed to be correct on average, that is,

$$\mathrm{E}[\widehat{\mu}(\boldsymbol{X})] = \mathrm{E}[\mu(\boldsymbol{X})] = \mathrm{E}[Y], \tag{6.3.1}$$

and both the response Y and predictor $\widehat{\mu}$ are assumed to be non-negative.

We denote as

$$F_{\widehat{\mu}}(t) = \mathrm{P}[\widehat{\mu}(\boldsymbol{X}) \leq t], \quad t \geq 0,$$

the distribution function of $\widehat{\mu}(\boldsymbol{X})$, as $f_{\widehat{\mu}}$ the corresponding probability density function,

$$F_{\widehat{\mu}}(t) = \int_0^t f_{\widehat{\mu}}(s)\mathrm{d}s, \quad t \geq 0,$$

and as $F_{\widehat{\mu}}^{-1}$ the associated quantile function (or Value-at-Risk) defined as the generalized inverse of $F_{\widehat{\mu}}$, i.e.

$$F_{\widehat{\mu}}^{-1}(\alpha) = \inf\{t | F_{\widehat{\mu}}(t) \geq \alpha\} \text{ for a probability level } \alpha.$$

Since the predictor is continuous, we have

$$F_{\widehat{\mu}}\Big(F_{\widehat{\mu}}^{-1}(\alpha)\Big) = \alpha \text{ for all probability level } \alpha.$$

To evaluate the performances of a predictor $\widehat{\mu}$, the following aspects appear to be relevant:

- the variability of $\widehat{\mu}$, as a more variable $\widehat{\mu}$ induces larger premium differentials, i.e. more lift.
- the ability of $\widehat{\mu}$ to match the true premium for increasing risk profiles.

The first objective can be formalized with the help of the convex order that can be characterized by means of the Lorenz curves. The convex order is often used in applied probability to compare the variability inherent to probability distributions. The second objective is assessed by means of concentration curves.

6.3.3 *Convex Order*

Clearly, the more $\widehat{\mu}(\boldsymbol{X})$ is dispersed, the more information it contains about the true premium. The constant predictor $\widehat{\mu}(\boldsymbol{X}) = \mathrm{E}[Y]$, the least dispersed one, does not bring any information about the relative riskiness of the different policies. Thus, comparing the underlying variability appears to be important in the problem under study.

The convex order is an effective probabilistic tool to assess the dispersion of random variables, beyond simple indicators such as standard deviations.

Definition 6.3.1 Consider two non-negative random variables Z_1 and Z_2. Then, Z_1 is said to be smaller than Z_2 in the convex order, henceforth denoted as $Z_1 \preceq_{\mathrm{cx}} Z_2$, if

$$\mathrm{E}[Z_1] = \mathrm{E}[Z_2] \text{ and } \mathrm{E}[(Z_1 - t)_+] \leq \mathrm{E}[(Z_2 - t)_+] \text{ for all } t \geq 0,$$

where $(z - t)_+$ denotes the positive part of $z - t$, i.e. $(z - t)_+ = \max\{z - t, 0\}$.

This means that the stop-loss premiums for Z_2 dominates the corresponding stop-loss premiums for Z_1 for all deductible levels t. The name convex order comes from the fact that $Z_1 \preceq_{cx} Z_2 \Leftrightarrow \mathrm{E}[g(Z_1)] \leq \mathrm{E}[g(Z_2)]$ for all the convex functions g for which the expectations exist. Moreover,

$$Z_1 \preceq_{cx} Z_2 \Rightarrow \mathrm{Var}[Z_1] \leq \mathrm{Var}[Z_2]. \tag{6.3.2}$$

This explains why $\preceq_{cx}$ is a variability order: it only applies to random variables with the same expected value and compares the dispersion of these variables. The convex order is a more sophisticated comparison than only focusing on the variances, yet (6.3.2) indicates that it agrees with this approach. Henceforth, we can interpret $Z_1 \preceq_{cx} Z_2$ as "Z_2 is more variable than Z_1", keeping in mind that the variability in question extends beyond the simple comparison of standard deviation. For a detailed presentation of the convex order, we refer the reader to Denuit et al. (2005).

From the identity

$$\mathrm{E}[(Z-t)_+] - \mathrm{E}\big[(t-Z)_+\big] = \mathrm{E}[Z] - t,$$

we find

$$Z_1 \preceq_{cx} Z_2 \Leftrightarrow \begin{cases} \mathrm{E}[Z_1] = \mathrm{E}[Z_2], \\ \mathrm{E}\big[(t-Z_1)_+\big] \leq \mathrm{E}\big[(t-Z_2)_+\big], \text{ for all } t \geq 0. \end{cases} \tag{6.3.3}$$

Note that partial integration leads to

$$\mathrm{E}[(Z-t)_+] = \int_t^\infty \big(1 - F_Z(z)\big)\mathrm{d}z \text{ and } \mathrm{E}[(t-Z)_+] = \int_{-\infty}^t F_Z(z)\mathrm{d}z. \tag{6.3.4}$$

Recall that for any random variables Z_1 and Z_2 with equal mean,

$$Z_1 \preceq_{cx} Z_2 \Leftrightarrow \int_\alpha^1 F_{Z_1}^{-1}(u)\mathrm{d}u \leq \int_\alpha^1 F_{Z_2}^{-1}(u)\mathrm{d}u \text{ for all } 0 < \alpha < 1.$$

As $\mathrm{E}[Z_k] = \int_0^1 F_{Z_k}^{-1}(u)\mathrm{d}u$, we also have

$$\begin{aligned} \int_0^\alpha F_{Z_1}^{-1}(u)\mathrm{d}u &= \mathrm{E}[Z_1] - \int_\alpha^1 F_{Z_1}^{-1}(u)\mathrm{d}u \\ &\geq \mathrm{E}[Z_2] - \int_\alpha^1 F_{Z_2}^{-1}(u)\mathrm{d}u \\ &= \int_0^\alpha F_{Z_2}^{-1}(u)\mathrm{d}u \end{aligned}$$

when $Z_1 \preceq_{cx} Z_2$. Notice that we also have

$$\int_0^\alpha F_{Z_k}^{-1}(u)\mathrm{d}u = \int_0^{F_{Z_k}^{-1}(\alpha)} z f_{Z_k}(z)\mathrm{d}z = \mathrm{E}\big[Z_k \mathrm{I}[Z_k \leq F_{Z_k}^{-1}(\alpha)]\big]$$

and

$$\frac{1}{\alpha}\int_0^\alpha F_{Z_k}^{-1}(u)du = \mathrm{E}\big[Z_k\big|Z_k \leq F_{Z_k}^{-1}(\alpha)\big].$$

An important characterization of the convex order is by construction on the same probability space using conditional expectations. Precisely, the random variables Z_1 and Z_2 satisfy $Z_1 \preceq_{\mathrm{cx}} Z_2$ if, and only if, there exist two random variables $\widetilde{Z}_1$ and $\widetilde{Z}_2$, defined on the same probability space, such that $\widetilde{Z}_1$ is distributed as Z_1, $\widetilde{Z}_2$ is distributed as Z_2, and $\{\widetilde{Z}_1, \widetilde{Z}_2\}$ is a martingale, that is, $\mathrm{E}[\widetilde{Z}_2|\widetilde{Z}_1] = \widetilde{Z}_1$ holds almost surely. This directly shows that increasing the number of features is beneficial as $\boldsymbol{X}_1 \subseteq \boldsymbol{X}_2 \Rightarrow \mu(\boldsymbol{X}_1) \preceq_{\mathrm{cx}} \mu(\boldsymbol{X}_2)$. Switching from $\boldsymbol{X}_1$ to the richer information $\boldsymbol{X}_2$ thus produces more dispersed premiums μ.

6.3.4 Concentration Curve

6.3.4.1 Definition

Given a binary response $Y \in \{0, 1\}$, where

$$Y = \begin{cases} 1 \text{ if there has been at least one claim filed against the insurer} \\ 0 \text{ otherwise} \end{cases},$$

Gourieroux and Jasiak (2007, Chap. 4) defined several performance measures for a predictor $\widehat{\mu}(\boldsymbol{X})$ of the mean response $\mathrm{E}[Y|\boldsymbol{X}] = \mathrm{P}[Y = 1|\boldsymbol{X}]$. These performance measures for the predictor $\widehat{\mu}$ for a binary response Y were based on the two curves

$$\alpha \mapsto \frac{\mathrm{P}[Y = 1|\widehat{\mu}(\boldsymbol{X}) \leq F_{\widehat{\mu}}^{-1}(\alpha)]}{\mathrm{P}[Y = 1]} = \frac{\mathrm{E}[Y|\widehat{\mu}(\boldsymbol{X}) \leq F_{\widehat{\mu}}^{-1}(\alpha)]}{\mathrm{E}[Y]}$$

and

$$\alpha \mapsto \mathrm{P}[\widehat{\mu}(\boldsymbol{X}) \leq F_{\widehat{\mu}}^{-1}(\alpha)|Y = 1].$$

The former curve gives the proportion of policies reporting at least one claim despite their associated prediction was low (precisely, among the $100\alpha\%$ smaller ones). The latter corresponds to the proportion of policies with small predictions among those with at least one claim reported. Notice that

$$\mathrm{P}[\widehat{\mu}(\boldsymbol{X}) \leq t|Y = 1] = \frac{\mathrm{P}[Y = 1|\widehat{\mu}(\boldsymbol{X}) \leq t]}{\mathrm{P}[Y = 1]}\mathrm{P}[\widehat{\mu}(\boldsymbol{X}) \leq t]$$

so that multiplying the first curve with α, we obtain the second one.

Considering that the identities

$$\frac{\mathrm{P}[Y=1|\widehat{\mu}(\boldsymbol{X})\leq t]}{\mathrm{P}[Y=1]}=\frac{\mathrm{E}[Y|\widehat{\mu}(\boldsymbol{X})\leq t]}{\mathrm{E}[Y]}$$

and

$$\mathrm{P}[\widehat{\mu}(\boldsymbol{X})\leq t|Y=1]=\frac{\mathrm{E}\big[Y\mathrm{I}[\widehat{\mu}(\boldsymbol{X})\leq t]\big]}{\mathrm{E}[Y]}$$

are both valid for binary responses, this suggests to base the performance measure of the predictor on the concentration curve of the response with respect to the predictor, whose definition is recalled next.

Definition 6.3.2 The concentration curve of the response Y with respect to the predictor $\widehat{\mu}$ based on the information contained in the vector $\boldsymbol{X}$ is defined as

$$\mathrm{CC}[Y,\widehat{\mu}(\boldsymbol{X});\alpha]=\frac{\mathrm{E}\big[Y\mathrm{I}[\widehat{\mu}(\boldsymbol{X})\leq F_{\widehat{\mu}}^{-1}(\alpha)]\big]}{\mathrm{E}[Y]}$$

for a probability level α.

In words, the concentration curve $\mathrm{CC}[Y,\widehat{\mu}(\boldsymbol{X});\alpha]$ measures the proportion of the total losses Y attributable to the sub-portfolio gathering a given percentage α of policies with the lower predictions. For an exhaustive review of the properties of the absolute concentration curve, we refer the interested reader to the book by Yitzhaki and Schechtman (2013).

The next result gives equivalent expressions for the concentration curve.

Property 6.3.3 *The concentration curve of the response Y with respect to the predictor $\widehat{\mu}$ based on the information contained in the vector $\boldsymbol{X}$ can be equivalently rewritten as*

$$\begin{aligned}
CC[Y,\widehat{\mu}(\boldsymbol{X});\alpha] &= \frac{1}{E[Y]}\int_0^{F_{\widehat{\mu}}^{-1}(\alpha)} E\big[Y|\widehat{\mu}(\boldsymbol{X})=t\big]f_{\widehat{\mu}}(t)dt && (6.3.5)\\
&= \frac{E\big[Y|\widehat{\mu}(\boldsymbol{X})\leq F_{\widehat{\mu}}^{-1}(\alpha)\big]}{E[Y]}\times\alpha && (6.3.6)\\
&= CC\big[\mu(\boldsymbol{X}),\widehat{\mu}(\boldsymbol{X});\alpha\big] && (6.3.7)\\
&= \frac{Cov\big[Y,I[\widehat{\mu}(\boldsymbol{X})\leq F_{\widehat{\mu}}^{-1}(\alpha)]\big]}{E[Y]}+\alpha && (6.3.8)
\end{aligned}$$

for every probability level α.

Proof Let $f_{(Y,\widehat{\mu})}$ denote the joint probability density function of the pair $(Y,\widehat{\mu}(\boldsymbol{X}))$. Then,

$$\mathrm{E}\big[Y|\widehat{\mu}(\boldsymbol{X})=t\big]=\int_0^{\infty}y\frac{f_{(Y,\widehat{\mu})}(y,t)}{f_{\widehat{\mu}}(t)}\mathrm{d}y$$

so that

$$\int_0^{F_{\widehat{\mu}}^{-1}(\alpha)} \mathrm{E}\big[Y|\widehat{\mu}(\boldsymbol{X})=t\big] f_{\widehat{\mu}}(t)\mathrm{d}t = \int_0^{F_{\widehat{\mu}}^{-1}(\alpha)} \int_0^{\infty} y f_{(Y,\widehat{\mu})}(y,t)\mathrm{d}y\mathrm{d}t$$
$$= \mathrm{E}\big[Y\mathrm{I}[\widehat{\mu}(\boldsymbol{X}) \leq F_{\widehat{\mu}}^{-1}(\alpha)]\big].$$

This establishes (6.3.5). To get (6.3.6), it suffices to note that

$$\mathrm{P}[Y \leq y|\widehat{\mu}(\boldsymbol{X}) \leq t] = \frac{\mathrm{P}[Y \leq y, \widehat{\mu}(\boldsymbol{X}) \leq t]}{\mathrm{P}[\widehat{\mu}(\boldsymbol{X}) \leq t]} = \frac{1}{\mathrm{P}[\widehat{\mu}(\boldsymbol{X}) \leq t]} \int_0^y \int_0^t f_{(Y,\widehat{\mu})}(z,s)\mathrm{d}z\mathrm{d}s$$

so that

$$\mathrm{E}\big[Y|\widehat{\mu}(\boldsymbol{X}) \leq t\big] = \int_0^{\infty} y\mathrm{dP}[Y \leq y, \widehat{\mu}(\boldsymbol{X}) \leq t]$$
$$= \frac{1}{\mathrm{P}[\widehat{\mu}(\boldsymbol{X}) \leq t]} \int_0^{\infty} \int_0^t y f_{(Y,\widehat{\mu})}(y,s)\mathrm{d}y\mathrm{d}s$$
$$= \frac{\mathrm{E}\big[Y\mathrm{I}[\widehat{\mu}(\boldsymbol{X}) \leq t]\big]}{\mathrm{P}[\widehat{\mu}(\boldsymbol{X}) \leq t]}.$$

Then, (6.3.7) follows from

$$\mathrm{E}\big[Y\mathrm{I}[\widehat{\mu}(\boldsymbol{X}) \leq t]\big] = \mathrm{E}\Big[\mathrm{E}\big[Y\mathrm{I}[\widehat{\mu}(\boldsymbol{X}) \leq t]\big|\boldsymbol{X}\big]\Big] = \mathrm{E}\Big[\mu(\boldsymbol{X})\mathrm{I}[\widehat{\mu}(\boldsymbol{X}) \leq t]\Big].$$

Finally, (6.3.8) is easily obtained from the definition of the concentration curve as

$$\mathrm{Cov}\big[Y, \mathrm{I}[\widehat{\mu}(\boldsymbol{X}) \leq F_{\widehat{\mu}}^{-1}(\alpha)]\big] = \mathrm{E}\big[Y\mathrm{I}[\widehat{\mu}(\boldsymbol{X}) \leq F_{\widehat{\mu}}^{-1}(\alpha)]\big] - \alpha\mathrm{E}[Y].$$

This ends the proof. □

Formula (6.3.7) shows that we can equivalently replace the response Y with the pure premium $\mu(\boldsymbol{X})$ in the concentration curve. This property is of utmost importance as the actuary is interested in the pure premium, which is unknown, but is allowed to replace it with the actual response values in the evaluation of the performances of the predictor under consideration.

6.3.4.2 Positive Dependence Structures

If the predictor brings a lot of information about the technical premium, or equivalently about the response, this means that these random variables are strongly correlated. The shape of the concentration curve of the response with respect to its predictor thus depends on the kind of positive relationship between the response and the predictor. This is why we recall the definition of the following positive dependence notions.

Definition 6.3.4 Consider two non-negative random variables Z_1 and Z_2.

(i) The random variable Z_1 is stochastically increasing on Z_2 if

$$z_2 \mapsto \mathrm{P}[Z_1 \leq z_1 | Z_2 = z_2] \text{ is non-increasing in } z_2,$$

for all $z_1 \geq 0$.

(ii) The random variable Z_1 is positively regression dependent on Z_2 if

$$z_2 \mapsto \mathrm{E}[Z_1 | Z_2 = z_2] \text{ is non-decreasing in } z_2.$$

(iii) The random variable Z_1 is left-tail decreasing on Z_2 if

$$z_2 \mapsto \mathrm{P}[Z_1 \leq z_1 | Z_2 \leq z_2] \text{ is non-increasing in } z_2,$$

for all $z_1 \geq 0$.

(iv) The random variable Z_1 is positively left-tail expectation dependent on Z_2 if

$$z_2 \mapsto \mathrm{E}[Z_1 | Z_2 \leq z_2] \text{ is non-decreasing in } z_2.$$

(v) The random variable Z_1 is positively expectation dependent on Z_2 if

$$\mathrm{E}[Z_1] \geq \mathrm{E}[Z_1 | Z_2 \leq z_2]$$

for all $z_2 \geq 0$.

Let us give some intuitive explanation about these different concepts. Considering stochastic increasingness, we see that the condition defining this positive dependence notion expresses the fact that the probability that Z_1 is small, in the sense that Z_1 falls below the threshold z_1, decreases as Z_2 gets larger. Intuitively speaking, Z_1 thus tends to become larger when Z_2 increases. Positive regression dependence is a weaker concept as

$$\mathrm{E}[Z_1 | Z_2 = z_2] = \int_0^\infty \mathrm{P}[Z_1 > z_1 | Z_2 = z_2] dz_1$$

is obviously increasing when Z_1 is stochastically increasing in Z_2. Positive regression dependence ensures that, on average, Z_1 gets larger when Z_2 increases. The next three concepts are defined by conditionings of the form "$Z_2 \leq z_2$" instead of "$Z_2 = z_2$". Intuitively speaking, "$Z_2 \leq z_2$" means that Z_2 is small. Left-tail decreasingness and left-tail expectation dependence are the counterparts to stochastic increasigness and positive regression dependence for these alternative conditionings. The last concept, called expectation dependence, expresses the fact that the knowledge that Z_2 is small, i.e. Z_2 is below the threshold z_2, decreases Z_1 on average compared to the situation where there is no information about Z_1.

6.3.4.3 Properties of the Concentration Curve

Monotonicity
The concentration curve is based on the function

$$t \mapsto \frac{\mathrm{E}\big[\mu(\boldsymbol{X})\mathrm{I}[\widehat{\mu}(\boldsymbol{X}) \leq t]\big]}{\mathrm{E}[Y]}$$

evaluated at quantiles of $\widehat{\mu}(\boldsymbol{X})$. This function is a distribution function, starting from $(0, 0)$ to reach $(1, 1)$, being non-decreasing and right-continuous. Therefore, $\alpha \mapsto \mathrm{CC}[\mu(\boldsymbol{X}), \widehat{\mu}(\boldsymbol{X}); \alpha]$ is non-decreasing and satisfies

$$\lim_{\alpha \to 0} \mathrm{CC}[\mu(\boldsymbol{X}), \widehat{\mu}(\boldsymbol{X}); \alpha] = 0 \text{ and } \lim_{\alpha \to 1} \mathrm{CC}[\mu(\boldsymbol{X}), \widehat{\mu}(\boldsymbol{X}); \alpha] = 1.$$

Line of Independence
In the particular case where the predictor brings no information about the response, in the sense that Y and $\widehat{\mu}(\boldsymbol{X})$ are mutually independent, then the concentration curve is the 45-degree line, often referred to as the line of independence in the literature. Formally, if Y and $\widehat{\mu}(\boldsymbol{X})$ are independent, then

$$\mathrm{CC}[\mu(\boldsymbol{X}), \widehat{\mu}(\boldsymbol{X}); \alpha] = \frac{\mathrm{E}[Y]\mathrm{P}\big[\widehat{\mu}(\boldsymbol{X}) \leq F_{\widehat{\mu}}^{-1}(\alpha)\big]}{\mathrm{E}[Y]} = \alpha.$$

Let us now study the position of the concentration curve with respect to the 45-degree line. If $\widehat{\mu}$ brings a lot of information about the true premium $\mu(\boldsymbol{X})$, this means that these random variables are strongly related and the concentration curve should be far from the line of independence. Furthermore, the shape of the concentration curve depends on the kind of relationship between $\mu(\boldsymbol{X})$ and $\widehat{\mu}(\boldsymbol{X})$. The next result shows that under weak positive dependence, every concentration curve lies below the independence line.

Property 6.3.5 *If $\mu(\boldsymbol{X})$ is positively expectation dependent on $\widehat{\mu}(\boldsymbol{X})$, that is, if the inequality*

$$E[\mu(\boldsymbol{X})] \geq E\big[\mu(\boldsymbol{X})\big|\widehat{\mu}(\boldsymbol{X}) \leq t\big]$$

holds for all t, then the concentration curve lies below the 45-degree line, i.e.

$$CC[\mu(\boldsymbol{X}), \widehat{\mu}(\boldsymbol{X}); \alpha] \leq \alpha \textit{ for all probability levels } \alpha.$$

Proof It suffices to write

$$\begin{aligned} \frac{\mathrm{E}\big[\mu(\boldsymbol{X})\mathrm{I}[\widehat{\mu}(\boldsymbol{X}) \leq t]\big]}{\mathrm{E}[Y]} &= \frac{\mathrm{P}[\widehat{\mu}(\boldsymbol{X}) \leq t]\mathrm{E}\big[\mu(\boldsymbol{X})\big|\widehat{\mu}(\boldsymbol{X}) \leq t\big]}{\mathrm{E}[Y]} \\ &\leq \mathrm{P}[\widehat{\mu}(\boldsymbol{X}) \leq t]. \end{aligned}$$

The announced then follows by replacing t with $F_{\widehat{\mu}}^{-1}(\alpha)$. □

Convexity

The next result states a positive dependence condition under which the concentration curve is convex. Again, the shape of the curve depends on the kind of relationship existing between the response and the predictor.

Property 6.3.6 *The concentration curve $\alpha \mapsto CC[\mu(\boldsymbol{X}), \widehat{\mu}(\boldsymbol{X}); \alpha]$ is convex if, and only if, $\mu(\boldsymbol{X})$ is positively regression dependent on $\widehat{\mu}(\boldsymbol{X})$, that is, if the function*

$$t \mapsto E\big[\mu(\boldsymbol{X})\big|\widehat{\mu}(\boldsymbol{X}) = t\big] \tag{6.3.9}$$

is non-decreasing.

Proof Let us start from the representation

$$\mathrm{E}\big[\mu(\boldsymbol{X})\mathrm{I}[\widehat{\mu}(\boldsymbol{X}) \leq t]\big] = \int_0^\infty \int_0^t u f_{(\mu,\widehat{\mu})}(u, s)\mathrm{d}u\mathrm{d}s,$$

where $f_{(\mu,\widehat{\mu})}$ denotes the joint probability density function of the pair $(\mu(\boldsymbol{X}), \widehat{\mu}(\boldsymbol{X}))$. Then, the first derivative of the selection curve is given by

$$\begin{aligned}\frac{\mathrm{d}}{\mathrm{d}\alpha}\mathrm{CC}[\mu(\boldsymbol{X}), \widehat{\mu}(\boldsymbol{X}); \alpha] &= \int_0^\infty u f_{(\mu,\widehat{\mu})}\big(u, F_{\widehat{\mu}}^{-1}(\alpha)\big)\mathrm{d}u \frac{1}{f_{\widehat{\mu}}\big(F_{\widehat{\mu}}^{-1}(\alpha)\big)} \\ &= \int_0^\infty u f_{\mu|\widehat{\mu}}\big(u|F_{\widehat{\mu}}^{-1}(\alpha)\big)\mathrm{d}u\end{aligned}$$

where $f_{\mu|\widehat{\mu}}(\cdot|t)$ denotes the conditional probability density function of the true premium $\mu(\boldsymbol{X})$, given $\widehat{\mu}(\boldsymbol{X}) = t$. Hence,

$$\frac{\mathrm{d}}{\mathrm{d}\alpha}\mathrm{CC}[\mu(\boldsymbol{X}), \widehat{\mu}(\boldsymbol{X}); \alpha] = \mathrm{E}[\mu(\boldsymbol{X})|\widehat{\mu}(\boldsymbol{X}) = F_{\widehat{\mu}}^{-1}(\alpha)].$$

We thus see that this derivative is increasing (and hence the primitive is convex) if, and only if, $\mu(\boldsymbol{X})$ is positively regression dependent on the score $\widehat{\mu}(\boldsymbol{X})$, as announced. This ends the proof. □

The convexity of the concentration curve ensures that the increments of the function

$$\mathrm{CC}[Y, \widehat{\mu}(\boldsymbol{X}); \alpha + \Delta] - \mathrm{CC}[Y, \widehat{\mu}(\boldsymbol{X}); \alpha] = \frac{\mathrm{E}\big[Y\mathrm{I}[F_{\widehat{\mu}}^{-1}(\alpha) < \widehat{\mu}(\boldsymbol{X}) \leq F_{\widehat{\mu}}^{-1}(\alpha + \Delta)]\big]}{\mathrm{E}[Y]}$$

are non-decreasing in α, for every positive Δ such that $\alpha + \Delta \leq 1$. This means that the sub-portfolios created by isolating a proportion Δ of policies with predictions comprised between $F_{\widehat{\mu}}^{-1}(\alpha)$ and $F_{\widehat{\mu}}^{-1}(\alpha + \Delta)$ brings an increasing share of the losses, on average as α increases.

This property is in relation with lift charts as described in Tevet (2013). To draw such graphs, the data set is sorted based on the values of $\widehat{\mu}(\boldsymbol{X})$. The data are then bucketed into equally populated classes based on quantiles. Within each bucket, the average predicted loss is calculated with the help of the predictor $\widehat{\mu}$ as well as the actual loss cost Y. The average predicted and average actual loss costs are then graphed for each class.

To assess the reasonableness of $\widehat{\mu}$, the analyst checks whether the actual loss costs monotonically increase as we move to higher buckets (by definition, this will be the case for the predicted loss costs). Property 6.3.6 precisely identifies the condition required to observe an increasing trend in such a lift chart.

The monotonicity condition imposed on $t \mapsto \mathrm{E}[\mu(\boldsymbol{X})|\widehat{\mu}(\boldsymbol{X}) = t]$ appears to be reasonable. However, this condition is not necessarily fulfilled for a given feature X_j, i.e. $t \mapsto \mathrm{E}[\mu(\boldsymbol{X})|X_j = t]$ is not necessarily non-decreasing. For instance, considering as response, the number of claims filed against the insurer in a motor third-party liability cover, the impact of policyholder's age on the expected claim frequency often exhibits a U-shape, which invalidates the monotonicity of the conditional expectation. Re-arranging the values of the feature is nevertheless possible, as explained in Shaked et al. (2012). But such a procedure makes the analysis less transparent. This is why we condition here on $\widehat{\mu}(\boldsymbol{X})$, which maps the vector of features to the prediction and induces a total order among risk profiles.

As positive regression dependence implies positive expectation dependence, the condition of Property 6.3.6 ensures that the concentration curve is non-decreasing and convex, starting from (0, 0) to end at (1, 1) with a graph everywhere below the 45-degree line.

6.3.4.4 Estimation

Assuming the observations $(Y_i, \boldsymbol{X}_i)$, $i = 1, \ldots, n$, to be independent and identically distributed, the concentration curve $\mathrm{CC}[\mu(\boldsymbol{X}), \widehat{\mu}(\boldsymbol{X}); \alpha]$ can be estimated as follows:

$$
\begin{aligned}
\widehat{\mathrm{CC}}\big[\mu(\boldsymbol{X}), \widehat{\mu}(\boldsymbol{X}); \alpha\big] &= \widehat{\mathrm{CC}}[Y, \widehat{\mu}(\boldsymbol{X}); \alpha] \\
&= \frac{1}{n\overline{Y}} \sum_{i|\widehat{\mu}(\boldsymbol{X}_i) \leq \widehat{F}_{\widehat{\mu}}^{-1}(\alpha)} Y_i \\
&= \frac{\sum_{i|\widehat{\mu}(\boldsymbol{X}_i) \leq \widehat{F}_{\widehat{\mu}}^{-1}(\alpha)} Y_i}{\sum_{i=1}^{n} Y_i},
\end{aligned}
$$

where

$$
\overline{Y} = \frac{\sum_{i=1}^{n} Y_i}{n}.
$$

Here, $\widehat{F_{\widehat{\mu}}}$ denotes the empirical distribution function of the resulting predictions, i.e.

$$\widehat{F}_{\widehat{\mu}}(t) = \frac{1}{n}\sum_{i=1}^{n} \mathrm{I}[\widehat{\mu}(\boldsymbol{X}_i) \le t].$$

The empirical counterpart $\widehat{\mathrm{CC}}$ to the population concentration curve CC can be interpreted as follows: a ratio of the total loss produced by those policies with predictor $\widehat{\mu}$ below its empirical quantile at level α and the aggregate loss of the entire portfolio. It means that $\widehat{\mathrm{CC}}$ expresses this total sub-portfolio loss in relative terms, as a percentage of the aggregate loss at the entire portfolio level.

6.3.5 *Assessing the Performances of a Given Predictor*

6.3.5.1 From Premiums to Ranks

Notice that

$$\widehat{\mu}(\boldsymbol{X}) \le F_{\widehat{\mu}}^{-1}(\alpha) \Leftrightarrow F_{\widehat{\mu}}\big(\widehat{\mu}(\boldsymbol{X})\big) = \alpha.$$

This means that it is enough to consider the ranking induced by the predictor, that is, we are free to replace every predictor $\widehat{\mu}(\boldsymbol{X})$ with the corresponding rank

$$M = F_{\widehat{\mu}}\big(\widehat{\mu}(\boldsymbol{X})\big)$$

obeying the unit uniform distribution. The intuitive meaning of M is as follows: M is the rank of a policyholder, once all contracts have been ordered according to their corresponding premiums (in ascending order).

Gourieroux and Jasiak (2007) were interested in credit scoring, that is, in the initial selection of applicants based on their propensity to reimburse their loan. Hence, the actual values of the predictor do no matter, only the rank they induce and the threshold defining acceptance or rejection of the application. In insurance pricing, however, the actual values $\widehat{\mu}(\boldsymbol{X})$ are also important and this information is captured by the Lorenz curve.

6.3.5.2 Lorenz Curve

The Lorenz curves are used in economics to measure the inequality of incomes. Intuitively speaking, the more incomes are variable in a given population, the less egalitarian it is. Lorenz curves are thus intimately related to the convex order. The Lorenz curve LC associated to the predictor $\widehat{\mu}(\boldsymbol{X})$ is defined by

$$\mathrm{LC}[\widehat{\mu}(\boldsymbol{X});\alpha] = \frac{1}{\mathrm{E}[\widehat{\mu}(\boldsymbol{X})]}\int_0^\alpha F_{\widehat{\mu}}^{-1}(u)\mathrm{d}u$$
$$= \frac{\mathrm{E}\big[\widehat{\mu}(\boldsymbol{X})\mathrm{I}[\widehat{\mu}(\boldsymbol{X}) \leq F_{\widehat{\mu}}^{-1}(\alpha)]\big]}{\mathrm{E}[\widehat{\mu}(\boldsymbol{X})]}$$

for a given probability level α.

The Lorenz curve is based on the function

$$t \mapsto \frac{\mathrm{E}\big[\widehat{\mu}(\boldsymbol{X})\mathrm{I}[\widehat{\mu}(\boldsymbol{X}) \leq t]\big]}{\mathrm{E}[\widehat{\mu}(\boldsymbol{X})]}$$

that can be seen as a distribution function. Hence, $\alpha \mapsto \mathrm{LC}[\widehat{\mu}(\boldsymbol{X});\alpha]$ is non-decreasing, starting from $(0, 0)$ to reach $(1, 1)$. Clearly,

$$\mathrm{LC}[\widehat{\mu}(\boldsymbol{X});\alpha] = \mathrm{CC}[\widehat{\mu}(\boldsymbol{X}), \widehat{\mu}(\boldsymbol{X});\alpha]$$

so that every Lorenz curve is convex by virtue of Property 6.3.6.

Assuming the observations $(Y_i, \boldsymbol{X}_i)$, $i = 1, \ldots, n$, to be independent and identically distributed, the empirical version of the Lorenz curve is obtained as

$$\widehat{\mathrm{LC}}\big(\widehat{\mu}(\boldsymbol{X});\alpha\big) = \frac{\sum_{i|\widehat{\mu}(\boldsymbol{X}_i)\leq \widehat{F}_{\widehat{\mu}}^{-1}(\alpha)} \widehat{\mu}(\boldsymbol{X}_i)}{\sum_{i=1}^n \widehat{\mu}(\boldsymbol{X}_i)}.$$

In words, $\widehat{\mathrm{LC}}$ is the percentage of the total premium income corresponding to the $100\alpha\%$ smaller premiums when the latter are computed using a predictor $\widehat{\mu}$. If the global balance $\sum_{i=1}^n \widehat{\mu}(\boldsymbol{X}_i) = \sum_{i=1}^n Y_i$ holds true (which is the case with GLMs as long as canonical link functions are used, for instance) then $\widehat{\mathrm{LC}}$ also expresses this proportion with respect to the aggregate loss at the entire portfolio level.

Remark 6.3.7 Considering $\mathrm{LC}[\widehat{\mu}(\boldsymbol{X});\alpha]$, the 45-degree line now refers to another limit case, called the line of equality. Assume that the predictor is constant, that is, $\widehat{\mu}(\boldsymbol{X}) = \mathrm{E}[Y]$. This may be due to the fact that none of the feature contained in $\boldsymbol{X}$ is related to the response Y (hence the link with the line of independence). Then, we have

$$\widehat{\mu}(\boldsymbol{X}) = \mathrm{E}[Y] \Rightarrow \mathrm{LC}[\widehat{\mu}(\boldsymbol{X});\alpha] = \alpha$$

so that this particular case corresponds to the 45-degree line. Notice that we must consider this limit case with some care because the constant predictor is not a continuous random variable (which invalidates some of the formulas derived earlier).

6.3.5.3 Canonical Predictors

If an actuary has access to the true premium $\mu(\boldsymbol{X})$ based on the information contained in $\boldsymbol{X}$ then there is no need to distinguish CC from LC. This is because if $\widehat{\mu}(\boldsymbol{X}) = \mu(\boldsymbol{X})$

then

$$\mathrm{LC}[\widehat{\mu}(\boldsymbol{X});\alpha]=\mathrm{CC}[\mu(\boldsymbol{X}),\widehat{\mu}(\boldsymbol{X});\alpha]$$

for all probability levels α.

In other words, the two performance curves reduce to the Lorenz curve of $\mu(\boldsymbol{X})$. This means that the sub-portfolio corresponding to $\widehat{\mu}(\boldsymbol{X})\leq F_{\widehat{\mu}}^{-1}(\alpha)$ is in equilibrium, as the premium matches the conditional expectation on average. A large difference between the two performance curves thus suggests that the predictor under consideration poorly approximates the true technical premium.

This explains why many empirical studies only use the Lorenz curve of $\widehat{\mu}(\boldsymbol{X})$ to evaluate performance of a predictive model. There is a confusion between the predictor $\widehat{\mu}$ and the conditional expectation whereas it is very likely that $\widehat{\mu}(\boldsymbol{X})$ only approximates $\mu(\boldsymbol{X})$, being different from it in reality. Because of this difference, we can resort to the pair of curves $\mathrm{CC}[\mu(\boldsymbol{X}),\widehat{\mu}(\boldsymbol{X});\alpha]$ and $\mathrm{LC}[\widehat{\mu}(\boldsymbol{X});\alpha]$ to evaluate the performance of a pricing model.

Remark 6.3.8 Many empirical studies also use Gini coefficients based on the Lorenz curve of the predictor. The reason for using this indicator is the following.

The Gini mean difference is one possible measure of variability defined for a non-negative continuous random variable Z as

$$\mathrm{Gini}[Z]=\mathrm{E}\big[|Z_1-Z_2|\big]=\mathrm{E}\big[\max\{Z_1,Z_2\}\big]-\mathrm{E}\big[\min\{Z_1,Z_2\}\big]$$

where Z_1 and Z_2 are independent and distributed as Z. It represents the average absolute difference between two observations distributed as Z. This is closely related to the variance, which can equivalently be expressed as

$$\mathrm{Var}[Z]=\frac{1}{2}\mathrm{E}\big[(Z_1-Z_2)^2\big].$$

If Z is continuous then it can be shown that

$$\mathrm{Gini}[Z]=4\mathrm{Cov}\big[Z,F_Z(Z)\big].$$

Thus, Gini mean difference measures the association between a random variable and its rank. In other words, considering a sequence of observations ranked in ascending order, Gini mean difference quantifies the relationship between the actual value of Z and its position in the sequence. Clearly, the more variability in Z, the larger its actual value when its rank is high (i.e. it appears among the largest observations) whereas a lower Gini mean difference indicates that the observations are more concentrated around their central value.

This formula relates the Gini mean difference to the Lorenz curve. Starting from the lower absolute deviation of Z_1, defined as $\mathrm{E}\big[(t-Z_1)\mathrm{I}[Z_1\leq t]\big]$, let us replace t with Z_2 to get

$$\mathrm{E}\big[(Z_2 - Z_1)\mathrm{I}[Z_1 \leq Z_2]\big] = \frac{1}{2}\mathrm{E}\big[|Z_1 - Z_2|\big] = \frac{1}{2}\mathrm{Gini}[Z] = 2\mathrm{Cov}\big[Z, F_Z(Z)\big].$$

Now, the area between the identity line and the Lorenz curve for Z is $\frac{1}{\mathrm{E}[Z]}\mathrm{Cov}[Z, F_Z(Z)]$. Therefore, the higher the Gini mean difference, the further the Lorenz curve of the predictor from the 45-degree line (i.e. from the Lorenz curve of the uninformative predictor constantly equal to $\mathrm{E}[Y]$). This is why candidate premiums with larger Gini mean difference tend to be preferred.

Notice that the Gini coefficient is the Gini mean difference divided by twice the mean. It is also known as the concentration ratio and represents the area between the 45-degree line and the actual Lorenz curve divided by the area between the 45-degree line and the Lorenz curve that yields the maximal value that this index can have.

6.3.5.4 Measuring Goodness-of-Lift

The performances of a predictor $\widehat{\mu}(\boldsymbol{X})$ can thus be assessed by means of the respective positions of the two curves

$$\alpha \mapsto \mathrm{LC}[\widehat{\mu}(\boldsymbol{X}); \alpha] \text{ and } \alpha \mapsto \mathrm{CC}[\mu(\boldsymbol{X}), \widehat{\mu}(\boldsymbol{X}); \alpha].$$

The first one represents the share of premiums collected from the $100\alpha\%$ of policies from the portfolio with the lowest $\widehat{\mu}(\boldsymbol{X})$ values. The second one gives the corresponding share of the true premium that should have been collected from this sub-portfolio.

As the total expected income of $\widehat{\mu}$ and μ match the total expected loss by (6.3.1), both ratios are directly comparable since we can compare two percentages: the one obtained with $\widehat{\mu}(\boldsymbol{X})$ with the true one corresponding to $\mu(\boldsymbol{X})$. As actuaries, we would like that the graph of the concentration curve is as close as possible to the graph of the Lorenz curve. In other words, the smaller the area between the two curves the better.

Gini mean difference measures the area between the 45-degree line and the Lorenz curve. As explained earlier, an alternative is to consider both the Lorenz curve of the predictor and the concentration curve of the true premium with respect to the predictor. More precisely, defining a distance between the two curves $\mathrm{CC}[\mu(\boldsymbol{X}), \widehat{\mu}(\boldsymbol{X}); \alpha]$ and $\mathrm{LC}[\widehat{\mu}(\boldsymbol{X}); \alpha]$ would be relevant, knowing that they coincide when $\widehat{\mu}(\boldsymbol{X}) = \mu(\boldsymbol{X})$. This is why the area between the concentration curve and the Lorenz curve turns out to be another good candidate for assessing the performance of a given predictor. This area between the curves, ABC in short, is given by

$$\begin{aligned}
\mathrm{ABC}[\widehat{\mu}(\boldsymbol{X})] &= \int_0^1 \Big(\mathrm{CC}[Y, \widehat{\mu}(\boldsymbol{X}); \alpha] - \mathrm{LC}[\widehat{\mu}(\boldsymbol{X}); \alpha]\Big)\mathrm{d}\alpha \\
&= \frac{1}{\mathrm{E}[\widehat{\mu}(\boldsymbol{X})]}\int_0^1 \Big(\mathrm{E}\big[Y\mathrm{I}[M \leq \alpha]\big] - \mathrm{E}\big[\widehat{\mu}(\boldsymbol{X})\mathrm{I}[M \leq \alpha]\big]\Big)\mathrm{d}\alpha \\
&= \frac{1}{\mathrm{E}[\widehat{\mu}(\boldsymbol{X})]}\int_0^1\int_0^\infty \Big(\mathrm{P}\big[\widehat{\mu}(\boldsymbol{X}) \leq y, M \leq \alpha\big] - \mathrm{P}\big[Y \leq y, M \leq \alpha\big]\Big)\mathrm{d}y\mathrm{d}\alpha
\end{aligned}$$

$$= \frac{1}{\mathrm{E}[\widehat{\mu}(\boldsymbol{X})]}\Big(\mathrm{Cov}\big[\widehat{\mu}(\boldsymbol{X}), M\big] - \mathrm{Cov}\big[Y, M\big]\Big) \tag{6.3.10}$$

where we recognize the difference between the Gini mean difference of the predictor $\widehat{\mu}(\boldsymbol{X})$ (up to the factor 4) and the Gini covariance of the response Y and the predictor $\widehat{\mu}(\boldsymbol{X})$. Let us notice that (6.3.10) can be rewritten as

$$\mathrm{ABC}[\widehat{\mu}(\boldsymbol{X})] = \frac{1}{\mathrm{E}[\widehat{\mu}(\boldsymbol{X})]}\mathrm{Cov}\big[\widehat{\mu}(\boldsymbol{X}) - Y, M\big].$$

Hence, if we think about $\widehat{\mu}(\boldsymbol{X}) - Y$ as the profit associated with a policy, then ABC may be interpreted to be proportional to the covariance between profits and the rank of premiums. This interpretation is similar to the one given in Frees et al. (2013) for the Gini index.

Remark 6.3.9 In the case where $\mu(\boldsymbol{X})$ and $\widehat{\mu}(\boldsymbol{X})$ are comonotonic, we have $\mathrm{CC}[\mu(\boldsymbol{X}), \widehat{\mu}(\boldsymbol{X}); \alpha] = \mathrm{LC}[\widehat{\mu}(\boldsymbol{X}); \alpha]$ for all $\alpha \in (0, 1)$ if and only if $\mu(\boldsymbol{X}) = \widehat{\mu}(\boldsymbol{X})$. Indeed, in such a case

$$\mu(\boldsymbol{X}) = F_{\mu}^{-1}(M)$$

where F_{μ}^{-1} is the quantile function associated to the distribution function F_{μ} of $\mu(\boldsymbol{X})$. Therefore, since $\mathrm{E}[\mu(\boldsymbol{X})] = \mathrm{E}[\widehat{\mu}(\boldsymbol{X})]$, we get $\mathrm{CC}[\mu(\boldsymbol{X}), \widehat{\mu}(\boldsymbol{X}); \alpha] = \mathrm{LC}[\widehat{\mu}(\boldsymbol{X}); \alpha]$ for all $\alpha \in (0, 1)$ if and only if

$$\begin{aligned}
&\mathrm{E}\big[\mu(\boldsymbol{X})\mathrm{I}[\widehat{\mu}(\boldsymbol{X}) \leq F_{\widehat{\mu}}^{-1}(\alpha)]\big] = \mathrm{E}\big[\widehat{\mu}(\boldsymbol{X})\mathrm{I}[\widehat{\mu}(\boldsymbol{X}) \leq F_{\widehat{\mu}}^{-1}(\alpha)]\big] \\
&\Leftrightarrow \mathrm{E}\big[F_{\mu}^{-1}(M)\mathrm{I}[F_{\widehat{\mu}}^{-1}(M) \leq F_{\widehat{\mu}}^{-1}(\alpha)]\big] = \mathrm{E}\big[F_{\widehat{\mu}}^{-1}(M)\mathrm{I}[F_{\widehat{\mu}}^{-1}(M) \leq F_{\widehat{\mu}}^{-1}(\alpha)]\big] \\
&\Leftrightarrow \int_0^{\alpha} F_{\mu}^{-1}(u)du = \int_0^{\alpha} F_{\widehat{\mu}}^{-1}(u)du.
\end{aligned} \tag{6.3.11}$$

Thus, since (6.3.11) must be fulfilled for all $\alpha \in (0, 1)$, it comes $F_{\mu}^{-1}(u) = F_{\widehat{\mu}}^{-1}(u)$ for all $u \in (0, 1)$ and so $\mu(\boldsymbol{X}) = \widehat{\mu}(\boldsymbol{X})$.

6.3.5.5 Assessing the Performances of Low-Risk Portfolios

Alternatively, we can also create a sub-portfolio gathering all the policies such that $\widehat{\mu}(\boldsymbol{X}) \leq F_{\widehat{\mu}}^{-1}(\alpha)$ and only consider this one, in isolation. The result of each portfolio is described by the performance curve

$$\Big(\mathrm{E}\big[\widehat{\mu}(\boldsymbol{X})\big|\widehat{\mu}(\boldsymbol{X}) \leq F_{\widehat{\mu}}^{-1}(\alpha)\big], \mathrm{E}\big[\mu(\boldsymbol{X})\big|\widehat{\mu}(\boldsymbol{X}) \leq F_{\widehat{\mu}}^{-1}(\alpha)\big]\Big).$$

Notice that we are allowed to replace the true premium $\mu(\boldsymbol{X})$ with the actual loss Y. The first component corresponds to the conditional lower-tail expectation of the predictor, defined as

$$\text{CLTE}[\widehat{\mu}(\boldsymbol{X}); \alpha] = \text{E}\big[\widehat{\mu}(\boldsymbol{X})\big|\widehat{\mu}(\boldsymbol{X}) \leq F_{\widehat{\mu}}^{-1}(\alpha)\big].$$

The second component looks like the conditional lower-tail expectation of the true premium

$$\text{CLTE}[\mu(\boldsymbol{X}); \alpha] = \text{E}\big[\mu(\boldsymbol{X})\big|\mu(\boldsymbol{X}) \leq F_{\mu}^{-1}(\alpha)\big]$$

except that the condition involves the predictor and not the true premium.

Clearly,

$$\alpha \mapsto \text{CLTE}[\widehat{\mu}(\boldsymbol{X}); \alpha]$$

is non-decreasing. But this is not necessarily the case for $\alpha \mapsto \text{E}\big[Y\big|\widehat{\mu}(\boldsymbol{X}) \leq F_{\widehat{\mu}}^{-1}(\alpha)\big]$ unless $\mu(\boldsymbol{X})$ is positively lower-tail expectation dependent on $\widehat{\mu}(\boldsymbol{X})$. Positive expectation dependence allows us to constrain the shape of this curve. If $\mu(\boldsymbol{X})$ and $\widehat{\mu}(\boldsymbol{X})$ are positively expectation dependent then

$$\lim_{t\to\infty} \text{E}[\mu(\boldsymbol{X})|\widehat{\mu}(\boldsymbol{X}) \leq t] = \text{E}[\mu(\boldsymbol{X})] \geq \text{E}[\mu(\boldsymbol{X})|\widehat{\mu}(\boldsymbol{X}) \leq t].$$

The next result gives the condition ensuring the monotonicity of $\alpha \mapsto \text{E}\big[\mu(\boldsymbol{X})\big|\widehat{\mu}(\boldsymbol{X}) \leq F_{\widehat{\mu}}^{-1}(\alpha)\big]$.

Property 6.3.10 *If $\mu(\boldsymbol{X})$ is positively lower-tail expectation dependent on $\widehat{\mu}(\boldsymbol{X})$ then*

$$\alpha \mapsto E\big[\mu(\boldsymbol{X})\big|\widehat{\mu}(\boldsymbol{X}) \leq F_{\widehat{\mu}}^{-1}(\alpha)\big]$$

is non-decreasing. A sufficient condition is provided by left-tail decreasingness.

Proof We see from its very definition that if the response is positively lower-tail expectation dependent on $\widehat{\mu}(\boldsymbol{X})$ then $t \mapsto \text{E}[\mu(\boldsymbol{X})|\widehat{\mu}(\boldsymbol{X}) \leq t]$ is non-decreasing. Furthermore, as

$$\begin{aligned}&\text{E}[\mu(\boldsymbol{X})|\widehat{\mu}(\boldsymbol{X}) \leq t] - \text{E}[\mu(\boldsymbol{X})|\widehat{\mu}(\boldsymbol{X}) \leq s]\\ &= \int_0^{\infty} \big(\text{P}[\mu(\boldsymbol{X}) \leq u|\widehat{\mu}(\boldsymbol{X}) \leq s] - \text{P}[\mu(\boldsymbol{X}) \leq u|\widehat{\mu}(\boldsymbol{X}) \leq t]\big)\text{d}u\end{aligned}$$

a sufficient condition for positive lower-tail expectation dependence is left-tail decreasingness. This dependence notion is stronger than positively quadrant dependence but weaker than stochastic increasingness. □

The next result establishes a lower bound for the second component.

Property 6.3.11 *For any predictor $\widehat{\mu}$, we have*

$$CLTE[\mu(\boldsymbol{X}); \alpha] \leq E\big[\mu(\boldsymbol{X})\big|\widehat{\mu}(\boldsymbol{X}) \leq F_{\widehat{\mu}}^{-1}(\alpha)\big] \textit{ for all } \alpha.$$

Proof Let us first establish that $\text{E}[\mu(\boldsymbol{X})|A] \geq \text{E}[\mu(\boldsymbol{X})|\mu(\boldsymbol{X}) \leq u]$ for any random event A such that $\text{P}[\mu(\boldsymbol{X}) \leq u] = \text{P}[A]$. This comes from

$$\begin{aligned}
\mathrm{E}[\mu(\boldsymbol{X})|\mu(\boldsymbol{X}) \leq u] &= u + \mathrm{E}[\mu(\boldsymbol{X}) - u|\mu(\boldsymbol{X}) \leq u, A]\mathrm{P}[A|\mu(\boldsymbol{X}) \leq u] \\
&\quad + \mathrm{E}[\mu(\boldsymbol{X}) - u|\mu(\boldsymbol{X}) \leq u, \overline{A}]\mathrm{P}[\overline{A}|\mu(\boldsymbol{X}) \leq u] \\
&\leq u + \mathrm{E}[\mu(\boldsymbol{X}) - u|\mu(\boldsymbol{X}) \leq u, A]\mathrm{P}[A|\mu(\boldsymbol{X}) \leq u] \\
&= u + \mathrm{E}[\mu(\boldsymbol{X}) - u|\mu(\boldsymbol{X}) \leq u, A]\mathrm{P}[\mu(\boldsymbol{X}) \leq u|A] \\
&\leq u + \mathrm{E}[\mu(\boldsymbol{X}) - u|\mu(\boldsymbol{X}) \leq u, A]\mathrm{P}[\mu(\boldsymbol{X}) \leq u|A] \\
&\quad + \mathrm{E}[\mu(\boldsymbol{X}) - u|\mu(\boldsymbol{X}) > u, A]\mathrm{P}[\mu(\boldsymbol{X}) > u|A] \\
&= \mathrm{E}[\mu(\boldsymbol{X})|A].
\end{aligned}$$

The announced result then follows by taking $\widehat{\mu}(\boldsymbol{X}) \leq F_{\widehat{\mu}}^{-1}(\alpha)$ for A and $u = F_{\mu}^{-1}(\alpha)$. □

As $\mu(\boldsymbol{X}) \preceq_{\mathrm{cx}} Y$ and $\widehat{\mu}(\boldsymbol{X})$ aims to predict $\mu(\boldsymbol{X})$, it is reasonable to assume that $\widehat{\mu}(\boldsymbol{X}) \preceq_{\mathrm{cx}} Y$ also holds true. Then, we know that the inequality

$$\mathrm{CLTE}[Y; \alpha] \leq \mathrm{CLTE}[\widehat{\mu}(\boldsymbol{X}); \alpha] \text{ holds for all } \alpha,$$

i.e. that we have

$$\mathrm{E}\big[\mu(\boldsymbol{X})\big|\mu(\boldsymbol{X}) \leq F_{\mu}^{-1}(\alpha)\big] \leq \mathrm{E}\big[\widehat{\mu}(\boldsymbol{X})\big|\widehat{\mu}(\boldsymbol{X}) \leq F_{\widehat{\mu}}^{-1}(\alpha)\big] \text{ for all } \alpha$$

since

$$\mathrm{CLTE}[\mu(\boldsymbol{X}); \alpha] = \mathrm{CLTE}[Y; \alpha].$$

Property 6.3.11 indicates that $\mathrm{E}\big[\mu(\boldsymbol{X})\big|\widehat{\mu}(\boldsymbol{X}) \leq F_{\widehat{\mu}}^{-1}(\alpha)\big]$ lies above $\mathrm{CLTE}[\mu(\boldsymbol{X}); \alpha]$ and thus it may cross $\mathrm{CLTE}[\widehat{\mu}(\boldsymbol{X}); \alpha]$. The corresponding sub-portfolio is therefore in disequilibrium, with true premiums above the actual premiums, on average.

Remark 6.3.12 If the information is so rich that $\mu(\boldsymbol{X})$ and $\widehat{\mu}(\boldsymbol{X})$ are comonotonic, i.e.

$$\mu(\boldsymbol{X}) = F_{\mu}^{-1}(M),$$

then we have equality in Property 6.3.11.

6.3.6 Comparison of the Performances of Two Predictors

Consider two predictors $\widehat{\mu}_1$ and $\widehat{\mu}_2$. Both attempt to predict the unknown pure premium $\mu(\boldsymbol{X})$. These predictors may differ in their functional form ($\widehat{\mu}_1$ instead of $\widehat{\mu}_2$) and/or in the information ($\boldsymbol{X}_1$ instead of $\boldsymbol{X}_2$) on which they are based. There are two important aspects when evaluating the performances of two predictors. First, their respective variability, i.e. their ability to identify different risk profiles. Second, their correlation with $\mu(\boldsymbol{X})$, i.e., the amount of information they bring in about the true premium.

6.3.6.1 Variability

The convex order appears to be the appropriate tool to measure the degree of lift induced by a predictor. This probabilistic tool indeed assesses the differentiation between the cheapest and costliest risk profiles identified by the model. In that respect, replacing the predictor with a more variable one, based on the convex order, appears to be a promising strategy.

Property 6.3.13 *If $\widehat{\mu}_2(\boldsymbol{X}_2) \preceq_{cx} \widehat{\mu}_1(\boldsymbol{X}_1)$ holds then*

(i) $\min[\widehat{\mu}_1(\boldsymbol{X}_1)] \leq \min[\widehat{\mu}_2(\boldsymbol{X}_2)]$ *and* $\max[\widehat{\mu}_1(\boldsymbol{X}_1)] \geq \max[\widehat{\mu}_2(\boldsymbol{X}_2)]$ *so that* $\widehat{\mu}_1(\boldsymbol{X}_1)$ *has a wider range than* $\widehat{\mu}_2(\boldsymbol{X}_2)$.
(ii) for any $\alpha < \beta$,

$$E\big[\widehat{\mu}_1(\boldsymbol{X}_1)\big|\widehat{\mu}_1(\boldsymbol{X}_1) > F_{\widehat{\mu}_1}^{-1}(\beta)\big] - E\big[\widehat{\mu}_1(\boldsymbol{X}_1)\big|\widehat{\mu}_1(\boldsymbol{X}_1) \leq F_{\widehat{\mu}_1}^{-1}(\alpha)\big]$$

$$\geq E\big[\widehat{\mu}_2(\boldsymbol{X}_2)\big|\widehat{\mu}_2(\boldsymbol{X}_2) > F_{\widehat{\mu}_2}^{-1}(\beta)\big] - E\big[\widehat{\mu}_2(\boldsymbol{X}_2)\big|\widehat{\mu}_2(\boldsymbol{X}_2) \leq F_{\widehat{\mu}_2}^{-1}(\alpha)\big].$$

Proof The proof of (i) is by contradiction. Suppose, for example, that $\max[\widehat{\mu}_1(\boldsymbol{X}_1)] < \max[\widehat{\mu}_2(\boldsymbol{X}_2)]$. Let t be such that $\max[\widehat{\mu}_1(\boldsymbol{X}_1)] < t < \max[\widehat{\mu}_2(\boldsymbol{X}_2)]$. Then

$$\mathrm{E}[(\widehat{\mu}_1(\boldsymbol{X}_1) - t)_+] = 0 < \mathrm{E}[(\widehat{\mu}_2(\boldsymbol{X}_2) - t)_+],$$

in contradiction to $\widehat{\mu}_2(\boldsymbol{X}_2) \preceq_{\mathrm{cx}} \widehat{\mu}_1(\boldsymbol{X}_1)$. Therefore we must have $\max[\widehat{\mu}_1(\boldsymbol{X}_1)] \geq \max[\widehat{\mu}_2(\boldsymbol{X}_2)]$. Similarly, it can be shown that $\min[\widehat{\mu}_1(\boldsymbol{X}_1)] \leq \min[\widehat{\mu}_2(\boldsymbol{X}_2)]$.

Considering (ii), we know that

$$\widehat{\mu}_2(\boldsymbol{X}_2) \preceq_{\mathrm{cx}} \widehat{\mu}_1(\boldsymbol{X}_1) \Leftrightarrow \begin{cases} \mathrm{E}\big[\widehat{\mu}_1(\boldsymbol{X}_1)\big|\widehat{\mu}_1(\boldsymbol{X}_1) > F_{\widehat{\mu}_1}^{-1}(\beta)\big] \geq \mathrm{E}\big[\widehat{\mu}_2(\boldsymbol{X}_2)\big|\widehat{\mu}_2(\boldsymbol{X}_2) > F_{\widehat{\mu}_2}^{-1}(\beta)\big] \\ \text{for all probability levels } \beta \\ \mathrm{E}\big[\widehat{\mu}_1(\boldsymbol{X}_1)\big|\widehat{\mu}_1(\boldsymbol{X}_1) \leq F_{\widehat{\mu}_1}^{-1}(\alpha)\big] \leq \mathrm{E}\big[\widehat{\mu}_2(\boldsymbol{X}_2)\big|\widehat{\mu}_2(\boldsymbol{X}_2) \leq F_{\widehat{\mu}_2}^{-1}(\alpha)\big] \\ \text{for all probability levels } \alpha \end{cases}$$

The announced result then follows by combining these two inequalities. □

6.3.6.2 More Positive Expectation Dependence

The respective distribution functions of the two predictors $\widehat{\mu}_1$ and $\widehat{\mu}_2$ are denoted as $F_{\widehat{\mu}_1}$ and $F_{\widehat{\mu}_2}$. Both $F_{\widehat{\mu}_k}$ are assumed to be continuous and strictly increasing, $k = 1, 2$. Define the scores $M_1 = F_{\widehat{\mu}_1}(\widehat{\mu}_1)$ and $M_2 = F_{\widehat{\mu}_2}(\widehat{\mu}_2)$ that are both uniformly distributed over the unit interval $[0, 1]$.

The more M_k is correlated to Y, the more information the corresponding predictor $\widehat{\mu}_k$ contains. More informative predictors thus lead to greater variability of the conditional expectation $\mathrm{E}[Y|M]$. This is formally stated in the next result established by Muliere and Petrone (1992) in their study of dependence orderings based on generalized Lorenz curves.

Property 6.3.14 *Assume that the functions $\alpha \mapsto E[Y|M_k = \alpha]$ are continuous and strictly increasing for $k \in \{1, 2\}$. Then*

$$E[Y|M_2] \preceq_{cx} E[Y|M_1] \Leftrightarrow E[Y|M_1 \geq \alpha] \geq E[Y|M_2 \geq \alpha] \text{ for all } \alpha.$$

Here, $\mathrm{E}[Y|M_k]$ measures how the rank M_k induced by the predictor $\widehat{\mu}_k$ explains the response Y. If all the ranks are equal, i.e. $\widehat{\mu}_k(\boldsymbol{X}) = \mathrm{E}[Y]$, then $\mathrm{E}[Y|M_k] = \mathrm{E}[Y]$ and the predictor does not bring any information about the response.

The next result shows that under the assumptions of Property 6.3.14 the mean square error of prediction (MSEP) is smaller with M_1 compared to M_2.

Property 6.3.15 *Under the conditions of Property (6.3.14), we have*

$$E\Big[\big(Y - E[Y|M_1]\big)^2\Big] \leq E\Big[\big(Y - E[Y|M_2]\big)^2\Big],$$

that is, Y is closer to $E[Y|M_1]$ in the L^2-norm.

Proof The announced result is a direct consequence of the convex inequality $\mathrm{E}[Y|M_2] \preceq_{\mathrm{cx}} \mathrm{E}[Y|M_1]$ since

$$\mathrm{Var}[Y] = \mathrm{E}\Big[\mathrm{Var}[Y|M_i]\Big] + \mathrm{Var}\Big[\mathrm{E}[Y|M_i]\Big] \text{ and } \mathrm{E}\Big[\mathrm{Var}[Y|M_i]\Big] = \mathrm{E}\Big[\big(Y - \mathrm{E}[Y|M_i]\big)^2\Big]$$

hold for $i = 1, 2$, so that

$$\mathrm{E}[Y|M_2] \preceq_{\mathrm{cx}} \mathrm{E}[Y|M_1] \Rightarrow \mathrm{Var}\Big[\mathrm{E}[Y|M_2]\Big] \leq \mathrm{Var}\Big[\mathrm{E}[Y|M_1]\Big].$$

This ends the proof. □

6.3.6.3 Discriminatory Power

The performance and selection curves are useful to evaluate the value of a given predictor $\widehat{\mu}(\boldsymbol{X})$. These curves are also helpful to compare the performances of different scores. Following Gourieroux and Jasiak (2007, Definition 4.5), we adopt the following comparison rule.

Definition 6.3.16 The predictor $\widehat{\mu}_1(\boldsymbol{X}_1)$ is more discriminatory than the predictor $\widehat{\mu}_2(\boldsymbol{X}_2)$ for a response Y if, and only if, the inequalities

$$\mathrm{E}[\widehat{\mu}_1(\boldsymbol{X}_1)|\widehat{\mu}_1(\boldsymbol{X}_1) \leq F^{-1}_{\widehat{\mu}_1}(\alpha)] \leq \mathrm{E}[\widehat{\mu}_2(\boldsymbol{X}_2)|\widehat{\mu}_2(\boldsymbol{X}_2) \leq F^{-1}_{\widehat{\mu}_2}(\alpha)]$$

and

$$\mathrm{E}[Y|\widehat{\mu}_1(\boldsymbol{X}_1) \leq F_{\widehat{\mu}_1}^{-1}(\alpha)] \leq \mathrm{E}[Y|\widehat{\mu}_2(\boldsymbol{X}_2) \leq F_{\widehat{\mu}_2}^{-1}(\alpha)]$$

both hold for all probability levels α.

The first condition is fulfilled if, and only if $\widehat{\mu}_2(\boldsymbol{X}_2) \preceq_{\mathrm{cx}} \widehat{\mu}_1(\boldsymbol{X}_1)$. The second condition can be rewritten as

$$\mathrm{E}[Y|\widehat{\mu}_1(\boldsymbol{X}_1) \leq F_{\widehat{\mu}_1}^{-1}(\alpha)] \leq \mathrm{E}[Y|\widehat{\mu}_2(\boldsymbol{X}_2) \leq F_{\widehat{\mu}_2}^{-1}(\alpha)]$$

$$\Leftrightarrow \mathrm{E}[Y|M_1 \leq \alpha] \leq \mathrm{E}[Y|M_2 \leq \alpha].$$

This amounts to requiring that Y is more positively expectation dependent on M_1 than on M_2, in the sense that the reduction in the expectation resulting from the knowledge that $M_k \leq \alpha$ is larger for M_1 compared to M_2. This can be seen from the inequality

$$\mathrm{E}[Y] - \mathrm{E}[Y|M_1 \leq \alpha] \geq \mathrm{E}[Y] - \mathrm{E}[Y|M_2 \leq \alpha].$$

This is equivalent to the corresponding condition appearing in Property 6.3.14 as the identity

$$\mathrm{E}[Y] = \alpha\mathrm{E}[Y|M_k \leq \alpha] + (1-\alpha)\mathrm{E}[Y|M_1 > \alpha]$$

holds for $k \in \{1, 2\}$.

The second condition in Definition 6.3.16 can be expressed in terms of the concordance order. Two random variables are said to be concordant if they tend to be all large together or small together. The concordance order expresses the idea that large and small values tend to be more often associated under the distribution that dominates the other one.

Definition 6.3.17 Let us consider two random couples (Z_1, Z_2) and (V_1, V_2) with the same marginal distributions, i.e. $\mathrm{P}[Z_k \leq t] = \mathrm{P}[V_k \leq t] = F_k(t)$ for $k \in \{1, 2\}$. If

$$\mathrm{P}[Z_1 \leq t_1, Z_2 \leq t_2] \leq \mathrm{P}[V_1 \leq t_1, V_2 \leq t_2] \text{ for all } t_1 \text{ and } t_2, \quad (6.3.12)$$

or, equivalently, if

$$\mathrm{P}[Z_1 > t_1, Z_2 > t_2] \leq \mathrm{P}[V_1 > t_1, V_2 > t_2] \text{ for all } t_1 \text{ and } t_2, \quad (6.3.13)$$

then (Z_1, Z_2) is said to be less concordant that (V_1, V_2). This is henceforth denoted as $(Z_1, Z_2) \preceq_{\mathrm{conc}} (V_1, V_2)$.

The intuitive meaning of a ranking with respect to $\preceq_{\mathrm{conc}}$ is clear from Definition 6.3.17. Indeed, $\mathrm{P}[Z_1 \leq t_1, Z_2 \leq t_2]$ and $\mathrm{P}[V_1 \leq t_1, V_2 \leq t_2]$ read as "Z_1 and Z_2 are both small" and "V_1 and V_2 are both small", respectively (small meaning that Z_1, resp. V_1, is smaller than the threshold t_1 and Z_2, resp. V_2, is smaller than the threshold t_2). So, (6.3.12) means that when $\boldsymbol{X} \preceq_{\mathrm{conc}} \boldsymbol{Y}$ holds, the probability that V_1 and V_2

are both small is larger than the corresponding probability for Z_1 and Z_2. Similarly from (6.3.13), $(Z_1, Z_2) \preceq_{\text{conc}} (V_1, V_2)$ also ensures that the probability that Z_1 and Z_2 are both large is smaller than the corresponding probability for V_1 and V_2. This corresponds to the intuitive content of "(V_1, V_2) being more positively dependent than (Z_1, Z_2)".

We have the following result in terms of covariances.

Proposition 6.3.18 *For random couples (Z_1, Z_2) and (V_1, V_2) with the same marginal distributions, we have*

$$(Z_1, Z_2) \preceq_{conc} (V_1, V_2) \Leftrightarrow Cov[g_1(Z_1), g_2(Z_2)] \leq Cov[g_1(V_1), g_2(V_2)]$$

for all the non-decreasing functions g_1 and g_2, provided the expectations exist.

Proposition 6.3.18 shows when $\preceq_{\text{conc}}$ holds, the correlations between $g_1(Z_1)$ and $g_2(Z_2)$ are less than between $g_1(V_1)$ and $g_2(V_2)$ for all increasing functions g_1 and g_2. Furthermore, Pearson's correlation coefficient as well as Kendall's and Spearman's rank correlation coefficients all agree with a ranking in the $\preceq_{\text{conc}}$-sense. This reinforces the intuitive meaning of $\preceq_{\text{conc}}$ as a tool to compare the strength of the dependence.

The next result gives a sufficient condition in terms of the concordance order.

Property 6.3.19 *If*

$$\widehat{\mu}_2(\boldsymbol{X}_2) \preceq_{cx} \widehat{\mu}_1(\boldsymbol{X}_1) \text{ and } (Y, M_2) \preceq_{conc} (Y, M_1)$$

then $\widehat{\mu}_1(\boldsymbol{X}_1)$ is more discriminatory than the predictor $\widehat{\mu}_2(\boldsymbol{X}_2)$ for the response Y.

Proof The result follows from

$$\mathrm{E}[Y|M_2 \leq \alpha] - \mathrm{E}[Y|M_1 \leq \alpha] = \frac{1}{\alpha}\int_0^\infty \Big(\mathrm{P}[Y \leq y, M_1 \leq \alpha] - \mathrm{P}[Y \leq y, M_2 \leq \alpha]\Big)\mathrm{d}y$$

which is indeed positive is (Y, M_1) is more concordant than (Y, M_2). □

Thus, we see that $\widehat{\mu}_1(\boldsymbol{X}_1)$ is more discriminatory than $\widehat{\mu}_2(\boldsymbol{X}_2)$ for the response Y if $\widehat{\mu}_1(\boldsymbol{X}_1)$ is simultaneously more variable (in the sense of the convex order) and more correlated (in the sense of positive expectation dependence or the stronger concordance order) with the response Y than $\widehat{\mu}_2(\boldsymbol{X}_2)$.

6.3.6.4 Integrated Concentration and Lorenz Curves

The preference relation proposed in Definition 6.3.16 only forms a partial ranking. Two predictors might well be incomparable because their respective concentration or Lorenz curves intersect: one predictor is better for low risks, and worse for high risks, for example. In such a case, we can base the comparison on the integral of the concentration curves. This amounts to considering the integrated concentration curve defined as

$$\begin{aligned}\mathrm{ICC}[\mu(\boldsymbol{X}), \widehat{\mu}(\boldsymbol{X}); \alpha] &= \int_0^\alpha \mathrm{CC}[\mu(\boldsymbol{X}), \widehat{\mu}(\boldsymbol{X}); \xi] \mathrm{d}\xi \\ &= \int_0^\alpha \frac{\mathrm{E}\big[\mu(\boldsymbol{X})\mathrm{I}[M \leq \xi]\big]}{\mathrm{E}[Y]} \mathrm{d}\xi \\ &= \frac{\mathrm{E}\big[\mu(\boldsymbol{X})(\alpha - M)_+\big]}{\mathrm{E}[Y]} \\ &= \frac{\mathrm{Cov}\big[\mu(\boldsymbol{X}), (\alpha - M)_+\big]}{\mathrm{E}[Y]} + \mathrm{E}\big[(\alpha - M)_+\big].\end{aligned}$$

The first term is driven by the correlation between the response and the predictor whereas the second one is just a constant as M is unit uniformly distributed:

$$\mathrm{E}\big[(\alpha - M)_+\big] = \int_0^\alpha (\alpha - \xi) \mathrm{d}\xi = \frac{\alpha^2}{2}.$$

The integral of the concentration curve over the whole interval [0, 1] is denoted ICC, i.e.

$$\begin{aligned}\mathrm{ICC} &= \mathrm{ICC}[\mu(\boldsymbol{X}), \widehat{\mu}(\boldsymbol{X}); 1] \\ &= \frac{\mathrm{Cov}\big[\mu(\boldsymbol{X}), 1 - M\big]}{\mathrm{E}[Y]} + \frac{1}{2} \\ &= \frac{1}{2} - \frac{\mathrm{Cov}\big[\mu(\boldsymbol{X}), M\big]}{\mathrm{E}[Y]}.\end{aligned}$$

Again, as

$$\begin{aligned}\mathrm{E}\big[Y(\alpha - M)_+\big] &= \mathrm{E}\big[\mathrm{E}[Y(\alpha - M)_+ | \boldsymbol{X}]\big] \\ &= \mathrm{E}\big[\mu(\boldsymbol{X})(\alpha - M)_+\big]\end{aligned}$$

we are allowed to replace $\mu(\boldsymbol{X})$ with Y in the definition of the integrated concentration curve. This means that we can use it to measure performance of $\widehat{\mu}(\boldsymbol{X})$ in predicting the unknown pure premium $\mu(\boldsymbol{X})$.

Let us now provide an intuitive interpretation for ICC. We still consider the $100\alpha\%$ of policies with the smallest $\widehat{\mu}$ values, as $(\alpha - M)_+ = 0$ for $\alpha \leq M$. Now, ICC is based on the covariance between $\mu(\boldsymbol{X})$ and $(\alpha - M)_+$. The idea is that, the smaller M with respect to α (i.e., the larger $(\alpha - M)_+$) the smaller the true premium should be. Hence, a positive relationship between $\widehat{\mu}(\boldsymbol{X})$ and $\mu(\boldsymbol{X})$ translates into a negative covariance between $\mu(\boldsymbol{X})$ and $(\alpha - M)_+$. And the more negative the covariance term entering the decomposition of ICC, the better the corresponding candidate premium.

Proceeding in a similar way with the Lorenz curve, we define the integrated Lorenz curve as

$$\begin{aligned}\text{ILC}[\widehat{\mu}(\boldsymbol{X});\alpha] &= \int_0^\alpha \frac{\text{E}\big[\widehat{\mu}(\boldsymbol{X})\text{I}[M \leq \xi]\big]}{\text{E}[\widehat{\mu}(\boldsymbol{X})]}d\xi \\ &= \frac{\text{E}\big[\widehat{\mu}(\boldsymbol{X})(\alpha - M)_+\big]}{\text{E}[\widehat{\mu}(\boldsymbol{X})]} \\ &= \frac{\text{Cov}\big[\widehat{\mu}(\boldsymbol{X}),(\alpha - M)_+\big]}{\text{E}[\widehat{\mu}(\boldsymbol{X})]} + \text{E}\big[(\alpha - M)_+\big].\end{aligned}$$

The integral of the Lorenz curve over the whole interval [0, 1] is denoted ILC.

The smaller the ILC metrics is, the better the corresponding candidate premium. Similarly, the smaller the ICC metrics, the better. Now, if we consider both metrics simultaneously, then one should prefer a predictor with smaller ICC and ILC metrics, or equivalently with smaller ICC and ABC values since

$$\text{ABC}[\widehat{\mu}(\boldsymbol{X})] = \text{ICC}[\mu(\boldsymbol{X}),\widehat{\mu}(\boldsymbol{X});1] - \text{ILC}[\widehat{\mu}(\boldsymbol{X});1].$$

It is worth noticing that one predictor can be better for ICC and worse for ABC, for instance.

6.3.7 *Ordered Lorenz Curve*

Let $\widehat{\mu}_1(\boldsymbol{X})$ and $\widehat{\mu}_2(\boldsymbol{X})$ be two predictors for a response Y. In ratemaking, these predictors are for the true technical premium $\mu(\boldsymbol{X})$. We can imagine that $\widehat{\mu}_1$ is the current predictor and that we consider replacing it with $\widehat{\mu}_2$ provided the latter's performances are better. In order to compare these two predictors, let us define the relativity as the ratio of the new to the old predictor, that is,

$$R = \frac{\widehat{\mu}_2(\boldsymbol{X})}{\widehat{\mu}_1(\boldsymbol{X})}.$$

If R is less than 1, this means that the risk profile $\boldsymbol{X}$ is overpriced with the current predictor. This profile is thus at risk of adverse selection: a competitor using the predictor $\widehat{\mu}_2$ could offer a better rate to such policyholders who could then leave the portfolio.

As before, both predictors are supposed to be balanced, i.e.

$$\text{E}[\widehat{\mu}_1(\boldsymbol{X})] = \text{E}[\widehat{\mu}_2(\boldsymbol{X})] = \text{E}[Y],$$

and to ease the explanation, we assume that $\widehat{\mu}_1(\boldsymbol{X})$, $\widehat{\mu}_2(\boldsymbol{X})$ and $\mu(\boldsymbol{X})$ are all continuous.

Following Frees et al. (2013), we define the ordered Lorenz curve as the set of points $(\text{CC}[\widehat{\mu}_1(\boldsymbol{X}), R(\boldsymbol{X}); \alpha], \text{CC}[Y, R(\boldsymbol{X}); \alpha])$

$$= \left(\frac{\text{E}\big[\widehat{\mu}_1(\boldsymbol{X})\text{I}[R(\boldsymbol{X}) \leq F_R^{-1}(\alpha)]\big]}{\text{E}[\widehat{\mu}_1(\boldsymbol{X})]}, \frac{\text{E}\big[Y\text{I}[R(\boldsymbol{X}) \leq F_R^{-1}(\alpha)]\big]}{\text{E}[Y]} \right)$$

where F_R denotes the distribution function of the relativity R, and F_R^{-1} the associated quantile function. Notice that

$$\begin{aligned} \text{CC}[Y, R(\boldsymbol{X}); \alpha] &= \text{E}\big[Y\text{I}[R(\boldsymbol{X}) \leq F_R^{-1}(\alpha)]\big] \\ &= \text{E}\Big[\text{E}\big[Y\text{I}[R(\boldsymbol{X}) \leq F_R^{-1}(\alpha)]|\boldsymbol{X}\big]\Big] \\ &= \text{E}\Big[\mu(\boldsymbol{X})\text{I}[R(\boldsymbol{X}) \leq F_R^{-1}(\alpha)]\Big] \\ &= \text{CC}\big[\mu(\boldsymbol{X}), R(\boldsymbol{X}); \alpha\big] \end{aligned}$$

so that we are allowed to replace the true premium $\mu(\boldsymbol{X})$ with the actual loss Y.

Both functions

$$s \mapsto \frac{\text{E}\big[\widehat{\mu}_1(\boldsymbol{X})\text{I}[R(\boldsymbol{X}) \leq s]\big]}{\text{E}[\widehat{\mu}_1(\boldsymbol{X})]}$$

and

$$s \mapsto \frac{\text{E}\big[Y\text{I}[R(\boldsymbol{X}) \leq s]\big]}{\text{E}[Y]}$$

can be interpreted as distribution functions. They give the proportion of total current premiums $\widehat{\mu}_1(\boldsymbol{X})$ and the proportion of total losses Y (or true premiums $\mu(\boldsymbol{X})$) in the sub-portfolio determined by the condition $R(\boldsymbol{X}) \leq s$. The approach is thus based on adverse selection against the insurer. Assume that a competitor attracts all profiles $\boldsymbol{X}$ that are overpriced under the current price list $\widehat{\mu}_0$, i.e. those such that $R(\boldsymbol{X}) \leq s$ for some s small enough. More precisely, in a sub-portfolio gathering all risk profiles such that $R(\boldsymbol{X}) \leq F_R^{-1}(\alpha)$, i.e. the $100\alpha\%$ of policies with the smaller relativities, we record the proportion

$$t_1 = \frac{\text{E}\big[Y\text{I}[R(\boldsymbol{X}) \leq F_R^{-1}(\alpha)]\big]}{\text{E}[Y]}$$

of total losses, for a proportion of

$$t_2 = \frac{\text{E}\big[\widehat{\mu}_1(\boldsymbol{X})\text{I}[R(\boldsymbol{X}) \leq F_R^{-1}(\alpha)]\big]}{\text{E}[\widehat{\mu}_1(\boldsymbol{X})]}$$

of premium income. Considering the point (t_1, t_2) of the ordered Lorenz curve, corresponding to the particular α, its meaning is as follows. By forming a portfolio with all policyholders whose relativities $R(\boldsymbol{X})$ are less than $F_R^{-1}(\alpha)$, i.e. all policies for which the new premium $\widehat{\mu}_2(\boldsymbol{X})$ is smaller than $F_R^{-1}(\alpha)$ times the old one $\widehat{\mu}_1(\boldsymbol{X})$, the corresponding premium income is t_1 and the corresponding losses is t_2, on average. If $t_1 < t_2$ then this is a profitable portfolio, one well worth retaining.

These graphical procedures can be supplemented with single numbers. Two coefficients have been proposed in the literature to measure the goodness-of-lift: the Value-of-Lift by Meyers and Cummings (2009) and the Gini index advocated by Frees et al. (2011).

6.3.8 Numerical Illustration

6.3.8.1 Assumptions

In this section, $\widehat{\mu}(\boldsymbol{X})$ is assumed to follow a Gamma distribution with unit mean $\mu = 1$ and variance σ^2, henceforth denoted as $\mathcal{G}am(\mu, \sigma^2)$. Such predictors are known to be ordered in the $\preceq_{\text{cx}}$-sense with σ.

Also, in this section, we consider two distributions for the true premium $\mu(\boldsymbol{X})$, namely

- a Gamma distribution with unit mean and variance σ_Y^2 (such true premiums are known to be ordered in the $\preceq_{\text{cx}}$-sense with σ_Y^2);
- a LogNormal distribution with unit mean, i.e. $\ln \mu(\boldsymbol{X})$ is Normally distributed with mean $-\sigma_Y^2/2$ and variance σ_Y^2, which is henceforth denoted as $\mathcal{LN}or(-\sigma_Y^2/2, \sigma_Y)$.

Notice that condition (6.3.1) is fulfilled in both cases since we have $\mathrm{E}[Y] = \mathrm{E}[\mu(\boldsymbol{X})] = \mathrm{E}[\widehat{\mu}(\boldsymbol{X})] = 1$. Also, it is worth noticing that the response may be discrete (such as the number of claims, for instance), the continuity assumption only concerns $\mu(\boldsymbol{X})$ and $\widehat{\mu}(\boldsymbol{X})$.

In addition to the parameters σ and σ_Y governing the variability of the predictor and the true premium, respectively, we consider different dependence structures for the random vector $(\mu(\boldsymbol{X}), \widehat{\mu}(\boldsymbol{X}))$. Specifically, we consider Frank and Clayton copulas, two copulas that are monotonically $\preceq_{\text{conc}}$-increasing with their parameter. Recall from Denuit et al. (2005) that the Clayton copula is given by

$$C_\theta(u, v) = \left(u^{-\theta} + v^{-\theta} - 1\right)^{-1/\theta}, \quad \theta > 0,$$

whereas the Frank copula is given by

$$C_\theta(u, v) = -\frac{1}{\theta} \ln\left(1 + \frac{(\exp(-\theta u) - 1)(\exp(-\theta v) - 1)}{\exp(-\theta) - 1}\right), \quad \theta \neq 0.$$

For positive values of θ in Frank's case, these two copulas express positive dependence. The parameter θ can be interpreted as a measure of strength of the dependence between $\mu(\boldsymbol{X})$ and $\widehat{\mu}(\boldsymbol{X})$. In order to make the dependence parameter more palatable, we rather use the corresponding Kendall's tau. For the Clayton copula, Kendall's tau is simply given by $\frac{\theta}{\theta+2}$. For the Frank copula, Kendall's tau, which also increases with θ, can only be expressed as a Debye function of the first kind.

6.3.8.2 Variability

The dependence structure between $\mu(\boldsymbol{X})$ and $\widehat{\mu}(\boldsymbol{X})$ is assumed to be fixed and modeled by means of the Clayton copula with Kendall's tau equal to 0.5. The predictor $\widehat{\mu}(\boldsymbol{X})$ is supposed to be Gamma distributed with mean and variance both equal to 1. In addition, the true premium $\mu(\boldsymbol{X})$ is supposed to be Gamma distributed with unit mean and variance σ_Y^2. We aim to assess the impact of σ_Y^2 on ABC values.

In that goal, we consider three values for σ_Y^2, that are 0.5, 1 and 2. The results are summarized in the following table and illustrated in Fig. 6.5.

Line type	$\widehat{\mu}(\boldsymbol{X})$	$\mu(\boldsymbol{X})$	Copula C	ABC
medium dash	$\mathcal{G}am(1,1)$	$\mathcal{G}am(1,2)$	Clayton($\tau = 0.5$)	6.33%
short dash	$\mathcal{G}am(1,1)$	$\mathcal{G}am(1,1)$	Clayton($\tau = 0.5$)	9.66%
dotted	$\mathcal{G}am(1,1)$	$\mathcal{G}am(1,0.5)$	Clayton($\tau = 0.5$)	13.08%

We observe that the concentration curves are non-crossing as a result of the convex order among the different distributions of $\mu(\boldsymbol{X})$. Furthermore, the smaller the variance of $\mu(\boldsymbol{X})$ the further away the concentration curve from the Lorenz curve which leads to a decreasing of the ABC value with the variance of $\mu(\boldsymbol{X})$. Notice that in this example, there is no need to complement ABC values with the ICC metrics, the Lorenz curve being the same in the three cases considered.

This example highlights the fact that when we have identically distributed predictors that perform similarly in terms of dependence with the true premium, the ABC metric will favor the case where the true premium is the most variable (in the convex order sense). Similarly, for a given true premium and predictors performing the same way in terms of dependence with the true premium, the ABC metric will favor the predictor that is the less variable in terms of the convex order.

The situation where $\mu(\boldsymbol{X}) \preceq_{\text{cx}} \widehat{\mu}(\boldsymbol{X})$ may be due to overfitting. This can happen when the predictor $\widehat{\mu}(\boldsymbol{X})$ integrates random noise. Indeed, assume that only the first q features, $q < p$, $X_1, \ldots, X_q$ matters and that $X_{q+1}, \ldots, X_p$ are independent, zero-mean random variables, independent of $X_1, \ldots, X_q$. Then, the true score $\beta_0 + \sum_{j=1}^{q} \beta_j X_j$ is dominated by $\beta_0 + \sum_{j=1}^{p} \beta_j X_j$ in the convex sense. On the contrary, the situation where $\widehat{\mu}(\boldsymbol{X}) \preceq_{\text{cx}} \mu(\boldsymbol{X})$ may be due to underfitting, which can be the case, for instance, when

$$\mu(\boldsymbol{X}) = \mathrm{E}[Y|X_1, \ldots, X_q, X_{q+1}, \ldots, X_p]$$

and

$$\widehat{\mu}(\boldsymbol{X}) = \mathrm{E}[Y|X_1, \ldots, X_q].$$

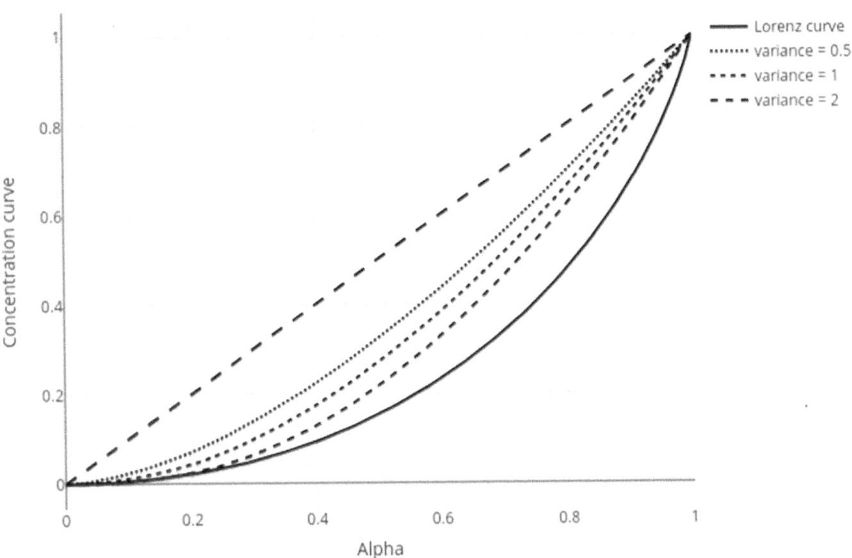

Fig. 6.5 Lorenz curve and several concentration curves for different variances of $\mu(\boldsymbol{X})$

Indeed, we have seen that increasing the number of features produce more dispersed premiums.

6.3.8.3 Dependence

Let us consider fixed distributions for $\mu(\boldsymbol{X})$ and $\widehat{\mu}(\boldsymbol{X})$, namely $\mu(\boldsymbol{X}) \sim \mathcal{G}am(1, 1)$ and $\widehat{\mu}(\boldsymbol{X}) \sim \mathcal{G}am(1, 1)$. We aim to assess the impact of the strength of the dependence between $\mu(\boldsymbol{X})$ and $\widehat{\mu}(\boldsymbol{X})$ on the ABC value. In that purpose, we consider the Clayton copula with three values for Kendall's tau. The results are summarized in the following table and illustrated in Fig. 6.6:

Line type	$\widehat{\mu}(\boldsymbol{X})$	$\mu(\boldsymbol{X})$	C	ABC
medium dash	$\mathcal{G}am(1, 1)$	$\mathcal{G}am(1, 1)$	Clayton($\tau = 0.75$)	3.46%
short dash	$\mathcal{G}am(1, 1)$	$\mathcal{G}am(1, 1)$	Clayton($\tau = 0.50$)	9.66%
dotted	$\mathcal{G}am(1, 1)$	$\mathcal{G}am(1, 1)$	Clayton($\tau = 0.25$)	17.04%

The weaker the dependence the further away the concentration curve from the Lorenz curve. This can be explained as follows. With the Clayton copula, increasing Kendall's tau results in a random pair $(\mu(\boldsymbol{X}), \widehat{\mu}(\boldsymbol{X}))$ larger in the sense of $\preceq_{\text{conc}}$. Therefore, from Property 6.3.19, we know that the concentration curve gets lower, and thus closer to the Lorenz curve.

We observe that the ABC value decreases with Kendall's tau, which is not surprising since increasing Kendall's tau means that $\widehat{\mu}(\boldsymbol{X})$ becomes more informative about the true premium $\mu(\boldsymbol{X})$.

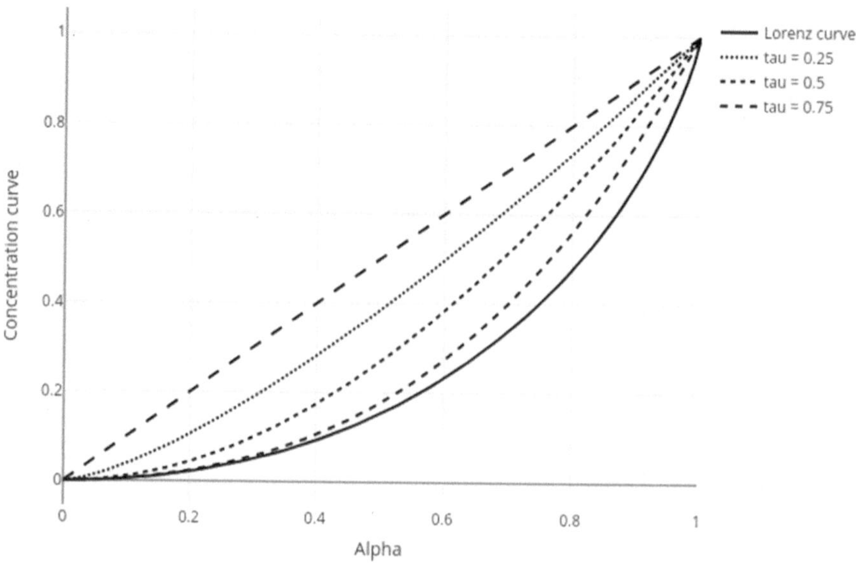

Fig. 6.6 Lorenz curve and several concentration curves for different values of Kendall's tau

Notice that there is no need here to consider ICC values. Indeed, the Lorenz curve being not impacted by the dependence between $\mu(\boldsymbol{X})$ and $\widehat{\mu}(\boldsymbol{X})$, ABC and ICC metrics behave the same way.

6.3.8.4 Distribution

Again, we suppose that the predictor $\widehat{\mu}(\boldsymbol{X})$ is Gamma distributed with mean and variance both equal to 1, and the dependence structure between $\mu(\boldsymbol{X})$ and $\widehat{\mu}(\boldsymbol{X})$ is assumed to be fixed and modeled by means of the Clayton copula with Kendall's tau equal to 0.5. The true premium $\mu(\boldsymbol{X})$ is assumed to be Gamma distributed, and this time, we also consider the LogNormal distribution for $\mu(\boldsymbol{X})$. Specifically, the following table summarizes the three cases considered here:

Line type	$\widehat{\mu}(\boldsymbol{X})$	$\mu(\boldsymbol{X})$	C	ABC
medium dash	$\mathcal{G}am(1, 1)$	$\mathcal{LN}or\left(-\frac{(1.25\sqrt{\ln 2})^2}{2}, 1.25\sqrt{\ln 2}\right)$	Clayton($\tau = 0.5$)	9.20%
short dash	$\mathcal{G}am(1, 1)$	$\mathcal{LN}or\left(-\frac{\ln 2}{2}, \sqrt{\ln 2}\right)$	Clayton($\tau = 0.5$)	11.58%
dotted	$\mathcal{G}am(1, 1)$	$\mathcal{G}am(1, 1)$	Clayton($\tau = 0.5$)	9.66%

In Fig. 6.7, we can see the corresponding concentration and Lorenz curves. In both cases where the variance of $\mu(\boldsymbol{X})$ is equal to 1, one sees that the LogNormal concentration curve (short dash) lies further away from the Lorenz curve than the Gamma one (dotted). One observes that the ABC value favors the case where the

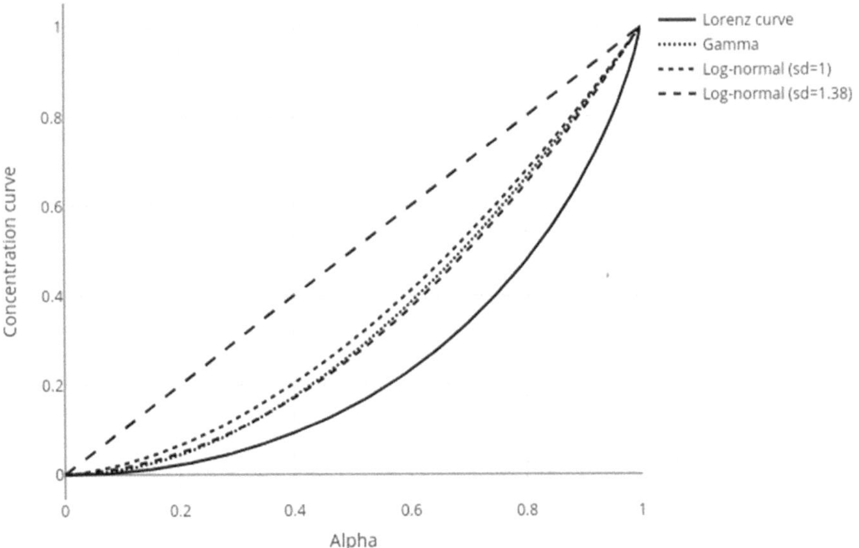

Fig. 6.7 Lorenz curve and several concentration curves for different distributions of $\mu(\boldsymbol{X})$

distributions of $\widehat{\mu}(\boldsymbol{X})$ and $\mu(\boldsymbol{X})$ are similar, the dependence structure being the same in both cases.

In the LogNormal case (short dash), increasing σ_Y by factor 1.25 (medium dash) yields concentration curves crossing around point 0.34. While in the previous examples the concentration curves were always ordered, we see that the use of different distributions can lead to crossing concentration curves.

Notice that in the latter case (medium dash), the ABC value is the smallest one, which is not surprising in light of Sect. 6.3.8.2 since it corresponds to the case where the variance of $\mu(\boldsymbol{X})$ is the largest one.

Again, there is no need to complement ABC values with the ICC metrics since the Lorenz curve remains the same across the three cases considered here.

6.3.8.5 Crossing Copulas

Similarly to the previous example, the use of different copulas can also lead to crossing concentration curves. Let us consider the Clayton copula C_1 and the Frank copula C_2 as in Example 2.3 of Denuit and Mesfioui (2013). In such a case, one can show that there exists a function f such that $C_1(u,v) - C_2(u,v) \leq 0$ if $v \leq f(u)$ and $C_1(u,v) - C_2(u,v) \geq 0$ if $v \geq f(u)$, so that these two copulas are not ordered according to the concordance order. We consider the two following cases:

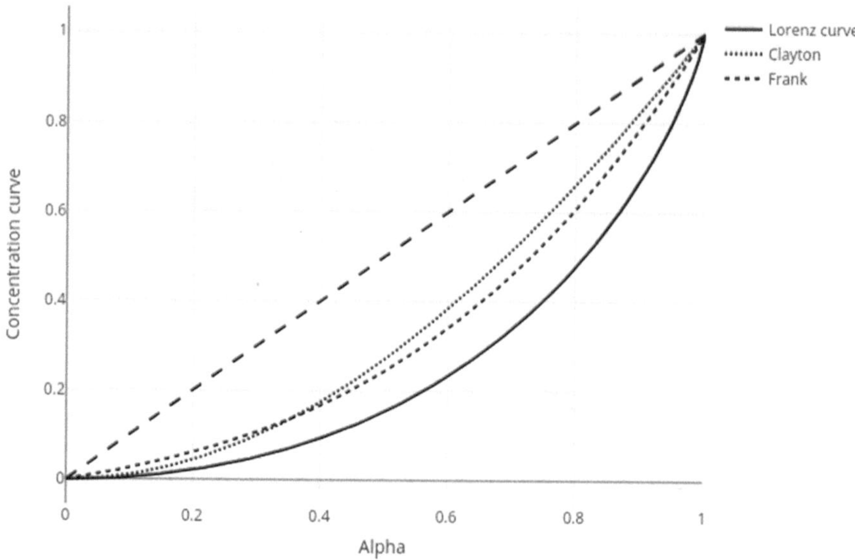

Fig. 6.8 Lorenz curve and two concentration curves for different copulas

Line type	$\widehat{\mu}(\boldsymbol{X})$	$\mu(\boldsymbol{X})$	C	ABC
short dash	$\mathcal{G}am(1,1)$	$\mathcal{G}am(1,1)$	Frank($\tau = 0.5$)	7.79%
dotted	$\mathcal{G}am(1,1)$	$\mathcal{G}am(1,1)$	Clayton($\tau = 0.5$)	9.66%

The corresponding concentration and Lorenz curves are depicted in Fig. 6.8. We observe that the concentration curves cross around point 0.35, which is an intuitive result. Indeed, Clayton copula has stronger dependence in the lower quadrant than Frank copula. Also, since the overall dependence is equal in both cases, the opposite holds in the upper quadrant. The stronger the dependence the closer the concentration curve is to the Lorenz curve. This is why the Clayton copula lies closer to the Lorenz curve for small values and further away for large values.

6.3.8.6 Non-regression Dependent Copula Impact

Finally, we can consider a copula that does not exhibit positive quadrant dependence but only positively expectation dependence. To this end, we can proceed as in Egozcue et al. (2011) by mixing two copulas expressing quadrant dependence of opposite signs. For instance, considering the Frechet–Hoeffding upper and lower bound copulas, we can use

$$C(u, v) = (1-\theta)\min\{u, v\} + \theta \max\{0, u+v-1\} \tag{6.3.14}$$

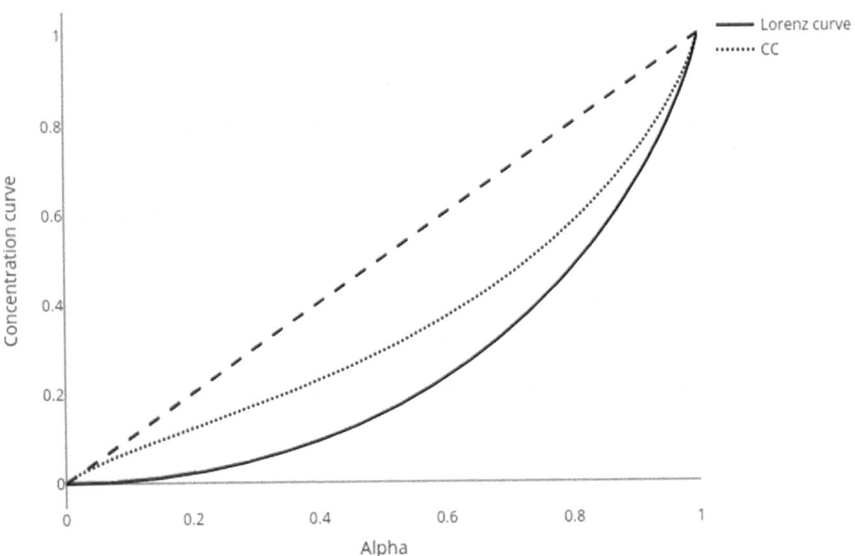

Fig. 6.9 Lorenz and concentration curves for a non-regression dependent copula

as in Example 2.1 of Denuit and Mesfioui (2017). We know from Egozcue et al. (2011) that this mixture expresses positive expectation dependence if, and only if, $\theta \leq \frac{1}{2}$. Alternatively, the Frechet–Hoeffding lower bound copula may be replaced with another copula expressing negative quadrant dependence (such as the Farlie-Gumbel-Morgenstern, or FGM copula with negative dependence parameter, for instance).

Considering the following setup

Line type	$\widehat{\mu}(\boldsymbol{X})$	$\mu(\boldsymbol{X})$	C	ABC
dotted	$\mathcal{G}am(1,1)$	$\mathcal{G}am(1,1)$	(6.3.14) with $\theta = 0.8$	10%

we see in Fig. 6.9 that the above mixture copula leads well to a non-convex concentration curve.

6.3.9 *Case Study*

We end this chapter by considering a French motor third-party liability insurance portfolio available in the `CASdatasets` package in `R`. Specifically, we investigate the dataset `freMTPL2freq` which contains 678 013 observations of the number of claims (response Y) together with nine features ($\boldsymbol{X} = (X_1, \ldots, X_9)$). The features correspond to several characteristics of the policyholder (age, density of inhabitants in the home city, region, area and bonus-malus) and the car (power, age, brand and fuel type). We refer the reader to Noll et al. (2018) for a broad description of the dataset.

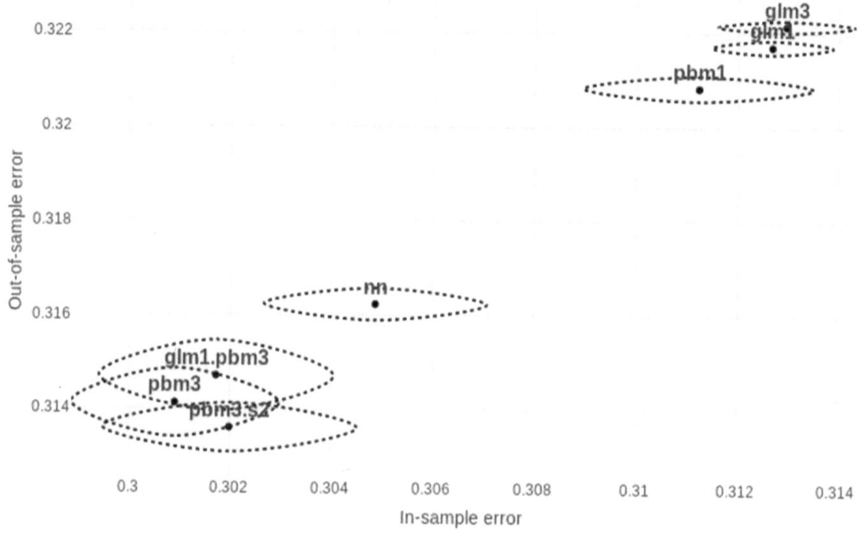

Fig. 6.10 In- and out-of-sample errors for models under consideration

In this section, we aim to compare some of the models investigated in Noll et al. (2018) by using ABC and ICC metrics discussed in this chapter. More specifically, we consider the following models of Noll et al. (2018) for the predictors $\widehat{\mu}_k(\boldsymbol{X}_k)$:

- glm1—Poisson GLM with a log-link function and all explanatory variables;
- glm3—same as glm1 but without area and region variables;
- pbm1—boosted SBS (Standardized Binary Splits) tree (depth = 1, iterations = 30);
- pbm3—boosted SBS tree (depth = 3, iterations = 50);
- pbm3.s2—boosted SBS tree (depth = 3, iterations = 50, shrinkage = 0.5);
- glm1.pbm3—boosted SBS tree starting from glm1 fit (depth = 3, iterations = 50);
- nn—shallow neural network (20 neurons with one hidden layer).

Models' implementation details can be found in Noll et al. (2018). We refer to Denuit et al. (2019a) for details on neural networks.

The dataset is partitioned into a training set of 610 000 observations and a validation set comprising the remaining observations.

Figure 6.10 shows the training sample estimate of the generalization error (in-sample error) and the validation sample estimate of the generalization error (out-of-sample error) for the models under study together with bootstrapped 95% confidence intervals. The bounds are derived for in- and out-of-sample errors individually, so only vertical and horizontal distances are meaningful. In particular, the oval shape is due to spline smoothing through the points (in-sample error, out-of-sample error): {(lower, observed), (observed,higher), (higher,observed), (observed,lower)}. Overall, in-sample error and out-of-sample error classify the models in a similar way, except the boosted tree model (pbm3) and its shrunken version. For the latter mod-

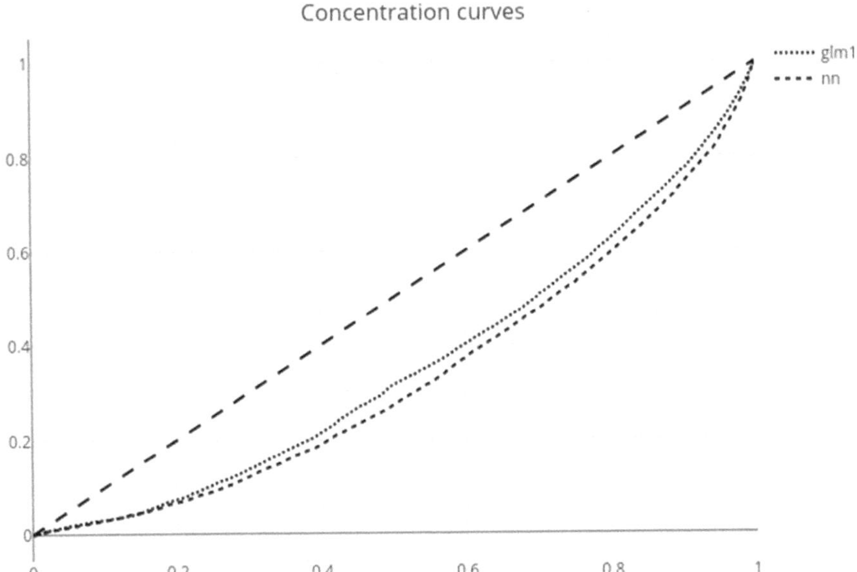

Fig. 6.11 $\widehat{\text{CC}}$ for models glm1 and nn

els, introducing a shrinkage factor increases the in-sample error while it reduces the out-of-sample error. This is not surprising as the introduction of a shrinkage factor aims to avoid overfitting issues. We also note that the boosted GLM model (glm1.pbm3) improves substantially over the original GLM model (glm1). However, it does not outperform the boosted SBS tree (pbm3). The latter observation indicates that the fixed structural form imposed to the expected claim frequency by the GLM model does not provide any additional explanatory insights compared to the boosted SBS tree. Finally, the optimal model with respect to the out-of-sample error metric is the boosted tree model with a shrinkage factor (pbm3.s2).

Looking at the bootstrapped confidence intervals, all the models except the ones based on (pbm3) are nicely separated. It also seems that boosted methods yield more varying results than GLMs or the neural network model (for out-of-sample error).

Let us now turn to the goodness-of-lift metrics discussed in this chapter. In the following, we use the empirical versions of the concentration curve $\widehat{\text{CC}}$ and the Lorenz curve $\widehat{\text{LC}}$ computed on the validation set in order to get ABC and ICC values.

In case the number of observations are insufficient, a smoothed version of the empirical concentration curve $\widehat{\text{CC}}$ could be used instead. Here, the size of the validation set is judged as sufficient to simply rely on $\widehat{\text{CC}}$, depicted in Fig. 6.11 for two of the considered models. The remaining models are close to these two models, forming two groups of curves for α larger than 0.15. The higher group of curves is related to models glm1, glm3 and pbm1, which are also the three worst models according to the out-of-sample errors.

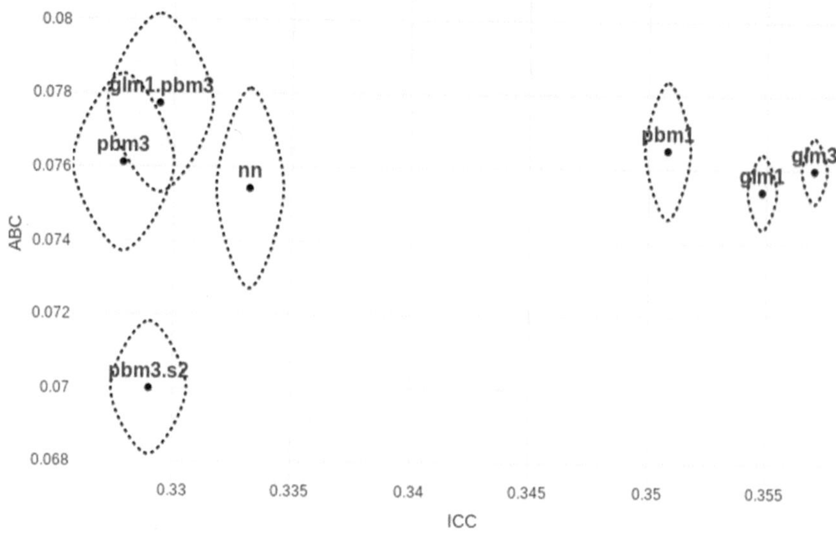

Fig. 6.12 Estimated ABC and ICC values for models under consideration

ABC and ICC values are displayed in Fig. 6.12 also with the same visualization of bootstrapped confidence intervals. We notice that ICC metric classifies the models as the out-of-sample error metric, except for models pbm3 and pbm3.s2. While both metrics agree that these two last models are the best ones, pbm3.s2 outperforms pbm3 according to the out-of-sample error while with ICC it is the other way around.

Regarding the ABC values, we observe that a model with a low ICC can either have low or large ABC. For instance, glm1.pbm3, which is one of the best model according to ICC, has the highest ABC, while pbm3.s2 has both low ICC and ABC. If we compare glm1.pbm3 and pbm3.s2 that have similar degrees of lift according to ICC, we notice that ABC metric favors pbm3.s2 that is less variable than glm1.pbm3, which is in line with Sect. 6.3.8.2. In the same way, while pbm3 and pbm3.s2 have similar ICC, pbm3.s2 outperforms pbm3 according to ABC, pbm3.s2 being less variable than pbm3 (since both models have the same number of trees while pbm3.s2 uses a shrinkage parameter). Finally, the optimal model with respect to ABC is pbm3.s2.

To end the case study, we display in Fig. 6.13 ICC and ABC as functions of α (i.e. integrating over the interval $[0, \alpha]$ instead of the whole interval $[0, 1]$). We present only curves for models pbm1 and glm1.pbm3 as the remaining curves look fairly similar. One sees that glm1.pbm3 has always lower ICC while the ABC values cross at around 91% quantile. Hence, from that quantile, one can say that pbm1 outperforms glm1.pbm3 according to ABC metric.

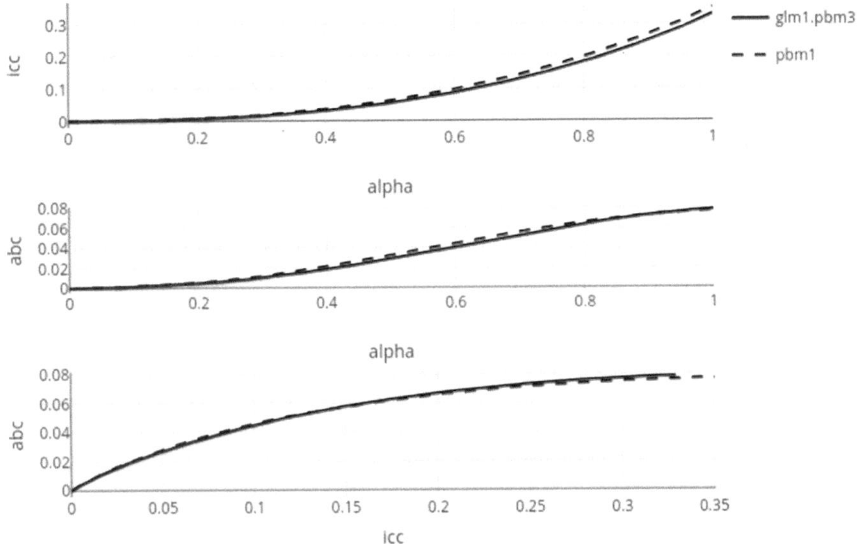

Fig. 6.13 Estimated ABC and ICC values for models under consideration

6.4 Bibliographic Notes and Further Reading

Denuit et al. (2019b) considered binary responses and derive the set of attainable values for concordance-based association measures so that the closeness to the best-possible fit can be properly assessed. Denuit et al. (2019c) and Mesfioui et al. (2020) obtained the best-possible upper bounds for Kendall's tau and Spearman's rho when the response is a discrete random variable. Section 6.2 is largely inspired from these two papers.

Several testing procedures have been proposed in the literature to detect dependence relations. For positive quadrant dependence, we refer the reader to Denuit and Scaillet (2004) and Scaillet (2005). Zhu et al. (2016) investigated hypothesis tests for first-degree and higher-degree expectation dependence. Testing procedures for the convex order have been proposed in economics (see e.g., Barrett and Donald 2003). Section 6.3 is strongly inspired from Denuit et al. (2019d), in which concentration curves and Lorenz curves are shown to provide actuaries with effective tools to evaluate whether a premium is appropriate or to compare two competing alternatives.

References

Barrett GF, Donald SG (2003) Consistent tests for stochastic dominance. Econometrica 71(1):71–104

Denuit M, Dhaene J, Goovaerts MJ, Kaas R (2005) Actuarial theory for dependent risks: measures, orders and models. Wiley, New York

Denuit M, Mesfioui M (2013) A sufficient condition of crossing-type for the bivariate orthant convex order. Stat Probab Lett 83(1):157–162

Denuit M, Mesfioui M (2017) Preserving the Rothschild-Stiglitz type increase in risk with background risk: a characterization. Insur: Math Econ 72:1–5

Denuit M, Hainaut D, Trufin J (2019a) Effective statistical learning methods for actuaries III: neural networks and extensions. Springer Actuarial Lecture Notes

Denuit M, Mesfioui M, Trufin J (2019b) Bounds on concordance-based validation statistics in regression models for binary responses. Methodol Comput Appl Probab 21(2):491–509

Denuit M, Mesfioui M, Trufin J (2019c) Concordance-based predictive measures in regression models for discrete responses. Scand Actuar J 10:824–836

Denuit M, Scaillet O (2004) J Financ Econ 2(3):422–450

Denuit M, Sznajder D, Trufin J (2019d) Model selection based on Lorenz and concentration curves, Gini indices and convex order. Insur: Math Econ 89:128–139

Egozcue M, Garcia L-F, Wong W-K, Zitikis R (2011) Grüss-type bounds for covariances and the notion of quadrant dependence in expectation. Cent Eur J Math 9(6):1288–1297

Frees E, Meyers G, Cummings A (2011) Summarizing insurance scores using a Gini index. J Amer Stat Asso 106(495):1085–1098

Frees EW, Meyers G, Cummings AD (2013) Insurance ratemaking and a Gini index. J Risk Insur 81(2):335–366

Gourieroux C, Jasiak J (2007) The econometrics of individual risk: credit, insurance, and marketing. Princeton University Press, Princeton

Mesfioui M, Tajar A (2005) On the properties of some nonparametric concordance measures in the discrete case. Nonparametric Stat 17(5):541–554

Mesfioui M, Trufin J, Zuyderhoff P (2020) Bounds on Spearman's rho when at least one random variable is discrete. Working paper

Meyers G, Cummings AD (2009) Goodness of Fit" vs. "Goodness of Lift. Actuar Rev 36–3:16–17

Muliere P, Petrone S (1992) Generalized Lorenz curve and monotone dependence orderings. Metron 50:19–38

Nešlehová J (2007) On rank correlation measures for non-continuous random variables. J Multivar Anal 98(3):544–567

Noll A, Salzmann R, Wüthrich M (2018) Case study: French motor third-party liability claims. Available at SSRN: https://ssrn.com/abstract=3164764

Scaillet O (2005) A Kolmogorov-Smirnov type test for Positive Quadrant Dependence. Can J Stat 33(3):415–427

Shaked M, Sordo MA, Suarez-Llorens A (2012) Global dependence stochastic orders. Methodol Comput Appl Probab 14(3):617–648

Tevet D (2013) Exploring model lift: is your model worth implementing. Actuar Rev 40(2):10–13

Yitzhaki S, Schechtman E (2013) The gini methodology: a primer on statistical methodology. Springer, Berlin

Zhu X, Guo X, Lin L, Zhu L (2016) Testing for positive expectation dependence. Ann Inst Stat Math 68:135–153